T0206108

Introduction to
Perturbation Theory
in Quantum Mechanics

Introduction to
Perturbation Theory
in Quantum Mechanics

Francisco M. Fernández, Ph.D.

CRC Press
Taylor & Francis Group
Boca Raton London New York

CRC Press is an imprint of the
Taylor & Francis Group, an **informa** business

CRC Press
Taylor & Francis Group
6000 Broken Sound Parkway NW, Suite 300
Boca Raton, FL 33487-2742

First issued in paperback 2020

ISBN 13: 978-0-367-57893-0 (pbk)
ISBN 13: 978-0-8493-1877-1 (hbk)

Visit the Taylor & Francis Web site at
http://www.taylorandfrancis.com

and the CRC Press Web site at
http://www.crcpress.com

Library of Congress Cataloging-in-Publication Data

Fernández, F.M. (Francisco M.), 1952-
    Introduction to perturbation theory in quantum mechanics/Francisco M. Fernández.
        p.   cm.
    Includes bibliographical references and index.
    ISBN 0-8493-1877-7 (alk. paper)
    1. Perturbation (Quantum dynamics) I. Title.

QC174.17.P45 F47 2000
530.12--dc21   00-042903

Library of Congress Card Number 00042903

Preface

Perturbation theory is an approximate method that enables one to solve a wide variety of problems in applied mathematics, and for this reason it has proved useful in theoretical physics and chemistry since long ago. Most textbooks on classical mechanics, quantum mechanics, and quantum chemistry exhibit a chapter, or at least a section, dedicated to that celebrated approach which is afterwards applied to several models.

In addition to the general view of perturbation theory offered by those textbooks, there is a wide variety of techniques that facilitate the application of the approach to particular problems in the fields mentioned above. Such implementations of perturbation theory are spread over many papers and specialized books. We believe that a single source collecting most of those methods may profit students of theoretical physics and chemistry.

For simplicity, in this book we concentrate on problems that allow exact analytical solutions of the perturbation equations and avoid those that require long and tedious numerical computation that may divert the reader's mind from the core of the problem. However, we also resort to numerical results when they are necessary to illustrate and complement important features of the theory.

In order to compare different methods, we apply them to the same models so that the reader may clearly understand why we prefer one or another. Sometimes, we also apply perturbation theory to exactly solvable models in order to illustrate the most relevant features of the approximate method and to disclose some of its limitations. This strategy is also suitable for clearly understanding the improvements in the perturbation series.

In this introductory book we try to keep the mathematics as simple as possible. Consequently, we avoid a thorough discussion of certain topics, such as the analytical properties of the eigenvalues of simple nontrivial quantum-mechanical models. The reader who is interested in going beyond the scope of this book will find the necessary references for that purpose.

Nowadays, there are many symbolic processors that greatly facilitate most analytical calculations, and this book would not be complete if it did not show how to apply them to perturbation theory. Here we choose Maple® because it is uncommonly powerful and simple at the same time. In addition, Maple offers a remarkably friendly interface that enables the user to organize his or her work in the form of useful worksheets which can be exported in several formats. For example, here we have chosen LaTeX® to produce some of the tables, thus avoiding unnecessary transcription of the results that may lead to misprints.

Maple allows one to do a great deal of calculation interactively, which is commonly useful to understand the main features of the problem, and when programming becomes necessary, Maple language is straightforward and easy to learn. Both modes of calculation have proved most useful for present work, and our programs reflect this fact in that they are not completely automatic or foolproof. In the program section we show several examples of the Maple procedures used to obtain the results discussed in this book, and we think that the hints given there are sufficient for their successful application. However, the reader who finds any difficulty is encouraged to contact the author via E-mail at: framfer@isis.unlp.edu.ar.

Biography

Francisco M. Fernández, Ph.D., is Professor at the University of La Plata, Buenos Aires, Argentina, where he graduated in 1977. Dr. Fernández has conducted research on theoretical chemistry and mathematical physics specializing in approximate methods in quantum mechanics and quantum chemistry. He has published more than 300 research papers and 3 books. The Ministry of Education and Culture of Argentina gave Dr. Fernández a physics and chemistry award for his research in the field from 1987 to 1990. Dr. Fernández is a Member of the Research Career at the National Research Council of Argentina (CONICET).

Contents

Chapter 1

Perturbation Theory in Quantum Mechanics

1.1 Introduction

It is well known that one cannot solve the Schrödinger equation in quantum mechanics except for some simple models. For that reason many authors have devoted considerable time and effort to develop efficient approximate methods. Among them, perturbation theory has been helpful since the earliest applications of quantum mechanics. One of the main advantages of this approach is that it provides analytical approximate solutions for many nontrivial simple problems which are suitable for subsequent discussion and interpretation of the physical phenomena. In fact, perturbation theory is probably one of the approximate methods that most appeals to intuition.

The standard textbook formulas for the perturbation corrections are somewhat cumbersome for a systematic calculation of sufficiently high order. In this book we show several alternative strategies that are easily programmable for numerical or algebraic calculation. We are mainly concerned with the derivation of exact perturbation corrections, and therefore concentrate on sufficiently simple nontrivial models having physical application. However, some of the algorithms discussed in this book are also suitable for numerical calculation.

Most of the methods discussed in this book lead to recurrence relations and other mathematical algorithms that are straightforward for hand calculation, and most suitable for computer algebra. The use of the latter is mandatory if one is interested in great perturbation orders. Among the many computer algebra packages, we have chosen Maple because it is easy to use, extremely powerful and reliable, and offers many facilities to write reports and convert the output into forms suitable for word processing [1].

In this chapter we briefly review those formulas of perturbation theory in quantum mechanics that we need in subsequent chapters. We assume that the reader is familiar with standard concepts and notation used in most textbooks on quantum mechanics. We are mainly concerned with perturbation theory for bound stationary states; however, in this chapter, we also outline time-dependent perturbation theory, and later in Chapter 8 we show simple applications of perturbation theory to stationary states in the continuum spectrum.

1.2 Bound States

We first consider bound states that are square-integrable solutions of the eigenvalue equation $\hat{H}\Psi = E\Psi$, where $\hat{H}$ is the Hamiltonian operator of the system and E is the energy of the state Ψ [2]. If Ψ is complex, then both its real Ψ_R and imaginary Ψ_I parts satisfy the eigenvalue equation (because

E is real) and are square integrable as follows from $< \Psi|\Psi > = < \Psi_R|\Psi_R > + < \Psi_I|\Psi_I > < \infty$. Therefore, without loss of generality we only consider real solutions of the eigenvalue equation. In principle, we apply perturbation theory to

$$\hat{H}\Psi_n = E_n\Psi_n, n = 1, 2, \ldots \tag{1.1}$$

provided that we can write

$$\hat{H} = \hat{H}_0 + \lambda\hat{H}' , \tag{1.2}$$

where $\hat{H}_0$ is a sufficiently close approximation to $\hat{H}$ so that $\hat{H}'$ may be considered to be a small perturbation, and λ is a perturbation parameter. In Chapter 6 we will discuss the meaning of the expression "small perturbation." We also assume that the eigenvalue equation for $\hat{H}_0$ is exactly solvable:

$$\hat{H}_0\Psi_{n,0} = E_{n,0}\Psi_{n,0}, \ n = 1, 2, \ldots . \tag{1.3}$$

The eigenvalues and eigenvectors of $\hat{H}$ given by equation (1.1) depend on the perturbation parameter λ and can be formally expanded in Taylor series about $\lambda = 0$:

$$E_n = \sum_{s=0}^{\infty} E_{n,s}\lambda^s, E_{n,s} = \frac{1}{s!} \frac{\partial^s E_n}{\partial \lambda^s}\bigg|_{\lambda=0} , \tag{1.4}$$

$$\Psi_n = \sum_{s=0}^{\infty} \Psi_{n,s}\lambda^s, \Psi_{n,s} = \frac{1}{s!} \frac{\partial^s \Psi_n}{\partial \lambda^s}\bigg|_{\lambda=0} . \tag{1.5}$$

From straightforward substitution of these series into the eigenvalue equation (1.1) we derive a system of equations for the perturbation coefficients:

$$\left[\hat{H}_0 - E_{n,0}\right]\Psi_{n,s} = \left[E_{n,1} - \hat{H}'\right]\Psi_{n,s-1} + \sum_{j=2}^{s} E_{n,j}\Psi_{n,s-j} . \tag{1.6}$$

For example the equation of first order is

$$\left[\hat{H}_0 - E_{n,0}\right]\Psi_{n,1} = \left[E_{n,1} - \hat{H}'\right]\Psi_{n,0} . \tag{1.7}$$

We say that the unperturbed states are nondegenerate if $E_n^{(0)} \neq E_m^{(0)}$ when $n \neq m$. In order to apply the method below we assume that the eigenfunctions $\Psi_{n,0}$ form a complete orthonormal set (a basis set) so that we can expand the perturbation corrections as follows:

$$\Psi_{n,s} = \sum_{m} C_{mn,s}\Psi_{m,0}, \ C_{mn,s} = \left\langle \Psi_{m,0}|\Psi_{n,s} \right\rangle . \tag{1.8}$$

Notice that $C_{mn,0} = \delta_{mn}$. For simplicity we write $|m>$ instead of $|\Psi_{m,0}>$ from now on.

On applying the bra vector $< m|$ to equation (1.6) we obtain

$$\left[E_{m,0} - E_{n,0}\right]C_{mn,s} = \sum_{j=1}^{s} E_{n,j}C_{mn,s-j} - \sum_{k} H'_{mk}C_{kn,s-1} , \tag{1.9}$$

where $H'_{mk} = < m|\hat{H}'|k >$. When $m = n$ we obtain an expression for the energy

$$E_{n,s} = \left\langle n\left|\hat{H}'\right|\Psi_{n,s-1}\right\rangle - \sum_{j=1}^{s-1} E_{n,j}C_{nn,s-j} . \tag{1.10}$$

Some authors choose the intermediate normalization condition $C_{nn,s} = \delta_{s0}$ because it leads to a simpler expression for the energy: $E_{n,s} = < n|\hat{H}'|\Psi_{n,s-1} >$ [3]. In that case one has to normalize the resulting approximate eigenfunction Ψ_n to unity. Here we choose the standard normalization condition $< \Psi_n|\Psi_n >= 1$ from which it follows that

$$\sum_{j=0}^{s} \langle \Psi_{n,j}|\Psi_{n,s-j} \rangle = \delta_{n0} \ . \tag{1.11}$$

When $m \neq n$ equation (1.9) gives us an expression for the expansion coefficients

$$C_{mn,s} = \left[E_{n,0} - E_{m,0}\right]^{-1} \left(\sum_{k} H'_{mk} C_{kn,s-1} - \sum_{j=1}^{s} E_{n,j} C_{mn,s-j} \right) \ . \tag{1.12}$$

The remaining coefficient $C_{nn,s}$ follows from equation (1.11):

$$C_{nn,1} = 0, C_{nn,s} = -\frac{1}{2} \sum_{j=1}^{s-1} \sum_{m} C_{mn,j} C_{mn,s-j}, s > 1 \ . \tag{1.13}$$

Equations (1.10) and (1.12) are the standard textbook perturbation expressions. For example, when $s = 1$ we obtain

$$E_{n,1} = H'_{nn} \tag{1.14}$$

from equation (1.10), and

$$C_{mn,1} = \frac{H'_{mn}}{E_{n,0} - E_{m,0}} \tag{1.15}$$

from equation (1.12).

For the second order we obtain

$$E_{n,2} = \left\langle n \left| \hat{H}' \right| \Psi_{n,1} \right\rangle = \sum_{m} \frac{H'^{\,2}_{mn}}{E_{n,0} - E_{m,0}} \tag{1.16}$$

from equation (1.10) and so on. We repeat this process as many times as needed. At each perturbation order we first calculate the energy and then the eigenfunction coefficients, both in terms of corrections already obtained in previous steps.

The recursion relations given by equations (1.10) and (1.12) yield analytical expressions provided that one is able to carry out the sums over intermediate states exactly. The simplest situation is that each such sum has a finite number of terms, which already happens if $H'_{mn} = 0$ for all $|m - n| > J$. Lie algebraic methods greatly facilitate the calculation of analytical matrix elements H_{mn} in certain cases [4].

Once we have the perturbation coefficients $C_{mn,s}$ we easily express matrix elements and between perturbed states in terms of matrix elements and between unperturbed states as follows:

$$\begin{aligned}
\left\langle \Psi_m \left| \hat{A} \right| \Psi_n \right\rangle &= \sum_{p=0}^{\infty} \lambda^p \sum_{s=0}^{p} \left\langle \Psi_{m,s} \left| \hat{A} \right| \Psi_{n,p-s} \right\rangle \\
&= \sum_{p=0}^{\infty} \lambda^p \sum_{s=0}^{p} \sum_{j} \sum_{k} C_{jm,s} C_{kn,p-s} \left\langle j \left| \hat{A} \right| k \right\rangle \ . \tag{1.17}
\end{aligned}$$

1.2.1   The $2s + 1$ Rule

The discussion above suggests that it is necessary to calculate the correction of order s to the eigenfunction in order to obtain the correction of order $s + 1$ to the energy. However, this is not the case; given all the corrections to the eigenfunction through order s we can obtain all the energy coefficients through order $2s + 1$. This calculation is based on more symmetric formulas that we briefly discuss in what follows. Consider a matrix element $< \Psi_{n,s}|[E_{n,1} - \hat{H}']|\Psi_{n,t} >$ with $s < t$. Using the general equation (1.6) we rewrite it as

$$
\begin{aligned}
\left\langle \Psi_{n,s} \left| \left[E_{n,1} - \hat{H}' \right] \right| \Psi_{n,t} \right\rangle &= \left\langle \left[E_{n,1} - \hat{H}' \right] \Psi_{n,s} | \Psi_{n,t} \right\rangle \\
&= \left\langle \left[\hat{H}_0 - E_{n,0} \right] \Psi_{n,s+1} | \Psi_{n,t} \right\rangle - \sum_{j=2}^{s+1} E_{n,j} \left\langle \Psi_{n,s+1-j} | \Psi_{n,t} \right\rangle \\
&= \left\langle \Psi_{n,s+1} \left| \left[\hat{H}_0 - E_{n,0} \right] \right| \Psi_{n,t} \right\rangle - \sum_{j=2}^{s+1} E_{n,j} \left\langle \Psi_{n,s+1-j} | \Psi_{n,t} \right\rangle \\
&= \left\langle \Psi_{n,s+1} \left| \left[E_{n,1} - \hat{H}' \right] \right| \Psi_{n,t-1} \right\rangle - \sum_{j=2}^{s+1} E_{n,j} \left\langle \Psi_{n,s+1-j} | \Psi_{n,t} \right\rangle \\
&\quad + \sum_{j=2}^{t} E_{n,j} \left\langle \Psi_{n,s+1} | \Psi_{n,t-j} \right\rangle .
\end{aligned}
\tag{1.18}
$$

The net result of this process is a reduction in the greater subscript and an increment in the smaller one making the matrix element more symmetric. We apply it to equation (1.10) as many times as required in order to obtain the most symmetric expression for the energy that consequently contains perturbation corrections of the smallest order to the eigenfunction. For example, the first energy coefficients are

$$
E_{n,3} = \left\langle \Psi_{n,1} \left| \left[\hat{H}' - E_{n,1} \right] \right| \Psi_{n,1} \right\rangle ,
\tag{1.19}
$$

$$
E_{n,4} = \left\langle \Psi_{n,2} \left| \left[\hat{H}' - E_{n,1} \right] \right| \Psi_{n,1} \right\rangle - E_{n,2} \left[\langle \Psi_{n,2} | n \rangle + \langle \Psi_{n,1} | \Psi_{n,1} \rangle \right] ,
\tag{1.20}
$$

$$
E_{n,5} = \left\langle \Psi_{n,2} \left| \left[\hat{H}' - E_{n,1} \right] \right| \Psi_{n,2} \right\rangle - E_{n,2} \left[\langle \Psi_{n,1} | \Psi_{n,2} \rangle + \langle \Psi_{n,2} | \Psi_{n,1} \rangle \right] ,
\tag{1.21}
$$

where we have used equation (1.11) to simplify the right-hand sides. Such symmetrized energy formulas and their generalizations are well known and have been discussed by other authors in more detail [4].

1.2.2   Degenerate States

When the unperturbed states are degenerate we cannot apply the perturbation equations given above in a straightforward way. If there are g_n linearly independent solutions to the unperturbed equation with the same eigenvalue:

$$
\hat{H}_0 \Phi_{n,a} = E_{n,0} \Phi_{n,a}, \ a = 1, 2, \ldots, g_n
\tag{1.22}
$$

we say that those states are g_n-fold degenerate. Any linear combination

$$
\Psi_{n,0} = \sum_{a=1}^{g_n} C_{a,n} \Phi_{n,a}
\tag{1.23}
$$

is an eigenfunction of $\hat{H}_0$ with eigenvalue $E_{n,0}$. Applying the bra $< \Phi_{n,a}|$ from the left to the equation of first order (1.7), we obtain an homogeneous system of g_n equations with g_n unknowns:

$$\sum_{b=1}^{g_n} \left(H'_{a,b} - E_{n,1}\delta_{a,b} \right) C_{b,n} = 0, \ a = 1, 2, \ldots, g_n \ . \tag{1.24}$$

As before we assume that $< \Phi_{n,a}|\Phi_{n,b} >= \delta_{ab}$ and write $H'_{a,b} =< \Phi_{n,a}|\hat{H}'|\Phi_{n,b} >$. Nontrivial solutions exist only if the secular determinant vanishes:

$$\left| H'_{a,b} - E_{n,1}\delta_{a,b} \right| = 0 \ . \tag{1.25}$$

The g_n real roots $E_{n,1,b}, b = 1, 2, \ldots, g_n$ are the corrections of first order for those states.

We may treat higher perturbation orders in the same way but the notation becomes increasingly awkward as the perturbation order increases. For this reason we do not proceed along these lines and will return to perturbation theory for degenerate states when we discuss a more systematic approach in Chapter 3.

In Chapters 5 and 7 we will show that it is sometimes convenient to choose a nonlinear perturbation parameter λ in the Hamiltonian operator and expand $\hat{H}(\lambda)$ in a Taylor series about $\lambda = 0$ as follows:

$$\hat{H}(\lambda) = \sum_{j=0}^{\infty} \hat{H}_j \lambda^j \ . \tag{1.26}$$

If we can solve the eigenvalue equation for $\hat{H}_0 = \hat{H}(0)$, then we can apply perturbation theory in the way outlined above. One easily proves that the perturbation equations for this case are

$$\left[\hat{H}_0 - E_{n,0} \right] \Psi_{n,s} = \sum_{j=1}^{s} \left[E_{n,j} - \hat{H}_j \right] \Psi_{n,s-j} \ , \tag{1.27}$$

and that the systematic calculation of the corrections is similar to that in preceding subsections.

1.3 Equations of Motion

In quantum mechanics one obtains the state $\Psi(t)$ of the system at time t from the state $\Psi(t_0)$ at time t_0 by means of a time-evolution operator $\hat{U}(t, t_0)$ [5]:

$$\Psi(t) = \hat{U}(t, t_0) \Psi(t_0) \ . \tag{1.28}$$

The time-evolution operator satisfies the differential equation

$$i\hbar \frac{d}{dt} \hat{U}(t, t_0) = \hat{H}\hat{U}(t, t_0) \tag{1.29}$$

with the initial condition

$$\hat{U}(t_0, t_0) = \hat{1} \ , \tag{1.30}$$

where $\hat{1}$ is the identity operator. It follows from the adjoint of equation (1.29)

$$i\hbar \frac{d}{dt} \hat{U}(t, t_0)^{\dagger} = -\hat{U}(t, t_0)^{\dagger} \hat{H} \tag{1.31}$$

that $\hat{U}(t, t_0)$ is unitary ($\hat{U}^\dagger = \hat{U}^{-1}$).

Other important properties of the time-evolution operator are

$$\hat{U}(t, t_0)^\dagger = \hat{U}(t, t_0)^{-1} = \hat{U}(t_0, t) \tag{1.32}$$

$$\hat{U}(t, t_0) = \hat{U}(t, t') \hat{U}(t', t_0) . \tag{1.33}$$

It follows from equation (1.33) that we can restrict ourselves to the case $t_0 = 0$ without loss of generality because $\hat{U}(t, t_0) = \hat{U}(t, 0)\hat{U}(0, t_0) = \hat{U}(t, 0)\hat{U}(t_0, 0)^\dagger$. Therefore, we consider $\hat{U} = \hat{U}(t) = \hat{U}(t, 0)$ from now on.

In the Schrödinger picture outlined above the states change with time; on the other hand, the states are time independent in the Heisenberg picture [5]. Given an observable $\hat{A}$ in the Schrödinger picture, we obtain its Heisenberg counterpart $\hat{A}_H$ as follows:

$$\hat{A}_H = \hat{U}^\dagger \hat{A} \hat{U} , \tag{1.34}$$

which satisfies the equation of motion

$$i\hbar \frac{d}{dt} \hat{A}_H = -\hat{U}^\dagger \hat{H} \hat{A} \hat{U} + \hat{U}^\dagger \hat{A} \hat{H} \hat{U} = \hat{U}^\dagger \left[\hat{A}, \hat{H} \right] \hat{U} = \left[\hat{A}_H, \hat{H}_H \right] , \tag{1.35}$$

where $[\hat{A}, \hat{B}] = \hat{A}\hat{B} - \hat{B}\hat{A}$ is the commutator between two linear operators $\hat{A}$ and $\hat{B}$. In order to derive equation (1.35) we have taken into account that $\hat{U}^\dagger \hat{A} \hat{B} \hat{U} = \hat{U}^\dagger \hat{A} \hat{U} \hat{U}^\dagger \hat{B} \hat{U}$.

If $\hat{H}$ is time independent then

$$\hat{U}(t, t_0) = \hat{U}(t - t_0) = \exp\left[-i(t - t_0) \hat{H}/\hbar \right] , \tag{1.36}$$

and $\hat{H}_H = \hat{H}$.

1.3.1 Time-Dependent Perturbation Theory

It is not possible to solve the Schrödinger equation (1.29) exactly, except for some simple models; for this reason one resorts to approximate methods. In order to apply perturbation theory we write the Hamiltonian operator as $\hat{H}_0 + \lambda \hat{H}'$, where, typically, $\hat{H}_0$ is time independent and $\hat{H}'$ may be time dependent. We further factorize the time-evolution operator as

$$\hat{U}(t) = \hat{U}_0(t)\hat{U}_I(t) \tag{1.37}$$

giving rise to the so-called interaction or intermediate picture [5]. The time-evolution operator in the interaction picture $\hat{U}_I$ is unitary and satisfies the differential equation

$$i\hbar \frac{d}{dt} \hat{U}_I = \lambda \hat{H}_I \hat{U}_I, \ \hat{H}_I = \hat{U}_0^\dagger \hat{H}' \hat{U}_0 . \tag{1.38}$$

The usual initial conditions are $\hat{U}_0(0) = \hat{1}$ and $\hat{U}_I(0) = \hat{1}$.

Expanding $\hat{U}_I$ in a Taylor series about $\lambda = 0$

$$\hat{U}_I = \sum_{j=0}^{\infty} \hat{U}_{I,j} \lambda^j \tag{1.39}$$

we obtain a recurrence relation for the coefficients [6]

$$\hat{U}_{I,j}(t) = -\frac{i}{\hbar} \int_0^t \hat{H}_I(t') \hat{U}_{I,j-1}(t') \, dt', \ \hat{U}_{I,0}(t) = \hat{1} . \tag{1.40}$$

Notice that any partial sum of the series (1.39) satisfies the initial condition $\hat{U}_I(0) = \hat{1}$, but it is not unitary.

In some cases we can choose $\hat{U}_0$ in such a way that equations (1.39) and (1.40) provide an approximate expression for $\hat{U}_I$ that may be suitable for the calculation of matrix elements and transition probabilities [6].

In order to illustrate the application of perturbation theory in the interaction picture we concentrate on the approximate calculation of operators in the Heisenberg picture when the Hamiltonian operator $\hat{H} = \hat{H}_0 + \lambda \hat{H}'$ is time independent.

If we expand a given Heisenberg operator $\hat{A}_H$ in a Taylor series about $\lambda = 0$

$$\hat{A}_H = \sum_{j=0}^{\infty} \hat{A}_{H,j} \lambda^j , \tag{1.41}$$

then equation (1.35) with $\hat{H}_H = \hat{H}$ gives us

$$i\hbar \frac{d}{dt} \hat{A}_{H,j} = \left[\hat{A}_{H,j}, \hat{H}_0\right] + \left[\hat{A}_{H,j-1}, \hat{H}'\right], \quad j = 1, 2, \dots . \tag{1.42}$$

We propose a solution to this operator differential equation of the form $\hat{A}_{H,j} = \hat{U}_0^\dagger \hat{B}_j \hat{U}_0$ and derive a differential equation for the time-dependent operator $\hat{B}_j$

$$i\hbar \frac{d\hat{B}_j}{dt} = \hat{U}_0 \left[\hat{A}_{H,j-1}, \hat{H}'\right] \hat{U}_0^\dagger \tag{1.43}$$

which we easily integrate:

$$\hat{B}_j(t) = -\frac{i}{\hbar} \int_0^t \hat{U}_0(t') \left[\hat{A}_{H,j-1}(t'), \hat{H}'\right] \hat{U}_0(t')^\dagger \, dt' . \tag{1.44}$$

Finally, we have

$$\hat{A}_{H,j}(t) = -\frac{i}{\hbar} \int_0^t \hat{U}_0^\dagger(t-t') \left[\hat{A}_{H,j-1}(t'), \hat{H}'\right] \hat{U}_0(t-t') \, dt' \tag{1.45}$$

where $j = 1, 2, \dots$, and $\hat{A}_{H,0} = \hat{U}_0^\dagger \hat{A} \hat{U}_0$.

If we define a dimensionless time variable $s = \omega t$ in terms of a frequency ω, and a dimensionless Hamiltonian operator

$$\hat{\mathcal{H}} = \frac{\hat{H}}{\hbar \omega} , \tag{1.46}$$

then we obtain a dimensionless Schrödinger equation

$$i\frac{d\hat{U}}{ds} = \hat{\mathcal{H}}\hat{U} . \tag{1.47}$$

Notice that we can derive equation (1.47) formally by setting $\hbar = 1$ in equation (1.29).

1.3.2 One-Particle Systems

Most of this book is devoted to one-particle models because they are convenient illustrative examples. More precisely, we consider a particle of mass m under the effect of a conservative force

$\mathbf{F}(\mathbf{r}) = -\nabla V(\mathbf{r})$, where $V(\mathbf{r})$ is a potential-energy function and $\mathbf{r}$ denotes the particle position. The Hamiltonian operator for this simple model reads

$$\hat{H} = \frac{|\hat{\mathbf{p}}|^2}{2m} + V(\mathbf{r}) \,, \tag{1.48}$$

where $\hat{\mathbf{p}}$ and $\hat{\mathbf{r}}$ are vector operators with components $(\hat{p}_x, \hat{p}_y, \hat{p}_z)$ and $(\hat{x}, \hat{y}, \hat{z})$, respectively. They satisfy the well-known commutation relations for coordinates and conjugate momenta

$$\left[\hat{u}, \hat{p}_v\right] = i\hbar\delta_{uv}, \ \left[\hat{u}, \hat{v}\right] = 0, \ \left[\hat{p}_u, \hat{p}_v\right] = 0, \ u, v = x, y, z \,. \tag{1.49}$$

The Hamiltonian operator (1.48) also applies to the relative motion of a pair of particles of masses m_1 and m_2. In this case $m = m_1 m_2/(m_1 + m_2)$ is the reduced mass, $\hat{\mathbf{r}} = \hat{\mathbf{r}}_2 - \hat{\mathbf{r}}_1$ is the relative position and $\hat{\mathbf{p}} = \hat{\mathbf{p}}_2 - \hat{\mathbf{p}}_1$ is the relative momentum.

Because mathematical equations are dimensionless, we believe it is appropriate to remove the dimensions from physical equations. The resulting equations are commonly simpler because they are free from most physical constants and parameters. Moreover, dimensionless equations clearly reveal the relevant parameters of the model. With that purpose in mind, we first define dimensionless coordinate $\mathbf{q} = \mathbf{r}/\gamma$ and momentum $\mathbf{p}' = \gamma\mathbf{p}/\hbar$, where γ is a yet undefined unit of length. The Hamiltonian operator reads

$$\hat{H} = \frac{\hbar^2}{m\gamma^2} \left[\frac{|\hat{\mathbf{p}}'|^2}{2} + v(\mathbf{q}) \right], \ v(\mathbf{q}) = \frac{m\gamma^2}{\hbar^2} V(\gamma\mathbf{q}) \,, \tag{1.50}$$

and we choose γ in such a way that the form of the dimensionless Hamiltonian operator $\hat{\mathcal{H}} = m\gamma^2\hat{H}/\hbar^2$ is as simple as possible.

In the case of the time-dependent Schrödinger equation one also defines a dimensionless time $s = \omega t$, as discussed earlier, and obtains

$$i\frac{d}{ds}\hat{U} = \hat{\mathcal{H}}\hat{U}, \ \hat{\mathcal{H}} = \left[\frac{|\hat{\mathbf{p}}'|^2}{2} + v(\mathbf{q}) \right] \tag{1.51}$$

provided that

$$\gamma^2 = \frac{\hbar}{m\omega} \,. \tag{1.52}$$

We obtain the dimensionless equation by formally setting $\hbar = m = 1$. For brevity we write $\mathbf{p}$ instead of $\mathbf{p}'$ when there is no room for confusion.

1.4   Examples

In what follows we illustrate the application of some of the general results derived above to simple one-dimensional models.

1.4.1   Stationary States of the Anharmonic Oscillator

As a first illustrative example we consider the anharmonic oscillator

$$\hat{H} = \frac{\hat{p}^2}{2m} + \frac{m\omega^2\hat{x}^2}{2} + k'\hat{x}^M, \ M = 4, 6, \ldots \tag{1.53}$$

which in dimensionless form reads

$$\hat{\mathcal{H}} = \frac{1}{2}\left(\hat{p}'^2 + \hat{q}^2\right) + \lambda\hat{q}^M, \ \lambda = \frac{k'\hbar^{M/2-1}}{m^{M/2}\omega^{M/2+1}} \ . \tag{1.54}$$

In particular we choose $M = 4$ and apply perturbation theory with $\hat{\mathcal{H}}_0 = (\hat{p}'^2 + \hat{q}^2)/2$, and $\hat{\mathcal{H}}' = \hat{q}^4$. The unperturbed problem $\hat{\mathcal{H}}_0|n> = (n+1/2)|n>$ is nondegenerate and we easily calculate the matrix elements $\mathcal{H}'_{mn}$ by means of the recurrence relation [7]

$$\left\langle m|\hat{q}^j|n\right\rangle = \sqrt{\frac{n}{2}}\left\langle m\left|\hat{q}^{j-1}\right|n-1\right\rangle + \sqrt{\frac{n+1}{2}}\left\langle m\left|\hat{q}^{j-1}\right|n+1\right\rangle \ . \tag{1.55}$$

Notice that in this case equations (1.10) and (1.12) yield exact analytical results because $< m|\hat{\mathcal{H}}'|n>$ $= 0$ if $|m-n| > 4$; consequently $C_{mn,s} = 0$ if $|m-n| > 4s$.

By means of the Maple procedures given in the program section we derived the results in Table 1.1. Notice that the matrix elements $< \Psi_0|\hat{q}|\Psi_3 >$ and $< \Psi_0|\hat{q}^2|\Psi_4 >$ vanish when $\lambda = 0$ because they are exactly zero for the harmonic oscillator and arise from the perturbation.

1.4.2 Harmonic Oscillator with a Time-Dependent Perturbation

In what follows we illustrate the application of time-dependent perturbation theory to a one-dimensional harmonic oscillator with a simple time-dependent perturbation. In the case of perturbed harmonic oscillators it is commonly convenient to express the dynamical variables in terms of the creation $\hat{a}^\dagger$ and annihilation $\hat{a}$ operators that satisfy the commutation relation $[\hat{a}, \hat{a}^\dagger] = \hat{1}$ (from now on we simply write 1 instead of $\hat{1}$). The model Hamiltonian operator is $\hat{H} = \hat{H}_0 + \lambda\hat{H}'$, where

$$\hat{H}_0 = \hbar\omega_0\left(\hat{a}^\dagger\hat{a} + 1/2\right), \ H' = f(t)\hat{a} + f(t)^*\hat{a}^\dagger \ , \tag{1.56}$$

$f(t)$ is a complex-valued function of time, and $f(t)^*$ its complex conjugate [8]. The dummy perturbation parameter λ is set equal to unity at the end of the calculation.

The dimensionless Schrödinger equation

$$i\frac{d\hat{U}}{ds} = \left[\hat{a}^\dagger\hat{a} + \frac{1}{2} + \frac{f(t)}{\hbar\omega_0}\hat{a} + \frac{f(t)^*}{\hbar\omega}\hat{a}^\dagger\right]\hat{U} \ , \tag{1.57}$$

where $s = \omega_0 t$, clearly reveals that the result will depend on the dimensionless function $f(t)/(\hbar\omega_0)$. In order to facilitate comparison with earlier results, in this case we prefer to work with the original Schrödinger equation.

Taking into account that

$$\hat{U}_0(t)^\dagger\hat{a}\hat{U}_0(t) = \hat{a}\exp\left(-i\omega_0 t\right) \tag{1.58}$$

we obtain [8]

$$\hat{H}_I = g(t)\hat{a} + g(t)^*\hat{a}^\dagger, \ g(t) = f(t)\exp\left(-i\omega_0 t\right) \ . \tag{1.59}$$

It is our purpose to write the perturbation corrections $\hat{H}_{I,j}$ in normal order (powers of $\hat{a}^\dagger$ to the left of powers of $\hat{a}$) because it facilitates the calculation of matrix elements.

According to equation (1.40) the perturbation correction of first order is

$$\hat{U}_{I,1}(t) = \beta_1(t)\hat{a}^\dagger + \beta_2(t)\hat{a} \ , \tag{1.60}$$

Table 1.1   Perturbation Corrections for the Dimensionless Anharmonic Oscillator $\hat{H} = \dfrac{\hat{p}^2 + \hat{q}^2}{2} + \lambda\,\hat{q}^4$

Perturbation corrections to the energy of the nth excited state

$$E_{n,1} = \tfrac{3}{2}\,n^2 + \tfrac{3}{2}\,n + \tfrac{3}{4}$$

$$E_{n,2} = -\tfrac{17}{4}\,n^3 - \tfrac{51}{8}\,n^2 - \tfrac{59}{8}\,n - \tfrac{21}{8}$$

$$E_{n,3} = \tfrac{1041}{16}\,n + \tfrac{177}{2}\,n^2 + \tfrac{375}{8}\,n^3 + \tfrac{333}{16} + \tfrac{375}{16}\,n^4$$

$$E_{n,4} = -\tfrac{111697}{128}\,n - \tfrac{80235}{64}\,n^2 - \tfrac{71305}{64}\,n^3 - \tfrac{30885}{128}$$

$$\qquad\quad -\tfrac{10689}{64}\,n^5 - \tfrac{53445}{128}\,n^4$$

First terms of the perturbation series for the ground state

$$E_0 = \tfrac{1}{2} + \tfrac{3}{4}\lambda - \tfrac{21}{8}\lambda^2 + \tfrac{333}{16}\lambda^3$$

$$\qquad -\tfrac{30885}{128}\lambda^4 + \tfrac{916731}{256}\lambda^5 - \tfrac{65518401}{1024}\lambda^6 + \tfrac{2723294673}{2048}\lambda^7$$

$$\qquad -\tfrac{1030495099053}{32768}\lambda^8 + \cdots$$

Some matrix elements $\langle \Psi_m | \hat{q}^k | \Psi_n \rangle$

$$\langle \Psi_0 | \hat{q} | \Psi_1 \rangle = \tfrac{\sqrt{2}}{2} - \tfrac{3\sqrt{2}}{4}\lambda + \tfrac{189\sqrt{2}}{32}\lambda^2 - \tfrac{4527\sqrt{2}}{64}\lambda^3 + \tfrac{1093701\sqrt{2}}{1024}\lambda^4 + \cdots$$

$$\langle \Psi_0 | \hat{q} | \Psi_3 \rangle = \tfrac{\sqrt{3}}{4}\lambda - \tfrac{39\sqrt{3}}{8}\lambda^2 + \tfrac{14041\sqrt{3}}{128}\lambda^3 - \tfrac{714681\sqrt{3}}{256}\lambda^4 + \cdots$$

$$\langle \Psi_0 | \hat{q}^2 | \Psi_0 \rangle = \tfrac{1}{2} - \tfrac{3}{2}\lambda + \tfrac{105}{8}\lambda^2 - \tfrac{333}{2}\lambda^3 + \tfrac{339735}{128}\lambda^4 + \cdots$$

$$\langle \Psi_0 | \hat{q}^2 | \Psi_2 \rangle = \tfrac{\sqrt{2}}{2} - \tfrac{15\sqrt{2}}{8}\lambda + \tfrac{1233\sqrt{2}}{64}\lambda^2 - \tfrac{68133\sqrt{2}}{256}\lambda^3 + \tfrac{16908219\sqrt{2}}{4096}\lambda^4 + \cdots$$

$$\langle \Psi_0 | \hat{q}^2 | \Psi_4 \rangle = \tfrac{\sqrt{6}}{2}\lambda - \tfrac{55\sqrt{6}}{4}\lambda^2 + \tfrac{6517\sqrt{6}}{16}\lambda^3 - \tfrac{212125\sqrt{6}}{16}\lambda^4 + \cdots$$

---

where

$$\beta_2(t) = -\beta_1(t)^* = -\frac{i}{\hbar}\int_0^t g(u)\,du\ . \tag{1.61}$$

A straightforward calculation shows that the correction of second order is

$$\hat{U}_{I,2}(t) = -|\beta_1|^2\,\hat{a}^\dagger\hat{a} + \beta_3 + \frac{\beta_1^2}{2}\left(\hat{a}^\dagger\right)^2 + \frac{\beta_2^2}{2}\,\hat{a}^2\ , \tag{1.62}$$

where

$$\beta_3(t) = \int_0^t \beta_1(u)\frac{d\beta_2(u)}{du}\,du\ . \tag{1.63}$$

In order to obtain equation (1.62) notice that $\Re(\beta_3) = -|\beta_1|^2/2$, where $\Re(z)$ stands for the real part of the complex number z.

It is not difficult to verify that the time-evolution operator for this simple model is exactly given by [8]

$$\hat{U}_I(t) = \exp\left(\beta_1 \hat{a}^\dagger\right) \exp\left(\beta_2 \hat{a}\right) \exp\left(\beta_3\right) . \tag{1.64}$$

Expanding the exponentials and keeping terms through second order in $\hat{H}'$ we obtain the results given above by perturbation theory.

On calculating the transition probabilities [8]

$$P_{mn} = \left|\left\langle m \left| \hat{U}(t) \right| n \right\rangle\right|^2 = \left|\left\langle m \left| \hat{U}_I(t) \right| n \right\rangle\right|^2 \tag{1.65}$$

by means of the approximate perturbation expression for $\hat{U}_I$, we conclude that at first order $P_{mn} = 0$ if $|m - n| > 1$, at second order $P_{mn} = 0$ if $|m - n| > 2$, and so on. The reader may easily obtain the nonzero transition probabilities in terms of $|\beta_1|$. If we keep only the perturbation correction of first order, we derive the usual approximate selection rule for the harmonic oscillator: $\Delta n = m - n = \pm 1$ [9]. If, on the other hand, we use the exact expression for $\hat{U}_I$, we realize that all the transition probabilities are nonzero [8]. However, at sufficiently short times, perturbation theory gives a reasonable approximation to the dynamics of the problem because the correction of order P is proportional to $|\beta_1|^P$ and $|\beta_1| \to 0$ as $t \to 0$. In order to have a deeper insight into this point we discuss a particular example below.

Consider the periodic interaction given by

$$f(t) = f_0 \cos(\omega t) , \tag{1.66}$$

where $|f_0| \ll \hbar\omega_0$ for a weak interaction. In the case of resonance $\omega = \omega_0$ we have

$$\beta_1(t) = \frac{f_0}{4\hbar\omega_0} \left[1 - 2i\omega_0 t - \exp\left(2i\omega_0 t\right)\right] . \tag{1.67}$$

As expected this result depends only on the dimensionless time variable $s = \omega_0 t$ and the ratio $f_0/(\hbar\omega_0)$. The absolute values of the first and third terms in the right-hand side of this equation are small for all values of t, while the absolute value of the second term increases linearly with time. Perturbation theory will give reasonable results provided that $|\beta_1|$ is sufficiently small; that is to say, when $|t| \ll \hbar/f_0 \ll 1/\omega_0$. In other words, perturbation theory is expected to be valid in a time interval sufficiently smaller than the period of the harmonic oscillator $2\pi/\omega_0$. Under such conditions $P_{02} = |\beta_1|^4/2 \ll P_{01} = |\beta_1|^2$ and the harmonic-oscillator selection rule is approximately valid.

1.4.3 Heisenberg Operators for Anharmonic Oscillators

In what follows we derive approximate expressions for Heisenberg operators in the particular case of anharmonic oscillators (1.53). It is not difficult to verify that the dimensionless time-evolution equation becomes

$$i\frac{\hat{U}}{ds} = \hat{\mathcal{H}}\hat{U} , \tag{1.68}$$

where $\hat{\mathcal{H}}$ is given by equation (1.54).

From now on we simply write $\hat{p}$ instead of $\hat{p}'$ and t instead of s to indicate the dimensionless momentum and time; one must keep in mind that it is necessary to substitute $\hat{x}/\gamma$ for $\hat{q}$, $\gamma\hat{p}/\hbar$ for $\hat{p}$, and ωt for t everywhere in order to recover the original units.

Notice that $\hat{\mathcal{H}}_0 = (\hat{p}^2 + \hat{q}^2)/2$, and $\hat{\mathcal{H}}' = \hat{q}^M$ play the role of $\hat{H}_0$ and $\hat{H}'$, respectively, in the perturbation equations developed earlier in this chapter. It is our purpose to obtain $\hat{q}_H$ for the cubic ($M = 3$) and quartic ($M = 4$) oscillators. In order to apply equation (1.45) recursively one must take into account the well-known canonical transformations

$$\hat{U}_0^\dagger \left(t - t'\right) \hat{q}\hat{U}_0 \left(t - t'\right) = \cos\left(t - t'\right)\hat{q} + \sin\left(t - t'\right)\hat{p} \tag{1.69}$$

$$\hat{U}_0^\dagger \left(t - t'\right) \hat{p}\hat{U}_0 \left(t - t'\right) = \cos\left(t - t'\right)\hat{p} - \sin\left(t - t'\right)\hat{q}. \tag{1.70}$$

Table 1.2 shows results through second order for $M = 3$ and of first order for $M = 4$. The calculation is straightforward but tedious. One carries out the commutators by hand and then uses Maple to calculate the necessary integrals. We should be careful with the order of the coordinate and momentum operators because they do not commute. It is convenient to choose an order for those operators and we have arbitrarily decided to write powers of $\hat{q}$ to the left of powers of $\hat{p}$ following the rule $\hat{p}\hat{q} = -i + \hat{q}\hat{p}$.

Table 1.2   Perturbation Corrections to the Heisenberg Operator $\hat{q}_H$ for Dimensionless Anharmonic Oscillators $\hat{H} = \dfrac{\hat{p}^2 + \hat{q}^2}{2} + \lambda \hat{q}^M$

$$M = 3$$

$$\hat{q}_{H,1} = \left(-2\hat{q}\,\hat{p} + i\right)\sin(t) + \left(\hat{q}\,\hat{p} - \tfrac{i}{2}\right)\sin(2t) + \left(2\hat{p}^2 + \hat{q}^2\right)\cos(t)$$

$$+ \left(-\tfrac{\hat{p}^2}{2} + \tfrac{\hat{q}^2}{2}\right)\cos(2t) - \tfrac{3\hat{p}^2}{2} - \tfrac{3\hat{q}^2}{2}$$

$$\hat{q}_{H,2} = \left(-\tfrac{9i\hat{q}}{16} + \tfrac{9\hat{q}^2\hat{p}}{16} - \tfrac{3\hat{p}^3}{16}\right)\sin(3t)$$

$$+ \left(\tfrac{65\hat{q}^2\hat{p}}{16} + \tfrac{5\hat{p}^3}{16} + \tfrac{15\hat{q}\hat{p}^2 t}{4} - \tfrac{15 i\hat{p}t}{4} + \tfrac{15\hat{q}^3 t}{4} - \tfrac{65 i\hat{q}}{16}\right)\sin(t)$$

$$+ \left(i\hat{q} - \hat{q}^2\hat{p} + 2\hat{p}^3\right)\sin(2t)$$

$$+ \left(-\tfrac{15\hat{p}^3 t}{4} + \tfrac{29\hat{q}^3}{16} - \tfrac{55\hat{q}\hat{p}^2}{16} - \tfrac{15\hat{q}^2\hat{p}t}{4} + \tfrac{55 i\hat{p}}{16} + \tfrac{15 i\hat{q}t}{4}\right)\cos(t)$$

$$+ \left(4\hat{q}\,\hat{p}^2 - 4i\,\hat{p} + \hat{q}^3\right)\cos(2t) + \left(-\tfrac{9\hat{q}\hat{p}^2}{16} + \tfrac{9 i\hat{p}}{16} + \tfrac{3\hat{q}^3}{16}\right)\cos(3t) - 3\hat{q}^3$$

$$M = 4$$

$$\hat{q}_{H,1} = \left(\tfrac{3\hat{q}^2\hat{p}}{8} - \tfrac{3 i\hat{q}}{8} - \tfrac{\hat{p}^3}{8}\right)\sin(3t)$$

$$+ \left(-\tfrac{3\hat{q}^3 t}{2} - \tfrac{21\hat{q}^2\hat{p}}{8} + \tfrac{3 i\hat{p}t}{2} - \tfrac{3\hat{q}\hat{p}^2 t}{2} - \tfrac{9\hat{p}^3}{8} + \tfrac{21 i\hat{q}}{8}\right)\sin(t)$$

$$+ \left(-\tfrac{3 i\hat{p}}{8} + \tfrac{3\hat{p}^3 t}{2} - \tfrac{3 i\hat{q}t}{2} + \tfrac{3\hat{q}^2\hat{p}t}{2} - \tfrac{\hat{q}^3}{8} + \tfrac{3\hat{q}\hat{p}^2}{8}\right)\cos(t)$$

$$+ \left(-\tfrac{3\hat{q}\hat{p}^2}{8} + \tfrac{\hat{q}^3}{8} + \tfrac{3 i\hat{p}}{8}\right)\cos(3t)$$

Chapter 2

Perturbation Theory in the Coordinate Representation

2.1 Introduction

In Chapter 1 we briefly showed how to solve the perturbation equations systematically in the number representation that is suitable for the calculation of the matrix elements necessary for the application of equations (1.10) and (1.12). Alternatively, if we write the Hamiltonian operator in the coordinate representation (substituting $-i\hbar\nabla$ for $\hat{\mathbf{p}}$ or $-i\nabla$ for $\hat{\mathbf{p}}'$), then the perturbation equations (1.6) become differential equations. The unperturbed equation is a solvable eigenvalue problem, and the perturbation corrections are solutions to inhomogeneous differential equations. In this chapter we discuss some widely used strategies for the solution of such equations.

2.2 The Method of Dalgarno and Stewart

Some time ago, Dalgarno and Stewart [10] developed a simple and practical method for the solution of perturbation equations, later adopted by many authors in the treatment of a variety of problems. For simplicity we apply this method to a one-particle model Hamiltonian operator, which in dimensionless form reads

$$\hat{H} = -\frac{1}{2}\nabla^2 + V(\mathbf{r}) \, , \tag{2.1}$$

where ∇^2 is the Laplacian operator and $V(\mathbf{r})$ is a dimensionless potential-energy function. Here $\mathbf{r}$ stands for the dimensionless coordinate introduced in Chapter 1. We assume that we can solve the eigenvalue equation for

$$\hat{H}_0 = -\frac{1}{2}\nabla^2 + V_0(\mathbf{r}) \, , \tag{2.2}$$

where $V_0(\mathbf{r})$ is a properly selected potential-energy function, and choose $\lambda V_1(\mathbf{r}) = V(\mathbf{r}) - V_0(\mathbf{r})$ to be the perturbation. We set the dummy perturbation parameter λ equal to unity at the end of the calculation.

For simplicity, in the following discussion we omit the label that indicates the selected stationary state and simply write Ψ_j and E_j for the perturbation corrections of order j to the eigenfunction and energy, respectively. That is to say, we expand a particular solution of $\hat{H}\Psi = E\Psi$ as

$$\Psi(\mathbf{r}) = \sum_{j=0}^{\infty} \Psi_j(\mathbf{r})\lambda^j, \ E = \sum_{j=0}^{\infty} E_j\lambda^j \ . \tag{2.3}$$

The method of Dalgarno and Stewart [10] consists of writing the perturbation corrections to the eigenfunction as

$$\Psi_j(\mathbf{r}) = F_j(\mathbf{r})\Psi_0(\mathbf{r}), \ j = 0, 1, \dots \tag{2.4}$$

and solving the resulting equations for the functions $F_j(\mathbf{r})$:

$$-\frac{1}{2}\nabla^2 F_j - \frac{1}{\Psi_0}\nabla\Psi_0 \cdot \nabla F_j + V_1 F_{j-1} - \sum_{i=1}^{j} E_i F_{j-i} = 0 \ . \tag{2.5}$$

In this equation ∇ is the gradient vector operator and the dot stands for the standard scalar product. These equations are easier to solve than the original differential equations for the perturbation corrections Ψ_j. In many cases the correction factors F_j are simple polynomial functions of the coordinates. Notice that $F_0 = 1$ is a suitable solution to the equation of order zero, and that E_0 does not appear in the perturbation equations (2.5).

In the following subsections we illustrate the application of the method of Dalgarno and Stewart to simple quantum-mechanical models.

2.2.1 The One-Dimensional Anharmonic Oscillator

As a first example we choose the widely discussed one-dimensional anharmonic oscillator

$$\hat{H} = -\frac{1}{2}\frac{d^2}{dx^2} + \frac{x^2}{2} + \lambda x^4 \tag{2.6}$$

that we split into a dimensionless harmonic oscillator and a quartic perturbation λx^4.

Upon substituting the unperturbed ground state normalized to unity

$$\Psi_0(x) = \pi^{-1/4} \exp\left(-x^2/2\right) \tag{2.7}$$

into equations (2.5) we have

$$-\frac{1}{2}F_j''' + xF_j' + x^4 F_{j-1} - \sum_{i=1}^{j} E_i F_{j-i} = 0 \ . \tag{2.8}$$

Straightforward inspection reveals that the solutions are polynomial functions of the form

$$F_j = \sum_{i=0}^{2j} c_{ji} x^{2i} \ . \tag{2.9}$$

Substitution of equation (2.9) into equation (2.8) for $j = 1$ leads to the polynomial equation

$$(4c_{12} + 1) x^4 + (2c_{11} - 6c_{12}) x^2 - c_{11} - E_1 = 0 \tag{2.10}$$

from which we obtain

$$c_{12} = -\frac{1}{4}, \; c_{11} = -\frac{3}{4}, \; E_1 = \frac{3}{4}. \tag{2.11}$$

The eigenvalue equation does not determine the coefficient c_{10} that we derive from the normalization condition (cf. equation (1.11))

$$\langle \Psi_0 | \Psi_1 \rangle = c_{10} - \frac{9}{16} = 0. \tag{2.12}$$

Finally, we have

$$F_1(x) = \frac{9}{16} - \frac{3x^2}{4} - \frac{x^4}{4}. \tag{2.13}$$

It is worth noticing that we have obtained the energy coefficient E_1 without having recourse to equation (1.14). The reason is that we have tacitly forced the solution to satisfy the boundary condition (that is to say, $F_1(x)\Psi_0(x)$ to be square integrable) and this requirement completely determines the energy.

By means of equations (1.16) and (1.19) we obtain the perturbation corrections of second and third order, respectively:

$$E_2 = \left\langle \Psi_0 | x^4 | \Psi_1 \right\rangle = -\frac{21}{8}, \tag{2.14}$$

$$E_3 = \left\langle \Psi_1 | x^4 - E_1 | \Psi_1 \right\rangle = \frac{333}{16} \tag{2.15}$$

that agree with the results of Table 1.1.

Proceeding along these lines, one easily obtains perturbation corrections of greater order. However, we prefer to illustrate such systematic calculation by means of more interesting, and slightly more complicated, quantum-mechanical models. The simple anharmonic oscillator discussed above serves just as an introductory example.

2.2.2 The Zeeman Effect in Hydrogen

Our second illustrative example is a spinless hydrogen atom in a uniform magnetic field. From a physical point of view this model is certainly more motivating than the anharmonic oscillator and has also been widely discussed in terms of perturbation theory [11, 12]. From a mathematical point of view this problem is more demanding because it is not separable and leads to perturbation equations in two variables. Arbitrarily choosing the z axis along the field the Hamiltonian operator reads

$$\hat{H} = -\frac{\hbar^2}{2m}\nabla^2 - \frac{e^2}{r} + \frac{eB}{2mc}\hat{L}_z + \frac{e^2 B^2}{8mc^2}(x^2 + y^2), \tag{2.16}$$

where m is the atomic reduced mass, e is the electron charge, c is the speed of light, $\hat{L}_z$ is the z-component of the angular-momentum operator, and B is the magnitude of the magnetic induction [13].

If we define units of length $\gamma = \hbar^2/(me^2)$ and energy $\hbar^2/(m\gamma^2) = e^2/\gamma$ we obtain a dimensionless Hamiltonian operator

$$\hat{H} = -\frac{1}{2}\nabla^2 - \frac{1}{r} + \sqrt{2\lambda}\hat{L}_z + \lambda\left(x^2 + y^2\right), \tag{2.17}$$

where $\hat{L}_z$ is given in units of $\hbar$ and $\lambda = B^2\hbar^6/(8m^4c^2e^6)$ is a perturbation parameter. We hope that the use of the same symbols for the original and dimensionless quantities will not be confusing. In order to recover the original units at the end of the calculation, one simply multiplies lengths, energy, linear momenta, angular momenta, and wavefunctions, respectively, by γ, e^2/γ, $\hbar/\gamma$, $\hbar$ and $\gamma^{-3/2}$.

Taking into account that $[\hat{H}, \hat{L}_z] = 0$ we omit the constant of the motion $\hat{L}_z$ in the perturbation calculation and then add the eigenvalue of $\sqrt{2\lambda}\hat{L}_z$ to the resulting energy. In other words, from now on, we consider $\hat{H} - \sqrt{2\lambda}\hat{L}_z$ instead of $\hat{H}$ and write

$$\hat{H} = \hat{H}_0 + \lambda\hat{H}', \quad \hat{H}_0 = -\frac{1}{2}\nabla^2 - \frac{1}{r}, \quad \hat{H}' = x^2 + y^2 . \tag{2.18}$$

We can solve the perturbation equations in several different coordinate systems. Here we choose a kind of modified spherical coordinates in which $u = \cos(\theta)$ takes the place of θ:

$$x = r\sqrt{1-u^2}\cos(\phi), \quad y = r\sqrt{1-u^2}\sin(\phi), \quad z = ru . \tag{2.19}$$

By straightforward application of the general method outlined in Appendix A, we obtain the form of the laplacian ∇^2 in terms of such coordinates, and the perturbation equations for the factor functions $F_j(r, u, \phi)$ become

$$-\frac{1}{2}\frac{\partial^2 F_j}{\partial r^2} - \frac{1}{r}\frac{\partial F_j}{\partial r} + \frac{u^2-1}{2r^2}\frac{\partial^2 F_j}{\partial u^2} + \frac{u}{r^2}\frac{\partial F_j}{\partial u} + \frac{1}{2r^2(u^2-1)}\frac{\partial^2 F_j}{\partial \phi^2}$$

$$-\frac{1}{\Psi_0}\left[\frac{\partial\Psi_0}{\partial r}\frac{\partial F_j}{\partial r} + \frac{1-u^2}{r^2}\frac{\partial\Psi_0}{\partial u}\frac{\partial F_j}{\partial u} + \frac{1}{r^2(1-u^2)}\frac{\partial\Psi_0}{\partial \phi}\frac{\partial F_j}{\partial \phi}\right]$$

$$+ r^2\left(1-u^2\right)F_{j-1} - \sum_{i=1}^{j}E_i F_{j-i} = 0 . \tag{2.20}$$

Notice that $\partial F_j/\partial\phi = 0$ because the perturbation is independent of ϕ, and the state depends on ϕ only through Ψ_0 which is an eigenfunction of $\hat{L}_z$.

In what follows we call ground state the one that correlates with the state $1s$ of the hydrogen atom as $B \to 0$ despite the well-known fact that states with negative values of the magnetic quantum number m may eventually have lower energy for sufficiently great values of B [14]. The dimensionless ground state energy and eigenfunction are

$$E_0 = -\frac{1}{2}, \quad \Psi_0(r) = 2\exp(-r) . \tag{2.21}$$

Because Ψ_0 is independent of u and ϕ, the perturbation equations (2.20) take a simpler form:

$$-\frac{1}{2}\frac{\partial^2 F_j}{\partial r^2} - \frac{1}{r}\frac{\partial F_j}{\partial r} + \frac{u^2-1}{2r^2}\frac{\partial^2 F_j}{\partial u^2} + \frac{u}{r^2}\frac{\partial F_j}{\partial u} + \frac{\partial F_j}{\partial r}$$

$$+ r^2\left(1-u^2\right)F_{j-1} - \sum_{i=1}^{j}E_i F_{j-i} = 0 . \tag{2.22}$$

Proceeding exactly as in the preceding example, at each step $j = 1, 2, \ldots$ we substitute a polynomial solution of the form

$$F_j(r, u) = \sum_{k=0}^{j}u^{2k}\sum_{i=0}^{3j}c_{jik}r^i \tag{2.23}$$

into equations (2.22), and solve for the coefficients c_{jik}. The calculation of the first perturbation corrections is straightforward and can be easily carried out by hand. However, as difficulty increases noticeably with the perturbation order, the use of computer algebra is recommended. In the program section we show a set of simple Maple procedures for the systematic calculation of the perturbation corrections to the ground state. Table 2.1 shows some results.

Table 2.1 Perturbation Corrections to the Ground State of a Hydrogen Atom in a Magnetic Field by Means of the Method of Dalgarno and Stewart

$$E_0 = -\frac{1}{2}$$

$$E_1 = 2$$

$$E_2 = -\frac{53}{3}$$

$$E_3 = \frac{5581}{9}$$

$$E_4 = -\frac{21577397}{540}$$

$$E_5 = \frac{31283298283}{8100}$$

$$E_6 = -\frac{13867513160861}{27000}$$

$$E_7 = \frac{53373333446078164463}{59535000}$$

$$E_8 = -\frac{99586066729159421 1123017}{50009400000}$$

$$F_1 = \frac{11}{3} - \frac{5}{6}r^2 - \frac{1}{3}r^3 + \left(\frac{1}{2}r^2 + \frac{1}{3}r^3\right)u^2$$

$$F_2 = -\frac{1489}{18} + \frac{703}{180}r^2 + \frac{131}{90}r^3 + \frac{11}{10}r^4 + \frac{17}{45}r^5 + \frac{1}{18}r^6$$
$$+ \left(-\frac{83}{60}r^2 - \frac{83}{90}r^3 - \frac{13}{10}r^4 - \frac{3}{5}r^5 - \frac{1}{9}r^6\right)u^2 + \left(\frac{5}{18}r^4 + \frac{2}{9}r^5 + \frac{1}{18}r^6\right)u^4$$

2.3 Logarithmic Perturbation Theory

Logarithmic perturbation theory is an alternative way of solving the perturbation equations in the coordinate representation. It was developed many years ago [15] and has lately been widely discussed and applied to many problems in quantum mechanics. Here we mention just some illustrative examples [16]–[20]. For simplicity, in what follows we restrict ourselves to the one-particle dimensionless Hamiltonian

$$\hat{H} = -\frac{1}{2}\nabla^2 + V_0(\mathbf{r}) + \lambda V_1(\mathbf{r}) . \tag{2.24}$$

Given the ground state Ψ we define the logarithmic derivative

$$\mathbf{f}(\mathbf{r}) = -\nabla\Psi(\mathbf{r})/\Psi(\mathbf{r}) \tag{2.25}$$

that satisfies the Riccati equation

$$\nabla \cdot \mathbf{f} - \mathbf{f} \cdot \mathbf{f} + 2\left(V_0 + \lambda V_1 - E\right) = 0 \,. \tag{2.26}$$

Expanding $\mathbf{f}$ in a Taylor series around $\lambda = 0$

$$\mathbf{f} = \sum_{j=0}^{\infty} \mathbf{f}_j \lambda^j \tag{2.27}$$

we obtain a linear differential equation for the perturbation correction of order j

$$\nabla \cdot \mathbf{f}_j - \sum_{i=0}^{j} \mathbf{f}_i \cdot \mathbf{f}_{j-i} + 2V_0 \delta_{j0} + 2V_1 \delta_{j1} - 2E_j = 0 \,. \tag{2.28}$$

Notice that perturbation theory enables us to transform the nonlinear Riccati equation (2.26) into a set of linear differential equations (2.28).

It is sometimes convenient to write $\Psi(\mathbf{r}) = \exp[-G(\mathbf{r})]$, so that $\mathbf{f} = \nabla G$, and the new function $G(\mathbf{r})$ satisfies

$$\nabla^2 G - \nabla G \cdot \nabla G + 2(V - E) = 0 \,. \tag{2.29}$$

The perturbation equations then read

$$\nabla^2 G_j - \sum_{i=0}^{j} \nabla G_i \cdot \nabla G_{j-i} + 2V_0 \delta_{j0} + 2V_1 \delta_{j1} - 2E_j = 0 \,. \tag{2.30}$$

The equations above are suitable for the nodeless ground state. In order to avoid singularities in the function $G(\mathbf{r})$, we commonly write excited states as $\Psi(\mathbf{r}) = N(\mathbf{r}) \exp[-G(\mathbf{r})]$, where $N(\mathbf{r})$ accounts for the state nodes. In this case we have to solve perturbation equations for both $N(\mathbf{r})$ and $G(\mathbf{r})$ [18]. For simplicity we only consider the ground state here.

The logarithmic perturbation theory is closely related to the method of Dalgarno and Stewart discussed earlier. To realize the connection between both approaches we simply expand the approximate state given by logarithmic perturbation theory

$$\exp\left(-G_0 - \lambda G_1 - \lambda^2 G_2 - \lambda^3 G_3 - \dots\right)$$
$$= \left[1 - \lambda G_1 + \lambda^2 \left(\frac{G_1^2}{2} - G_2\right) + \lambda^3 \left(G_1 G_2 - G_3 - \frac{G_1^3}{6}\right) + \cdots\right] \exp\left(-G_0\right) \tag{2.31}$$

and take into account that $\Psi_0 \propto \exp(-G_0)$. In order to obtain exactly the same results, order by order, by means of both methods, one may have to add appropriate normalization constants $c_1, c_2, \dots$ as follows:

$$F_1 = c_1 - G_1, \quad F_2 = c_2 - G_2 + \frac{(c_1 - G_1)^2}{2}, \dots \,. \tag{2.32}$$

2.3.1 The One-Dimensional Anharmonic Oscillator

As a first simple illustrative example we choose the anharmonic oscillator (2.6). The logarithmic derivative for this one-dimensional problem is a scalar. By inspection of the perturbation equations

$$f_j' - \sum_{i=0}^{j} f_i f_{j-i} + x^2 \delta_{j0} + 2x^4 \delta_{j1} - 2E_j = 0 \tag{2.33}$$

we conclude that

$$f_j = \sum_{i=0}^{j} c_{ji} x^{2i+1} \,. \tag{2.34}$$

For example, an appropriate solution to the unperturbed equation

$$f_0' - f_0^2 + x^2 - 2E_0 = 0 \tag{2.35}$$

is $f_0 = x$, $E_0 = 1/2$. At first order

$$f_1' - 2f_0 f_1 + 2x^4 - 2E_1 = 0 \tag{2.36}$$

we try

$$f_1 = c_{10} x + c_{11} x^3 \tag{2.37}$$

and obtain

$$f_1 = \frac{3x}{2} + x^3, \ E_1 = \frac{3}{4} \,. \tag{2.38}$$

Notice that

$$c_1 - G_1(x) = c_1 - \int f_1(x) \, dx = c_1 - \frac{3}{4}x^2 - \frac{1}{4}x^4 \tag{2.39}$$

agrees with the function $F_1(x)$, equation (2.13), given by the method of Dalgarno and Stewart if $c_1 = 9/16$.

We do not proceed with the discussion of higher perturbation corrections for this simple model and turn our attention to the Zeeman effect in hydrogen.

2.3.2 The Zeeman Effect in Hydrogen

Here we consider the reduced dimensionless Hamiltonian operator (2.18) for a hydrogen atom in a magnetic field. Using the same coordinates chosen earlier for the method of Dalgarno and Stewart we derive the perturbation equations

$$\frac{\partial^2 G_j}{\partial r^2} + \frac{2}{r} \frac{\partial G_j}{\partial r} + \frac{1-u^2}{r^2} \frac{\partial^2 G_j}{\partial u^2} - \frac{2u}{r^2} \frac{\partial G_j}{\partial u}$$

$$- \sum_{i=0}^{j} \left[\frac{\partial G_i}{\partial r} \frac{\partial G_{j-i}}{\partial r} + \frac{1-u^2}{r^2} \frac{\partial G_i}{\partial u} \frac{\partial G_{j-i}}{\partial u} \right]$$

$$- \frac{2\delta_{j0}}{r} + 2r^2 \left(1 - u^2 \right) \delta_{j1} - 2E_j = 0 \,. \tag{2.40}$$

By inspection we conclude that

$$G_j(r, u) = \sum_{i=0}^{j} \sum_{k=0}^{2j+1} c_{jki} r^k u^{2i} \tag{2.41}$$

for the ground state. For simplicity we arbitrarily choose the normalization constants c_{j00} equal to zero.

The algorithm for the calculation of perturbation corrections is so similar to that for the method of Dalgarno and Stewart that we do not think it is necessary to show the Maple program here. Suffice to say that the program for logarithmic perturbation theory runs faster and requires less computer memory, so that we could calculate more perturbation corrections. The first eight perturbation corrections to the energy omitted in Table 2.2 agree with those in Table 2.1. Table 2.2 also shows the normalization constants c_1 and c_2 that enable one to obtain F_1 and F_2 from G_1 and G_2 according to equations (2.32).

2.4   The Method of Fernández and Castro

The method of Dalgarno and Stewart discussed earlier in this chapter is powerful and straightforward. However, the explicit occurrence of the unperturbed eigenfunction $\Psi_0(\mathbf{r})$ in the perturbation equations for the factor functions $F_j(\mathbf{r})$ is a disadvantage as it forces us to treat just one particular state at a time. Consequently, it is not easy to obtain analytical expressions in terms of the unperturbed quantum numbers for all the states simultaneously. One can certainly perform such a calculation for separable problems by means of other methods. Logarithmic perturbation theory exhibits the same limitation, and even the treatment of particular excited states by means of this approach is rather awkward [18].

Fernández and Castro developed an alternative approach that in principle overcomes the above-mentioned limitation retaining the simplicity of the method of Dalgarno and Stewart and logarithmic perturbation theory [21]– [23]. This implementation of perturbation theory is particularly suitable for separable problems to which we restrict here.

We say that an eigenvalue equation is separable when it is possible to split it into a set of one-dimensional differential equations in terms of an appropriate set of coordinates. Typical examples are central-field problems in spherical coordinates, and the hydrogen atom in spherical and parabolic coordinates [24], among others. After separation we are left with one-dimensional equations with additional unknowns called separation constants that play the role of eigenvalues. Such equations are typically of the form

$$P(x)\Phi''(x) + Q(x)\Phi'(x) + R(x)\Phi(x) = 0 \,, \tag{2.42}$$

where we assume that $P(x)$, $Q(x)$, and $R(x)$ are differentiable functions. If we cannot solve equation (2.42) exactly we resort to an approximate method.

In order to apply perturbation theory to equation (2.42), we need a closely related and exactly solvable equation of the form

$$P(x)\Phi_0''(x) + Q(x)\Phi_0'(x) + R_0(x)\Phi_0(x) = 0 \,, \tag{2.43}$$

which we call the unperturbed or reference equation. We then write a solution to equation (2.42) in terms of a solution to equation (2.43) as follows

$$\Phi(x) = A(x)\Phi_0(x) + P(x)B(x)\Phi_0'(x) \,, \tag{2.44}$$

where $A(x)$ and $B(x)$ are two functions to be determined.

In order to obtain the master equation for $A(x)$ and $B(x)$, we first substitute equation (2.44) into equation (2.42) and obtain

$$\left[PA'' + QA' + RA\right]\Phi_0 + \left[2PA' + P(PB)'' + QA + Q(PB)' + PRB\right]\Phi_0'$$
$$+ \left[PA + 2P(PB)' + PQB\right]\Phi_0'' + P^2B\Phi_0''' = 0 \,. \tag{2.45}$$

Table 2.2 Logarithmic Perturbation Theory for the Ground State of Hydrogen in a Magnetic Field *(Continued)*

$$E_9 = \frac{8662946342386597559274204742 3}{15752961000000}$$

$$E_{10} = -\frac{612787354461355179309164710303303 3}{3308121810000000}$$

$$E_{11} = \frac{28609067916890546438864135875927117890 49}{3820880690550000000}$$

$$E_{12} = -\frac{632122787704554404111616455430558208289840606 3}{1765246879034100000000 0}$$

$$E_{13} = \frac{648359227541411978018618130859580430426860226276373 03}{32395222308407458500000000 0}$$

$$E_{14} = -\frac{3396736985021053214938088952478747232823198749628247631265499 3}{2626637019987985142638500000000000}$$

$$E_{15} = \frac{22642641094823076310600000641858301878713803408913224993147208305977287}{23663372913071758150030246500000000000}$$

$E_{16} = -89127611194002329289600505325358500669879758411133183993848 4\backslash$
$02093665288991091 41/1108553298184090039702836951736200000000000 0$

$E_{17} = 1846826604847204664087127274868002680909705982968538931488046\backslash$
$0028549287902636512351170 89/$
$2425403761096970597865836966703631980000000000000 0$

$E_{18} = -4199133085850427980415379969247341465130443502288757572559 15\backslash$
$0063632969281602151622760088675873526 3/$
$5200410071125980731649251405461858880861160000000000000 0$

$E_{19} = 1518092578288527437732059802137479306225379703305765542184136\backslash$
$0394984575893556663362174955165225112073380955 3/$
$1592916807246314653988555601001400146361058474960000000000000 00$

$E_{20} = -2193369107020357669508037479593090426045555878846432000681 80\backslash$
$1369766912678321174612627293587091686383720143432304444677 2\backslash$
$3\big/1761391111773006659620672828032408892364886061804262 9\backslash$
13600000000000000000

$E_{21} = 1837484722422899428119801480748793174932836600343205332902060\backslash$
$2286004788907024859824151462804311508733184520456961407529 8\backslash$
$4068073054 7\big/1025096865177210897975362641400260572550965\backslash$
$7010893314564250070400000000000000000 0$

$E_{22} = -4100286553650066053429462195485176310062414194908562116295 73\backslash$
$7423258085675978113801733166474140226824423168214903106091 9\backslash$
$920597529670139688493 73\big/14488548926335079414974711663 35\backslash$
$28104891803057588902299760892829521913600000000000000000000$

$$G_0 = r$$

$$G_1 = \tfrac{5}{6}r^2 + \tfrac{1}{3}r^3 + \left(-\tfrac{1}{2}r^2 - \tfrac{1}{3}r^3\right)u^2$$

$$G_2 = -\tfrac{1253}{180}r^2 - \tfrac{241}{90}r^3 - \tfrac{271}{360}r^4 - \tfrac{1}{10}r^5 + \left(\tfrac{193}{60}r^2 + \tfrac{193}{90}r^3 + \tfrac{53}{60}r^4 + \tfrac{7}{45}r^5\right)u^2$$
$$+\left(-\tfrac{11}{72}r^4 - \tfrac{1}{18}r^5\right)u^4$$

Table 2.2 *(Cont.)* Logarithmic Perturbation Theory for the Ground State of Hydrogen in a Magnetic Field

$$G_3 = \frac{648977}{2700} r^2 + \frac{4567}{50} r^3 + \frac{530879}{18900} r^4 + \frac{29572}{4725} r^5 + \frac{10757}{11340} r^6 + \frac{1}{14} r^7$$

$$+ \left(-\frac{90877}{900} r^2 - \frac{90877}{1350} r^3 - \frac{94727}{3150} r^4 - \frac{41746}{4725} r^5 - \frac{6347}{3780} r^6 - \frac{289}{1890} r^7 \right) u^2$$

$$+ \left(\frac{803}{180} r^4 + \frac{122}{45} r^5 + \frac{431}{540} r^6 + \frac{1}{10} r^7 \right) u^4 + \left(-\frac{7}{108} r^6 - \frac{1}{54} r^7 \right) u^6$$

$$c_1 = \frac{11}{3}$$

$$c_2 = -\frac{805}{9}$$

By means of equation (2.43) and its first derivative we remove Φ_0'' and Φ_0''' from equation (2.45) to derive an expression only in terms of Φ_0 and Φ_0':

$$\left[PA'' + QA' + (R - R_0) A - 2PR_0B' - (R_0P' + R_0'P)B \right] \Phi_0$$
$$+ P \left[PB'' + (2P' - Q)B' + (P'' - Q')B + (R - R_0) B + 2A' \right] \Phi_0' = 0 . \quad (2.46)$$

In order to solve this equation, it is sufficient to choose the arbitrary functions $A(x)$ and $B(x)$ to satisfy the following set of coupled differential equations of second order

$$\begin{aligned} PA'' + QA' + (R - R_0) A - 2PR_0B' - (R_0P' + R_0'P)B &= 0 \\ PB'' + (2P' - Q) B' + (P'' - Q') B + (R - R_0) B + 2A' &= 0 . \end{aligned} \quad (2.47)$$

To facilitate the application of perturbation theory, we introduce a perturbation parameter λ into the function R in such a way that $R(\lambda, x)$ satisfies $R(0, x) = R_0(x)$ and $R(1, x) = R(x)$. Assuming that we can expand the difference $R - R_0$ in a Taylor series about $\lambda = 0$

$$R(\lambda, x) - R_0(x) = \sum_{j=1}^{\infty} r_j(x)\lambda^j \quad (2.48)$$

we look for a solution to the set of coupled equations (2.47) in the form of λ-power series

$$A(\lambda, x) = \sum_{k=0}^{\infty} A_k(x)\lambda^k, \quad B(\lambda, x) = \sum_{k=0}^{\infty} B_k(x)\lambda^k , \quad (2.49)$$

where $A(0, x) = A_0(x) \equiv 1$ and $B(0, x) = B_0(x) \equiv 0$.

The coefficients $A_k(x)$ and $B_k(x)$ satisfy

$$PA_k'' + QA_k' - 2PR_0B_k' - (R_0P' + R_0'P) B_k + \sum_{j=1}^{k} r_j A_{k-j} = 0$$

$$PB_k'' + (2P' - Q) B_k' + (P'' - Q') B_k + 2A_k' + \sum_{j=1}^{k} r_j B_{k-j} = 0 \quad (2.50)$$

and the perturbation corrections to the eigenfunction

$$\Phi(x) = \sum_{k=0}^{\infty} \Phi_k(x)\lambda^k \quad (2.51)$$

are given by

$$\Phi_k(x) = A_k(x)\Phi_0(x) + B_k(x)\Phi_0'(x) . \tag{2.52}$$

The perturbation equations (2.50) look more complicated than the ones in the method of Dalgarno and Stewart and in the logarithmic perturbation theory discussed above. This is the price we have to pay for the advantage that Φ_0 does not appear explicitly in them. If we are able to solve them, then we obtain general perturbation corrections in terms of the quantum numbers of the unperturbed model as we will shortly see.

2.4.1 The One-Dimensional Anharmonic Oscillator

The time-independent Schrödinger equation for a dimensionless one-dimensional quantum-mechanical model

$$\Phi''(q) + 2[E - V(q)]\Phi(q) = 0 \tag{2.53}$$

is a particular case of (2.42) with $P = 1$, $Q = 0$, and $R = 2(E - V)$. Given a closely related exactly solvable problem of the form

$$\Phi_0''(q) + 2[E_0 - V_0(q)]\Phi_0(q) = 0 , \tag{2.54}$$

the perturbation equations (2.50) become

$$A_k'' + 4\left(V_0 - E_0\right) B_k' + 2V_0' B_k + 2\left(E_1 - V_1\right) A_{k-1} + 2\sum_{j=2}^{k} E_j A_{k-j} = 0$$

$$B_k'' + 2A_k' + 2\left(E_1 - V_1\right) B_{k-1} + 2\sum_{j=2}^{k} E_j B_{k-j} = 0 . \tag{2.55}$$

As a simple nontrivial example we choose the anharmonic oscillator

$$V_0(q) = \frac{q^2}{2}, \quad V_1(q) = q^4 \tag{2.56}$$

for which equations (2.55) become

$$A_k'' + 2\left(q^2 - 2E_0\right) B_k' + 2q B_k + 2\left(E_1 - q^4\right) A_{k-1} + 2\sum_{j=2}^{k} E_j A_{k-j} = 0$$

$$B_k'' + 2A_k' + 2\left(E_1 - q^4\right) B_{k-1} + 2\sum_{j=2}^{k} E_j B_{k-j} = 0 . \tag{2.57}$$

We first note that Φ_0 and Φ have definite parity (either even or odd) because both $V_0(q)$ and $V(q)$ are parity-invariant (even). It therefore follows that $A(q)$ is even and $B(q)$ is odd in order that Φ_0 and Φ have the same parity. By straightforward inspection of equations (2.57) we conclude that

$$A_k(q) = \sum_{i=0}^{\alpha_k} a_{ki} q^{2i}, \ B_k(q) = \sum_{i=0}^{\beta_k} b_{ki} q^{2i+1} , \tag{2.58}$$

where $\alpha_1 = \beta_1 = 1$, $\alpha_k = \beta_{k-1} + 3$, and $\beta_k = \alpha_{k-1} + 1$ for all $k > 1$.

The perturbation equations (2.57) do not determine the coefficients a_{k0} which we obtain by normalization. For simplicity we choose an intermediate normalization such that $a_{k0} = 0$. If necessary we normalize the eigenfunction afterwards. Table 2.3 shows some results in terms of $E_0 = n + 1/2$, where $n = 0, 1, \ldots$ is the harmonic oscillator quantum number. We have obtained many more perturbation corrections by means of the simple program shown in the program section.

Table 2.3 Method of Fernández and Castro for the Anharmonic Oscillator

$$\hat{H} = \frac{\hat{p}^2 + \hat{q}^2}{2} + \lambda \hat{q}^4$$

$$E_1 = \tfrac{3}{2} E_0^2 + \tfrac{3}{8}$$

$$E_2 = -\tfrac{17}{4} E_0^3 - \tfrac{67}{16} E_0$$

$$E_3 = \tfrac{375}{16} E_0^4 + \tfrac{1707}{32} E_0^2 + \tfrac{1539}{256}$$

$$E_4 = -\tfrac{10689}{64} E_0^5 - \tfrac{89165}{128} E_0^3 - \tfrac{305141}{1024} E_0$$

$$E_5 = \tfrac{87549}{64} E_0^6 + \tfrac{587265}{64} E_0^4 + \tfrac{9317949}{1024} E_0^2 + \tfrac{1456569}{2048}$$

$$E_6 = -\tfrac{3132399}{256} E_0^7 - \tfrac{124269873}{1024} E_0^5 - \tfrac{912774217}{4096} E_0^3 - \tfrac{1056412343}{16384} E_0$$

$$E_7 = \tfrac{238225977}{2048} E_0^8 + \tfrac{3294251289}{2048} E_0^6 + \tfrac{78698260599}{16384} E_0^4 + \tfrac{104934994197}{32768} E_0^2$$
$$\quad + \tfrac{106611707169}{524288}$$

$$E_8 = -\tfrac{18945961925}{16384} E_0^9 - \tfrac{349771437429}{16384} E_0^7 - \tfrac{12484329668811}{131072} E_0^5$$
$$\quad - \tfrac{30818393633217}{262144} E_0^3 - \tfrac{114146479775437}{4194304} E_0$$

$$E_9 = \tfrac{194904116847}{16384} E_0^{10} + \tfrac{4646230497315}{16384} E_0^8 + \tfrac{233413756830177}{131072} E_0^6$$
$$\quad + \tfrac{938878529637915}{262144} E_0^4 + \tfrac{7968922195994439}{4194304} E_0^2 + \tfrac{110903991788745}{1048576}$$

$$E_{10} = -\tfrac{8240234242929}{65536} E_0^{11} - \tfrac{988099602430105}{262144} E_0^9 - \tfrac{16709415020420169}{524288} E_0^7$$
$$\quad - \tfrac{201401788392932181}{2097152} E_0^5 - \tfrac{1579588053085388253}{16777216} E_0^3$$
$$\quad - \tfrac{1268701005225618653}{67108864} E_0$$

$$A_1 = -\tfrac{3}{8} q^2$$

$$B_1 = \tfrac{3}{4} E_0 q + \tfrac{1}{4} q^3$$

$$A_2 = \left(\tfrac{43}{32} E_0 - \tfrac{9}{16} E_0^3\right) q^2 + \left(\tfrac{191}{384} - \tfrac{3}{32} E_0^2\right) q^4 + \tfrac{1}{8} E_0 q^6 + \tfrac{1}{32} q^8$$

$$B_2 = \left(-\tfrac{91}{64} - \tfrac{77}{32} E_0^2\right) q - \tfrac{95}{96} E_0 q^3 - \tfrac{5}{24} q^5$$

$$A_3 = \left(-\tfrac{1539}{256} - \tfrac{735}{128} E_0^2 + \tfrac{231}{64} E_0^4\right) q^2 + \left(-\tfrac{131}{32} E_0 + \tfrac{13}{16} E_0^3\right) q^4$$
$$\quad + \left(-\tfrac{331}{768} E_0^2 - \tfrac{3467}{3072}\right) q^6 - \tfrac{79}{384} E_0 q^8 - \tfrac{9}{256} q^{10}$$

$$B_3 = \left(\tfrac{6093}{256} E_0 + \tfrac{1731}{128} E_0^3\right) q + \left(\tfrac{2143}{512} + \tfrac{1507}{256} E_0^2 - \tfrac{9}{64} E_0^4\right) q^3$$
$$\quad + \left(-\tfrac{9}{128} E_0^3 + \tfrac{911}{512} E_0\right) q^5 + \left(\tfrac{397}{1536} + \tfrac{3}{128} E_0^2\right) q^7 + \tfrac{7}{384} E_0 q^9 + \tfrac{1}{384} q^{11}$$

In addition to giving general expressions for the eigenfunctions and eigenvalues in terms of the quantum numbers of the unperturbed problem, the method of Fernández and Castro is faster and requires less computer memory than the method of Dalgarno and Stewart and the logarithmic perturbation theory. The reason is that the polynomial functions $A_k(x)$ and $B_k(x)$ are remarkably simple because $\Phi_0(x)$ and $\Phi_0'(x)$ carry out most of the details of the perturbed eigenfunction $\Phi(x)$ (such as the oscillation in the classical region). The disadvantage of the method of Fernández and Castro is that its application to nonseparable models is not so obvious.

The method of Fernández and Castro has already been applied to several separable quantum-mechanical models [21]–[23]. We will mention some of them later in this book.

Chapter 3

Perturbation Theories without Wavefunction

3.1 Introduction

In the preceding chapters we solved the time-independent Schrödinger equation approximately by calculation of perturbation corrections to both the eigenvalue and eigenfunction at each perturbation order. Typically, the calculation of the perturbation corrections to the eigenfunction is more tedious, lengthy, and time consuming. For this reason, if one is primarily interested in the energy, then it will be desirable to bypass the explicit treatment of the eigenfunction. In what follows we discuss some well-known strategies for this purpose. Two approaches simply substitute expectation values and moments for the eigenfunction, whereas the third approach is a true perturbation theory without wavefunction as it only takes the Hamiltonian operator into consideration.

3.2 Hypervirial and Hellmann–Feynman Theorems

An hermitian operator $\hat{A}$ satisfies

$$\left\langle \Psi | \hat{A} | \Phi \right\rangle = \left\langle \hat{A} \Psi | \Phi \right\rangle \tag{3.1}$$

for any pair Ψ, Φ of vectors of the state space. In particular, the Hamiltonian operator satisfies equation (3.1), and in the case that Ψ is an eigenfunction of $\hat{H}$ with eigenvalue E, $\hat{H}\Psi = E\Psi$, we have

$$\left\langle \Psi \left| \left(\hat{H} - E \right) \right| \Phi \right\rangle = \left\langle \left(\hat{H} - E \right) \Psi | \Phi \right\rangle = 0 \,. \tag{3.2}$$

If $\hat{W}$ is an arbitrary linear operator such that $\Phi = \hat{W}\Psi$ then $(\hat{H} - E)\Phi = \hat{W}(H - E)\Psi + [\hat{H}, \hat{W}]\Psi$, and equation (3.2) leads to the hypervirial theorem [25]

$$\left\langle \Psi \left| \left[\hat{H}, \hat{W} \right] \right| \Psi \right\rangle = 0 \,. \tag{3.3}$$

If Ψ is normalized to unity $< \Psi | \Psi >\, = 1$ we simply write

$$\left\langle \left[\hat{H}, \hat{W} \right] \right\rangle = 0 \tag{3.4}$$

where $< \ldots >$ denotes the quantum-mechanical expectation value.

If the Hamiltonian operator depends on a parameter λ, which may be a particle charge or mass, a field strength, or simply a dummy parameter artificially introduced to apply perturbation theory, then the eigenvalues and eigenfunctions of $\hat{H}$ will also depend on that parameter. If we differentiate $< \Psi|(\hat{H} - E)|\Psi >= 0$ with respect to λ

$$\left\langle \frac{\partial \Psi}{\partial \lambda} \left| \left(\hat{H} - E \right) \right| \Psi \right\rangle + \left\langle \Psi \left| \frac{\partial (\hat{H} - E)}{\partial \lambda} \right| \Psi \right\rangle + \left\langle \Psi \left| \left(\hat{H} - E \right) \right| \frac{\partial \Psi}{\partial \lambda} \right\rangle = 0 \qquad (3.5)$$

and take into account equation (3.2), then we obtain the Hellmann–Feynman theorem [25]

$$\frac{\partial E}{\partial \lambda} < \Psi|\Psi >= \left\langle \Psi \left| \frac{\partial \hat{H}}{\partial \lambda} \right| \Psi \right\rangle , \qquad (3.6)$$

or

$$\frac{\partial E}{\partial \lambda} = \left\langle \frac{\partial \hat{H}}{\partial \lambda} \right\rangle \qquad (3.7)$$

if Ψ is normalized to unity.

In what follows we show that the general results just derived facilitate the application of perturbation theory.

3.3 The Method of Swenson and Danforth

The first method that we discuss here was developed long ago [26], but it had to wait until its main equations were rewritten in a simpler way [27] to become popular [28, 29]. The approach is based on the combination of the hypervirial and Hellmann–Feynman theorems with perturbation theory, and for this reason we will also call it hypervirial-Hellmann–Feynman method or hypervirial perturbative method [28].

The first step is the derivation of a recurrence relation for the expectation values of properly chosen functions of the coordinate by means of the hypervirial theorem. Such recurrence relations have only been obtained for separable quantum-mechanical models as shown in the examples below.

3.3.1 One-Dimensional Models

We illustrate the main ideas behind the method of Swenson and Danforth by means of a one-dimensional model

$$\hat{H} = \frac{\hat{p}^2}{2m} + \hat{V}, \ \hat{V} = V\left(\hat{x}\right) , \qquad (3.8)$$

where $[\hat{x}, \hat{p}] = i\hbar$.

First, rewrite the commutator $[\hat{H}, \hat{W}]$ for the linear operator

$$\hat{W} = W\left(\hat{x}, \hat{p}\right) = \hat{f}\hat{p} + \hat{g} , \qquad (3.9)$$

where $\hat{f} = f(\hat{x})$, and $\hat{g} = g(\hat{x})$, in the following way:

$$\begin{aligned}
\left[\hat{H}, \hat{W}\right] &= \frac{1}{m}\left[\hat{p}, \hat{f}\right]\hat{p}^2 + \frac{1}{m}\left(\frac{1}{2}\left[\hat{p}, \left[\hat{p}, \hat{f}\right]\right] + \left[\hat{p}, \hat{g}\right]\right)\hat{p} \\
&\quad + \frac{1}{2m}\left[\hat{p}, \left[\hat{p}, \hat{g}\right]\right] + \hat{f}\left[\hat{V}, \hat{p}\right] .
\end{aligned} \qquad (3.10)$$

Second, choose the arbitrary functions $f(\hat{x})$ and $g(\hat{x})$ so that the coefficient of $\hat{p}$ in equation (3.10) vanishes:

$$\hat{g} = \frac{1}{2}\left[\hat{f}, \hat{p}\right] , \qquad (3.11)$$

and write $\hat{p}^2$ in terms of $\hat{H}$ and $\hat{V}$. Equation (3.10) becomes

$$\left[\hat{H}, \hat{W}\right] = 2\left[\hat{p}, \hat{f}\right]\left(\hat{H} - \hat{V}\right) + \frac{1}{4m}\left[\hat{p}, \left[\hat{p}, \left[\hat{f}, \hat{p}\right]\right]\right] + \hat{f}\left[\hat{V}, \hat{p}\right] . \qquad (3.12)$$

Finally, the hypervirial theorem (3.4) gives us an expression for expectation values of operators that commute with the coordinate operator:

$$2E\left\langle\left[\hat{p}, \hat{f}\right]\right\rangle + \frac{1}{4m}\left\langle\left[\hat{p}, \left[\hat{p}, \left[\hat{f}, \hat{p}\right]\right]\right]\right\rangle + \left\langle\hat{f}\left[\hat{V}, \hat{p}\right]\right\rangle + 2\left\langle\left[\hat{f}, \hat{p}\right]\hat{V}\right\rangle = 0 . \qquad (3.13)$$

Notice, for example, that $[\hat{p}, \hat{f}]$ is a function only of the coordinate operator as it follows from the Jacobi identity

$$\left[\hat{p}, \left[\hat{x}, \hat{f}\right]\right] + \left[\hat{f}, \left[\hat{p}, \hat{x}\right]\right] + \left[\hat{x}, \left[\hat{f}, \hat{p}\right]\right] = \left[\hat{x}, \left[\hat{f}, \hat{p}\right]\right] = 0 . \qquad (3.14)$$

Equation (3.13) reduces to the well-known virial theorem for the particular case $\hat{f} = \hat{x}$:

$$2E - 2\left\langle\hat{V}\right\rangle + \frac{i}{\hbar}\left\langle\hat{x}\left[\hat{V}, \hat{p}\right]\right\rangle = 0 . \qquad (3.15)$$

In the coordinate representation $\hat{p} = -i\hbar d/dx$, and equation (3.13) reduces to

$$2E\left\langle f'\right\rangle + \frac{\hbar^2}{4m}\left\langle f'''\right\rangle - \left\langle fV'\right\rangle - 2\left\langle f'V\right\rangle = 0 , \qquad (3.16)$$

where the prime denotes the derivative with respect to x. To simplify the notation we drop the caret on functions of the coordinate operator reflecting the fact that $\hat{x}\Psi(x) = x\Psi(x)$. Although it is simpler to derive equation (3.16) directly in the coordinate representation, we have followed a more lengthy way with the purpose of obtaining the main equation (3.13) independent of any particular representation.

Equation (3.16) enables one to obtain expectation values in terms of the energy of a stationary state of a given exactly solvable model. As an example consider the dimensionless harmonic oscillator

$$\hat{H} = \frac{1}{2}\left(-\frac{d^2}{dx^2} + x^2\right) . \qquad (3.17)$$

Choosing $f(x) = x^{2j+1}$, $j = 0, 1, \ldots$, equation (3.16) with $\hbar = m = 1$ becomes a three-term recurrence relation that we conveniently rewrite as

$$X_{j+1} = \frac{1}{2(j+1)}\left[2(2j+1)EX_j + \frac{j}{2}\left(4j^2 - 1\right)X_{j-1}\right] , \qquad (3.18)$$

where

$$X_j = \left\langle x^{2j}\right\rangle . \qquad (3.19)$$

Notice that we do not consider the expectation values $< x^{2j+1} >$ that vanish because the eigenfunctions are either even or odd. We obtain all the expectation values X_j, $j = 1, 2, \ldots$ by simply setting $j = 0, 1, \ldots$ in equation (3.18) and taking into account that $X_0 = 1$. The expectation value X_j is a polynomial function of E of degree j, where $E = n + 1/2$, and $n = 0, 1, \ldots$ is the quantum number.

If the quantum-mechanical model is not exactly solvable we apply perturbation theory to the recurrence relation for the expectation values. To illustrate this approach we consider the simple anharmonic oscillator

$$\hat{H} = \frac{1}{2}\left(-\frac{d^2}{dx^2} + x^2\right) + \lambda x^{2K} , \tag{3.20}$$

where $K = 2, 3, \ldots$ and λ is a perturbation parameter. The recurrence relation for the expectation values becomes

$$2(2j + 1)EX_j + \frac{j}{2}\left(4j^2 - 1\right)X_{j-1} - 2(j + 1)X_{j+1} - 2\lambda(2j + K + 1)X_{j+K} = 0 . \tag{3.21}$$

If we expand the expectation values and the energy in a Taylor series around $\lambda = 0$

$$E = \sum_{i=0}^{\infty} E_i\lambda^i, \; X_j = \sum_{i=0}^{\infty} X_{j,i}\lambda^i , \tag{3.22}$$

then equation (3.21) gives us a recurrence relation for the coefficients E_i and $X_{j,i}$. Substituting $j - 1$ for j it reads

$$X_{j,i} = \frac{1}{2j}\left\{\frac{j-1}{2}[4(j-1)^2 - 1]X_{j-2,i} + 2(2j-1)\sum_{m=0}^{i} E_m X_{j-1,i-m}\right.$$

$$\left. - 2(2j + K - 1)X_{j+K-1,i-1}\right\} , \tag{3.23}$$

where

$$X_{0,i} = \delta_{0i} \tag{3.24}$$

as follows from the normalization condition $X_0 = 1$.

The recurrence relation (3.23) gives us the perturbation corrections $X_{j,i}$ in terms of the yet unknown energy coefficients E_i. The Hellmann–Feynman theorem

$$\frac{\partial E}{\partial \lambda} = X_K \tag{3.25}$$

provides an additional equation that enables us to solve the problem completely. Expanding equation (3.25) in a Taylor series about $\lambda = 0$ we have

$$E_i = \frac{1}{i}X_{K,i-1} \tag{3.26}$$

for all $i > 0$. By straightforward inspection of the recurrence relation (3.23) one easily convinces oneself that it is sufficient to calculate the coefficients $X_{j,i}$, for all $i = 1, 2, \ldots, p - 1$ and $j = 1, 2, \ldots, (p - i)(K - 1) + 1$ in order to obtain E_p.

Table 3.1   Method of Swenson and Danforth for the Anharmonic Oscillator

$$\hat{H} = -\frac{1}{2}\frac{d^2}{dx^2} + \frac{\hat{x}^2}{2} + \lambda\hat{x}^4$$

$X_{1,0} = E_0$

$X_{1,1} = -\frac{3}{4} - 3\,E_0{}^2$

$X_{1,2} = \frac{335}{16}\,E_0 + \frac{85}{4}\,E_0{}^3$

$X_{1,3} = -\frac{1539}{32} - \frac{1707}{4}\,E_0{}^2 - \frac{375}{2}\,E_0{}^4$

$X_{1,4} = \frac{3356551}{1024}\,E_0 + \frac{980815}{128}\,E_0{}^3 + \frac{117579}{64}\,E_0{}^5$

$X_{1,5} = -\frac{10195983}{1024} - \frac{65225643}{512}\,E_0{}^2 - \frac{4110855}{32}\,E_0{}^4 - \frac{612843}{32}\,E_0{}^6$

$X_{1,6} = \frac{17959009831}{16384}\,E_0 + \frac{15517161689}{4096}\,E_0{}^3 + \frac{2112587841}{1024}\,E_0{}^5 + \frac{53250783}{256}\,E_0{}^7$

$X_{1,7} = -\frac{533058535845}{131072} - \frac{524674970985}{8192}\,E_0{}^2 - \frac{393491302995}{4096}\,E_0{}^4$

$\qquad - \frac{16471256445}{512}\,E_0{}^6 - \frac{1191129885}{512}\,E_0{}^8$

$X_{1,8} = \frac{2625369034835051}{4194304}\,E_0 + \frac{708823053563991}{262144}\,E_0{}^3 + \frac{287139582382653}{131072}\,E_0{}^5$

$\qquad + \frac{8044743060867}{16384}\,E_0{}^7 + \frac{435757124275}{16384}\,E_0{}^9$

$X_{1,9} = -\frac{1441751893253685}{524288} - \frac{103595988547927707}{2097152}\,E_0{}^2 - \frac{12205420885292895}{131072}\,E_0{}^4$

$\qquad - \frac{3034378838792301}{65536}\,E_0{}^6 - \frac{60400996465095}{8192}\,E_0{}^8 - \frac{2533753519011}{8192}\,E_0{}^{10}$

Note: The energy coefficients are identical to those in Table 2.3.

In the program section we show a set of simple Maple procedures that facilitates the systematic calculation of the perturbation corrections for the anharmonic oscillator (3.20). Table 3.1 shows perturbation corrections E_i and $X_{1,i}$ for $K = 4$. The former agree with our previous calculation in Chapter 2. From the perturbation corrections to the energy and equation (3.26), one easily obtains the coefficients $X_{2,i}$.

The method of Swenson and Danforth also applies to potentials that are not parity invariant. Consider, for example, the cubic-quartic anharmonic oscillator

$$\hat{H} = \frac{1}{2}\left(-\frac{d^2}{dx^2} + x^2\right) + \lambda\left(\alpha x^3 + x^4\right)\,. \tag{3.27}$$

In this case we choose $f(x) = x^j$, $j = 0, 1, \ldots$, and define

$$X_j = \left\langle x^j \right\rangle\,. \tag{3.28}$$

Repeating the algebraic steps in the discussion of the preceding example we obtain the recurrence relation

$$X_{j,i} = \frac{1}{j}\left[\frac{(j-1)(j-2)(j-3)}{4}X_{j-4,i} + 2(j-1)\sum_{m=0}^{i}E_m X_{j-2,i-m}\right.$$

$$\left. - \alpha(2j+1)X_{j+1,i-1} - 2(j+1)X_{j+2,i-1}\right] \tag{3.29}$$

from the hypervirial theorem, and

$$E_i = \frac{1}{i} \left(\alpha X_{3,i-1} + X_{4,i-1} \right) \tag{3.30}$$

from the Hellmann–Feynman theorem. The initial conditions are also given by equation (3.24). In order to obtain E_p we need the perturbation coefficients $X_{j,i}$ for all $i = 0, 1, \ldots, p - 1$, and $j = 1, 2, \ldots, 2(p - i + 1)$.

Table 3.2 shows perturbation corrections to the energy and some moments. Notice that when $\alpha = 0$ the energy coefficients reduce to those in Table 2.3, and that all the $X_{1,i}$ vanish because the resulting potential-energy function is parity-invariant.

3.3.2 Central-Field Models

In Chapters 1 and 2 we briefly outlined the Hamiltonian operator for one- and two-particle systems. In this chapter we specialize in central conservative forces that are given by the gradient of a potential-energy function $V(r)$, where $r = |\mathbf{r}|$ is the distance between the particles. It is well known that in such a case the Schrödinger equation is separable in spherical coordinates r, θ, ϕ [30]. We can therefore write the eigenfunctions of the Hamiltonian operator as $\Psi_{nlm}(r, \theta, \phi) = R_{nl}(r)Y_{lm}(\theta, \phi)$, where $n = 1, 2, \ldots, l = 0, 1, \ldots$, and $m = -l, -l + 1, \ldots, l$ are three quantum numbers that completely specify the state, and Y_{lm} are the spherical harmonics [30, 31]. From now on we consider the dimensionless form of the Schrödinger equation derived in Chapter 1. It only remains to obtain the radial factor $R(r)$ of the eigenfunction. For convenience we consider $\Phi(r) = r R(r)$ that is an eigenfunction of the one-dimensional-like Hamiltonian operator [30]

$$\hat{H} = -\frac{1}{2}\frac{d^2}{dr^2} + \frac{l(l+1)}{2r^2} + V(r) . \tag{3.31}$$

The bound states satisfy the boundary conditions

$$\Phi(0) = 0, \ \lim_{r \to \infty} \Phi(r) = 0 . \tag{3.32}$$

If we naively follow the steps of the discussion above regarding the hypervirial theorem for actual one-dimensional models, and choose the function $f(r) = r^j$, we obtain

$$2jE\left\langle r^{j-1} \right\rangle + (j - 1)\left[\frac{j(j-2)}{4} - l(l+1) \right]\left\langle r^{j-3} \right\rangle$$
$$- \left\langle r^j V' \right\rangle - 2j\left\langle r^{j-1}V \right\rangle = 0 . \tag{3.33}$$

However, this equation is not valid for all values of j because of the boundary condition at $r = 0$. Although the form of the hypervirial theorem under arbitrary boundary conditions is well known [32], we briefly address this point here for completeness.

Consider the scalar product

$$< \Phi | \chi > = \int_0^\infty \Phi(r)^* \chi(r)dr , \tag{3.34}$$

where $\Phi(0) = 0$. The radial Hamiltonian (3.31) satisfies

$$\left\langle \Phi \left| \hat{H} \right| \chi \right\rangle = \left\langle \hat{H}\Phi | \chi \right\rangle - \frac{1}{2}\Phi'(0)^* \chi(0) , \tag{3.35}$$

Table 3.2 Method of Swenson and Danforth for the Anharmonic Oscillator

$$\hat{H} = -\frac{1}{2}\frac{d^2}{dx^2} + \frac{\hat{x}^2}{2} + \lambda\left(\alpha\hat{x}^3 + \hat{x}^4\right)$$

$$E_1 = \frac{3}{8} + \frac{3}{2}E_0{}^2$$

$$E_2 = \left(-\frac{15}{4}E_0{}^2 - \frac{7}{16}\right)\alpha^2 - \frac{67}{16}E_0 - \frac{17}{4}E_0{}^3$$

$$E_3 = \left(\frac{225}{4}E_0{}^3 + \frac{459}{16}E_0\right)\alpha^2 + \frac{1539}{256} + \frac{1707}{32}E_0{}^2 + \frac{375}{16}E_0{}^4$$

$$E_4 = \left(-\frac{705}{16}E_0{}^3 - \frac{1155}{64}E_0\right)\alpha^4 + \left(-\frac{40261}{512} - \frac{62013}{64}E_0{}^2 - \frac{24945}{32}E_0{}^4\right)\alpha^2$$
$$- \frac{305141}{1024}E_0 - \frac{89165}{128}E_0{}^3 - \frac{10689}{64}E_0{}^5$$

$$E_5 = \left(\frac{131817}{1024} + \frac{116325}{64}E_0{}^4 + \frac{239985}{128}E_0{}^2\right)\alpha^4$$
$$+ \left(\frac{338625}{32}E_0{}^5 + \frac{1597845}{64}E_0{}^3 + \frac{3909285}{512}E_0\right)\alpha^2 + \frac{1456569}{2048} + \frac{9317949}{1024}E_0{}^2$$
$$+ \frac{587265}{64}E_0{}^4 + \frac{87549}{64}E_0{}^6$$

$$E_6 = \left(-\frac{101479}{2048} - \frac{115755}{128}E_0{}^4 - \frac{209055}{256}E_0{}^2\right)\alpha^6$$
$$+ \left(-\frac{53000175}{2048}E_0 - \frac{25428615}{256}E_0{}^3 - \frac{6383475}{128}E_0{}^5\right)\alpha^4$$
$$+ \left(-\frac{70997745}{128}E_0{}^4 - \frac{18237765}{128}E_0{}^6 - \frac{104283313}{4096} - \frac{809619141}{2048}E_0{}^2\right)\alpha^2$$
$$- \frac{1056412343}{16384}E_0 - \frac{912774217}{4096}E_0{}^3 - \frac{124269873}{1024}E_0{}^5 - \frac{3132399}{256}E_0{}^7$$

$$X_{1,0} = 0$$

$$X_{1,1} = -3\,\alpha\,E_0$$

$$X_{1,2} = \left(\frac{23}{4} + 39\,E_0{}^2\right)\alpha$$

$$X_{1,3} = \left(-45\,E_0{}^2 - \frac{21}{4}\right)\alpha^3 + \left(-\frac{4677}{16}E_0 - \frac{2055}{4}E_0{}^3\right)\alpha$$

$$X_{1,4} = \left(\frac{7545}{4}E_0{}^3 + \frac{14259}{16}E_0\right)\alpha^3 + \left(\frac{49495}{64} + \frac{72255}{8}E_0{}^2 + \frac{27195}{4}E_0{}^4\right)\alpha$$

$$X_{1,5} = \left(-\frac{19035}{16}E_0{}^3 - \frac{31185}{64}E_0\right)\alpha^5 + \left(-\frac{839475}{16}E_0{}^4 - \frac{1131183}{256} - \frac{1890135}{32}E_0{}^2\right)\alpha^3$$
$$+ \left(-\frac{72618213}{1024}E_0 - \frac{28507965}{128}E_0{}^3 - \frac{5770737}{64}E_0{}^5\right)\alpha$$

$$X_{2,0} = E_0$$

$$X_{2,1} = -\frac{3}{4} - 3\,E_0{}^2$$

$$X_{2,2} = \left(15\,E_0{}^2 + \frac{7}{4}\right)\alpha^2 + \frac{335}{16}E_0 + \frac{85}{4}E_0{}^3$$

$$X_{2,3} = \left(-\frac{1575}{4}E_0{}^3 - \frac{3213}{16}E_0\right)\alpha^2 - \frac{1539}{32} - \frac{1707}{4}E_0{}^2 - \frac{375}{2}E_0{}^4$$

$$X_{2,4} = \left(\frac{6345}{16}E_0{}^3 + \frac{10395}{64}E_0\right)\alpha^4 + \left(\frac{124725}{16}E_0{}^4 + \frac{201305}{256} + \frac{310065}{32}E_0{}^2\right)\alpha^2$$
$$+ \frac{3356551}{1024}E_0 + \frac{980815}{128}E_0{}^3 + \frac{117579}{64}E_0{}^5$$

$$X_{2,5} = \left(-\frac{348975}{16}E_0{}^4 - \frac{719955}{32}E_0{}^2 - \frac{395451}{256}\right)\alpha^4$$
$$+ \left(-\frac{50820705}{512}E_0 - \frac{20771985}{64}E_0{}^3 - \frac{4402125}{32}E_0{}^5\right)\alpha^2 - \frac{10195983}{1024}$$
$$- \frac{65225643}{512}E_0{}^2 - \frac{4110855}{32}E_0{}^4 - \frac{612843}{32}E_0{}^6$$

as follows from integration by parts. For the particular case $\chi = \hat{D}\Phi$, $\hat{D} = d/dr$, we have

$$\left\langle \left[\hat{H}, \hat{D} \right] \right\rangle = l(l+1)\left\langle r^{-3} \right\rangle - \left\langle V' \right\rangle = -\frac{1}{2}\left| \Phi'(0) \right|^2 . \tag{3.36}$$

If

$$\lim_{r \to 0} r^2 V(r) = 0 , \tag{3.37}$$

then the eigenfunctions of $\hat{H}$ behave as Cr^{l+1} close to $r = 0$, where C is a nonzero constant. We clearly see that the master equation (3.33) does not hold when $j = 0$ for those states with $l = 0$ because $\Phi'(0) \neq 0$. This apparent complication does not take place when $l > 0$.

Another important point to notice is that the integrand in the expectation value

$$\left\langle r^j \right\rangle = \int_0^\infty |\Phi(r)|^2 r^j dr \tag{3.38}$$

behaves as $|C|^2 r^{2l+j+2}$ close to origin. For this reason the expectation value (3.38) does not exist if $j < -2l - 2$. It is most important to take into account this fact in the case of perturbations that are singular at origin.

In what follows we consider two exactly solvable models as illustrative examples. The master equation (3.33) for the harmonic oscillator in three dimensions

$$V(r) = \frac{r^2}{2} \tag{3.39}$$

becomes

$$2jER_{j-1} + (j-1)\left[\frac{j(j-2)}{4} - l(l+1) \right] R_{j-3} - (j+1)R_{j+1} = 0 \tag{3.40}$$

for all $j > 0$, where

$$R_j = \left\langle r^j \right\rangle . \tag{3.41}$$

According to equation (3.36), for $j = 0$ we have

$$l(l+1)R_{-3} - R_1 = -\frac{1}{2}\left| \Phi'(0) \right|^2 , \tag{3.42}$$

and for $j = 2$ equation (3.40) reduces to

$$4ER_1 - l(l+1)R_{-1} - 3R_3 = 0 . \tag{3.43}$$

It is clear that we cannot obtain the expectation values of odd powers (either positive or negative) of the radial coordinate, unless we have additional information about the solutions of the Schrödinger equation. On the other hand, equation (3.40) is sufficient for the calculation of R_{2j} for all $j > 0$ in terms of the energy; for example:

$$R_2 = E, \quad R_4 = \frac{3}{8} - \frac{l(l+1)}{2} + \frac{3E^2}{2}, \dots . \tag{3.44}$$

Substituting the energy eigenvalues [33]

$$E = 2v + l + \frac{3}{2}, \quad v = 0, 1, \dots \tag{3.45}$$

into the Hellmann–Feynman theorem

$$\frac{\partial E}{\partial l} = \left(l + \frac{1}{2}\right) R_{-2} \tag{3.46}$$

we obtain

$$R_{-2} = \frac{2}{2l + 1} . \tag{3.47}$$

Taking into account this expression and the master equation (3.40), we easily calculate the expectation values R_{2j} for all $-(l + 1) \le j < 0$.

We are entitled to apply the Hellmann–Feynman theorem as in equation (3.46) because the expression (3.45) for the energy is also valid for noninteger values of l [33]. We substitute the actual value of the angular momentum quantum number l into the resulting expression.

It follows from the master equation (3.40) with $j = -1$ that

$$R_{-4} = \frac{8E}{(2l + 1)[4l(l + 1) - 3]} . \tag{3.48}$$

Since $R_j > 0$ for all j we clearly see that this expression is not valid when $l = 0$ in agreement with the discussion above. Analogously, it follows from equation (3.40) with $j = -3$ that

$$R_{-6} = \frac{8[24\nu^2 + 24\nu l + 36\nu + 4l^2 + 16l + 15]}{(2l + 1)[4l(l + 1) - 3][4l(l + 1) - 15]} \tag{3.49}$$

provided that $l \ge 2$.

We now consider anharmonic oscillators with potential-energy functions

$$V(r) = \frac{r^2}{2} + \lambda r^{2K}, \quad K = 2, 3, \ldots \tag{3.50}$$

as illustrative examples. In order to apply the method of Swenson and Danforth, we take into account the perturbation series for the expectation values

$$R_j = \left\langle r^{2j} \right\rangle = \sum_{i=0}^{\infty} R_{j,i} \lambda^i \tag{3.51}$$

and proceed as in the one-dimensional case obtaining the recurrence relation

$$R_{j,i} = \frac{1}{j} \left\{ (j - 1) \left[\frac{(2j - 1)(2j - 3)}{4} - l(l + 1) \right] R_{j-2,i} \right.$$
$$\left. + (2j - 1) \sum_{m=0}^{i} E_m R_{j-1,i-m} - (2j + K - 1) R_{j+K-1,i-1} \right\} . \tag{3.52}$$

The normalization condition and the Hellmann–Feynman theorem give us $R_{0,j} = \delta_{0j}$ and

$$E_i = \frac{1}{i} R_{K,i-1} , \tag{3.53}$$

respectively. In order to obtain the perturbation correction to the energy E_p, we need all the coefficients $R_{j,i}$ with $i = 0, 1, \ldots, p - 1$ and $j = 1, 2, \ldots, (p - i)(K - 1) + 1$.

Table 3.3 Method of Swenson and Danforth for the Anharmonic Oscillator

$$\hat{H} = -\frac{\nabla^2}{2} + \frac{\hat{r}^2}{2} + \lambda \hat{r}^4$$

$E_1 = \frac{3}{8} - \frac{1}{2} l(l+1) + \frac{3}{2} E_0^2$

$E_2 = -\frac{67}{16} E_0 + \frac{9}{4} E_0 l(l+1) - \frac{17}{4} E_0^3$

$E_3 = \frac{1539}{256} + \frac{1707}{32} E_0^2 - \frac{129}{8} l(l+1) E_0^2 + \frac{375}{16} E_0^4 - \frac{273}{32} l(l+1)$

$\quad + \frac{11}{16} l^2 (l+1)^2$

$E_4 = \frac{28647}{128} E_0 l(l+1) - \frac{909}{64} E_0 l^2 (l+1)^2 + \frac{4455}{32} l(l+1) E_0^3 - \frac{305141}{1024} E_0$

$\quad - \frac{89165}{128} E_0^3 - \frac{10689}{64} E_0^5$

$R_{3,1} = -\frac{945}{128} + \frac{1711 l(l+1)}{16} - \frac{885 E_0^2}{16} - \frac{9 l^2 (l+1)^2}{8} + \frac{63 l(l+1) E_0^2}{4} - \frac{165 E_0^4}{8}$

$R_{3,2} = -\frac{10707}{32} E_0 l(l+1) + \frac{393}{16} E_0 l^2 (l+1)^2 - \frac{1425}{8} l(l+1) E_0^3 + \frac{117281}{256} E_0$

$\quad + \frac{29555}{32} E_0^3 + \frac{3129}{16} E_0^5$

$R_{3,3} = \frac{136923}{64} l(l+1) - \frac{4259307}{256} E_0^2 - \frac{4907}{16} l^2 (l+1)^2 + \frac{246465}{32} l(l+1) E_0^2$

$\quad - \frac{367605}{256} - \frac{473985}{32} E_0^4 - \frac{6915}{16} l^2 (l+1)^2 E_0^2 + \frac{16815}{8} l(l+1) E_0^4$

$\quad + \frac{37}{4} l^3 (l+1)^3 - \frac{31983}{16} E_0^6$

$R_{3,4} = -\frac{262575777}{2048} E_0 l(l+1) + \frac{7429035}{512} E_0 l^2 (l+1)^2 - \frac{39645255}{256} l(l+1) E_0^3$

$\quad + \frac{1230998721}{8192} E_0 + \frac{965258733}{2048} E_0^3 + \frac{118198899}{512} E_0^5 - \frac{46515}{128} E_0 l^3 (l+1)^3$

$\quad + \frac{895125}{128} l^2 (l+1)^2 E_0^3 - \frac{3261825}{128} l(l+1) E_0^5 + \frac{2742687}{128} E_0^7$

Table 3.3 shows perturbation corrections to the energy and R_3. We do not give more results because the length of the corrections rapidly increases with the perturbation order. One easily obtains many more perturbation corrections by means of a simple Maple program which we do not show here because it is similar to that discussed earlier for the one-dimensional model.

Another interesting exactly-solvable model is the nonrelativistic hydrogen atom. Upon solving the dimensionless Schrödinger equation with the dimensionless Coulomb interaction between electron and nucleus

$$V(r) = -\frac{1}{r} \tag{3.54}$$

one obtains the dimensionless energy eigenvalues

$$E = -\frac{1}{2n^2}, \; n = v + l + 1, \tag{3.55}$$

where $v = 0, 1, \ldots, n - l - 1$. The master equation (3.33) reduces to

$$2jER_{j-1} + (j-1)\left[\frac{j(j-2)}{4} - l(l+1)\right] R_{j-3} + (2j-1)R_{j-2} = 0. \tag{3.56}$$

When $j = 1$ we obtain the well-known virial theorem

$$R_{-1} = -2E. \tag{3.57}$$

The Hellmann–Feynman theorem (3.46) gives us

$$R_{-2} = \frac{2}{(2l+1)n^3} \, . \tag{3.58}$$

From the master equation (3.56) with $j = 0$ we obtain

$$R_{-3} = \frac{R_{-2}}{l(l+1)}, \; l > 0 \, . \tag{3.59}$$

In what follows we consider simple perturbations of the form λr^K, $K > 0$. The method also applies to singular perturbations ($K < 0$) with the reservations mentioned above regarding the expectation values of negative powers of the radial coordinate. Substituting the potential-energy function

$$V(r) = -\frac{1}{r} + \lambda r^K \tag{3.60}$$

into the master equation (3.33) we obtain the recurrence relation

$$2jER_{j-1} \;+\; (j-1)\left[\frac{j(j-2)}{4} - l(l+1)\right]R_{j-3}$$
$$+\; (2j-1)R_{j-2} - (2j+K)\lambda R_{j+K-1} = 0 \, . \tag{3.61}$$

Expanding the energy and expectation values in Taylor series about $\lambda = 0$ as in preceding examples, equation (3.61) gives us the recurrence relation

$$R_{j,i} \;=\; \frac{1}{2(j+1)E_0}\left\{ j\left[l(l+1) - \frac{j^2-1}{4}\right]R_{j-2,i} - (2j+1)R_{j-1,i}\right.$$
$$\left. - 2(j+1)\sum_{m=1}^{i} E_m R_{j,i-m} + (2j+K+2)R_{j+K,i-1}\right\} \tag{3.62}$$

where we have substituted $j + 1$ for j for convenience. It follows from the Hellmann–Feynman theorem that equation (3.53) also applies to this example. The recurrence relation (3.62) is unsuitable for the calculation of the perturbation corrections to R_{-1} because the denominator vanishes. In order to obtain them we resort to the virial theorem that follows from equation (3.61) with $j = 1$. Expanding the resulting expression in a Taylor series we have

$$R_{-1,i} = -2E_i + (K+2)R_{K,i-1} \, . \tag{3.63}$$

Table 3.4 shows sample results: one easily obtains more perturbation corrections by means of a simple Maple program written according to equations (3.53), (3.62), (3.63), and the normalization condition $R_{0,j} = \delta_{0j}$.

3.3.3 More General Polynomial Perturbations

The method of Swenson and Danforth applies to more general polynomial perturbations than the ones discussed above. For example, the reader may easily derive suitable perturbation equations from the hypervirial and Hellmann–Feynman theorems for the following cases:

$$V(x) \;=\; \frac{x^2}{2} + \sum_{j=1}^{\infty} \lambda^j \sum_{i=1}^{n_j} c_{ji} x^i \, , \tag{3.64}$$

$$V(r) \;=\; \frac{r^2}{2} + \sum_{j=1}^{\infty} \lambda^j \sum_{i=1}^{n_j} c_{ji} r^{2i} \, , \tag{3.65}$$

Table 3.4   Method of Swenson and Danforth for $\hat{H} = -\dfrac{\nabla^2}{2} - \dfrac{1}{\hat{r}} + \lambda \hat{r}$

$E_1 = -\dfrac{1}{2} l\,(l+1) - \dfrac{3}{4} \dfrac{1}{E_0}$

$E_2 = -\dfrac{3}{16} \dfrac{l^2\,(l+1)^2}{E_0} + \dfrac{7}{64} \dfrac{1}{E_0{}^3} - \dfrac{5}{32} \dfrac{1}{E_0{}^2}$

$E_3 = -\dfrac{5}{32} \dfrac{l^3\,(l+1)^3}{E_0{}^2} + \dfrac{7}{128} \dfrac{l^2\,(l+1)^2}{E_0{}^3} - \dfrac{33}{512} \dfrac{1}{E_0{}^5} + \dfrac{75}{256} \dfrac{1}{E_0{}^4}$

$E_4 = \dfrac{45}{512} \dfrac{l^3\,(l+1)^3}{E_0{}^4} + \dfrac{45}{256} \dfrac{l^2\,(l+1)^2}{E_0{}^4} - \dfrac{21}{128} \dfrac{l^4\,(l+1)^4}{E_0{}^3} - \dfrac{2275}{4096} \dfrac{1}{E_0{}^6} + \dfrac{55}{256} \dfrac{1}{E_0{}^5}$

$\qquad - \dfrac{99}{2048} \dfrac{l^2\,(l+1)^2}{E_0{}^5} + \dfrac{465}{8192} \dfrac{1}{E_0{}^7}$

$E_5 = -\dfrac{91}{1024} \dfrac{l^3\,(l+1)^3}{E_0{}^6} - \dfrac{2093}{4096} \dfrac{l^2\,(l+1)^2}{E_0{}^6} + \dfrac{33}{256} \dfrac{l^4\,(l+1)^4}{E_0{}^5} + \dfrac{55}{128} \dfrac{l^3\,(l+1)^3}{E_0{}^5}$

$\qquad - \dfrac{99}{512} \dfrac{l^5\,(l+1)^5}{E_0{}^4} - \dfrac{1995}{32768} \dfrac{1}{E_0{}^9} + \dfrac{4335}{4096} \dfrac{1}{E_0{}^8} - \dfrac{11409}{8192} \dfrac{1}{E_0{}^7}$

$\qquad + \dfrac{465}{8192} \dfrac{l^2\,(l+1)^2}{E_0{}^7}$

$R_{2,0} = \dfrac{3}{4} \dfrac{l\,(l+1)}{E_0} - \dfrac{1}{4} \dfrac{1}{E_0} + \dfrac{5}{8} \dfrac{1}{E_0{}^2}$

$R_{2,1} = \dfrac{15}{32} \dfrac{l^2\,(l+1)^2}{E_0{}^2} - \dfrac{45}{128} \dfrac{1}{E_0{}^4} + \dfrac{63}{64} \dfrac{1}{E_0{}^3} + \dfrac{5}{16} \dfrac{l\,(l+1)}{E_0{}^2} - \dfrac{7}{32} \dfrac{l\,(l+1)}{E_0{}^3}$

$R_{2,2} = \dfrac{63}{128} \dfrac{l^3\,(l+1)^3}{E_0{}^3} - \dfrac{45}{128} \dfrac{l^2\,(l+1)^2}{E_0{}^4} + \dfrac{91}{256} \dfrac{1}{E_0{}^6} - \dfrac{649}{256} \dfrac{1}{E_0{}^5} + \dfrac{99}{512} \dfrac{l\,(l+1)}{E_0{}^5}$

$\qquad - \dfrac{225}{256} \dfrac{l\,(l+1)}{E_0{}^4} + \dfrac{21}{32} \dfrac{l^2\,(l+1)^2}{E_0{}^3} + \dfrac{93}{128} \dfrac{1}{E_0{}^4}$

$R_{2,3} = -\dfrac{253}{512} \dfrac{l^3\,(l+1)^3}{E_0{}^5} - \dfrac{2893}{1024} \dfrac{l^2\,(l+1)^2}{E_0{}^5} - \dfrac{55}{64} \dfrac{l\,(l+1)}{E_0{}^5} + \dfrac{315}{512} \dfrac{l^4\,(l+1)^4}{E_0{}^4}$

$\qquad + \dfrac{285}{256} \dfrac{l^3\,(l+1)^3}{E_0{}^4} + \dfrac{24855}{4096} \dfrac{1}{E_0{}^7} - \dfrac{465}{2048} \dfrac{l\,(l+1)}{E_0{}^7} - \dfrac{819}{128} \dfrac{1}{E_0{}^6}$

$\qquad + \dfrac{455}{1024} \dfrac{l^2\,(l+1)^2}{E_0{}^6} + \dfrac{2275}{1024} \dfrac{l\,(l+1)}{E_0{}^6} - \dfrac{3621}{8192} \dfrac{1}{E_0{}^8}$

---

or

$$V(r) = -\dfrac{1}{r} + \sum_{j=1}^{\infty} \lambda^j \sum_{i=1}^{n_j} c_{ji} r^i \,. \qquad (3.66)$$

3.4   Moment Method

In what follows we call moments of a given vector Ψ of the state space the inner products $< f|\Psi >$ between Ψ and other vectors f. In particular, we choose Ψ to be an eigenvector of the Hamiltonian operator $\hat{H}$. It is known since long ago that the moments of an eigenvector of $\hat{H}$ prove suitable for nonperturbative [34] and perturbative [35] treatments of the Schrödinger equation. The application of the moment method to perturbation theory has since been considerably generalized and extended [36]–[39] (and references therein).

The main idea behind the moment method is that if $(\hat{H} - E)\Psi$ is orthogonal to a complete set of (not necessarily orthogonal) vectors $\{f_j\}$, then Ψ is eigenvector of Ψ with eigenvalue E. In other words,

$$\left\langle f_j \left| \hat{H} - E \right| \Psi \right\rangle = \left\langle \left(\hat{H} - E \right) f_j | \Psi \right\rangle = 0 \tag{3.67}$$

for all j is equivalent to the Schrödinger equation $\hat{H}\Psi = E\Psi$.

Since the set of vectors $\{f_j\}$ is complete we have

$$\hat{H} f_j = \sum_{i=0}^{\infty} H_{i,j} f_i \,, \tag{3.68}$$

where $H_{i,j}$ are coefficients. It follows from equations (3.67) and (3.68) that

$$\sum_{i=0}^{\infty} \left(H_{i,j} - E\delta_{ij} \right) F_i = 0 \,, \tag{3.69}$$

where

$$F_j = \left\langle f_j | \Psi \right\rangle \tag{3.70}$$

is a moment of Ψ.

Notice that if the set $\{f_j\}$ is orthonormal, then $H_{i,j} = < f_i | \hat{H} | f_j >$ is the usual matrix element of the Hamiltonian operator and F_j is a coefficient of the expansion of Ψ in the orthonormal basis set. In what follows we do not assume that the set of vectors $\{f_j\}$ is orthonormal.

3.4.1 Exactly Solvable Cases

When $H_{i,j} = 0$ for all $i > j$ the eigenvalue equation (3.69) becomes an exactly solvable triangular system of homogeneous linear equations

$$
\begin{aligned}
\left(H_{0,0} - E \right) F_0 &= 0 \\
H_{0,1} F_0 + \left(H_{1,1} - E \right) F_1 &= 0 \\
&\cdots \\
H_{0,j} F_0 + H_{1,j} F_1 + \cdots + (H_{j,j} - E) F_j &= 0 \\
&\cdots
\end{aligned}
\tag{3.71}
$$

We first consider the nondegenerate case $H_{i,i} \neq H_{j,j}$ for all $i \neq j$. If $E \neq H_{j,j}$ for all j, then all the moments F_j vanish and Ψ is the null vector because the set $\{f_j\}$ is complete. Consequently, E has to be one of the diagonal coefficients in order for a nontrivial solution to exist. Suppose that $E = H_{n,n}$; then $F_j = 0$ for all $j < n$, and

$$F_j = \frac{1}{H_{n,n} - H_{j,j}} \sum_{i=n}^{j-1} H_{i,j} F_i, \quad j = n+1, n+2, \ldots \,. \tag{3.72}$$

All the nonzero moments are proportional to F_n which we may arbitrarily choose equal to unity as an intermediate normalization condition.

As an illustrative example consider the dimensionless harmonic oscillator

$$\hat{H} = \frac{1}{2} \left(-\frac{d^2}{dx^2} + x^2 \right) \tag{3.73}$$

and the set of functions

$$f_j = x^{2j+s} \exp\left(-x^2/2\right), \ s = 0, 1 . \tag{3.74}$$

In this case we have

$$\hat{H} f_j = -\frac{(2j+s)(2j+s-1)}{2} f_{j-1} + \left(2j+s+\frac{1}{2}\right) f_j \tag{3.75}$$

so that the only nonzero coefficients $H_{i,j}$ are

$$H_{j-1,j} = -\frac{(2j+s)(2j+s-1)}{2}, \ H_{j,j} = \left(2j+s+\frac{1}{2}\right) . \tag{3.76}$$

The argument above leads to the well-known energy eigenvalues $E = (2j+s+1/2)$, where $s = 0$ and $s = 1$ apply to even and odd states, respectively. The recurrence relation for the moments is

$$F_j = -\frac{(2j+s)(2j+s-1)}{4(j-n)} F_{j-1}, \ j = n+1, n+2, \dots . \tag{3.77}$$

The fully degenerate case $H_{i,i} = H_{j,j}$ for all i, j is also interesting. Choosing $E = H_{j,j}$, the eigenvalue equation (3.71) becomes

$$\begin{aligned} H_{0,1} F_0 &= 0 \\ H_{0,2} F_0 + H_{1,2} F_1 &= 0 \\ &\cdots \\ H_{0,j} F_0 + H_{1,j} F_1 + \cdots + H_{j-1,j} F_{j-1} &= 0 . \end{aligned} \tag{3.78}$$

If all the coefficients $H_{j-1,j}$ are nonzero, then all the moments F_j are zero, and Ψ is the null vector. We must therefore assume that $H_{n-1,n} = 0$. The nonzero moments given by the recurrence relation

$$F_j = -\frac{1}{H_{j,j+1}} \sum_{i=n-1}^{j-1} H_{i,j+1} F_i, \ j = n, n+1, \dots \tag{3.79}$$

are proportional to F_{n-1} which we may arbitrarily choose equal to unity.

A simple example of full degeneration is given by the radial Hamiltonian operator for the Coulomb problem

$$\hat{H} = -\frac{1}{2} \frac{d^2}{dr^2} + \frac{l(l+1)}{2r^2} - \frac{1}{r} . \tag{3.80}$$

Choosing the functions

$$f_j = r^{j+l+1} \exp(-\alpha r), \ j = 0, 1, \dots \tag{3.81}$$

the only nonzero Hamiltonian coefficients are

$$H_{j-2,j} = -\frac{j(j+2l+1)}{2}, \ H_{j-1,j} = \alpha(j+l+1) - 1, \ H_{j,j} = -\frac{\alpha^2}{2} . \tag{3.82}$$

Therefore, from $E = H_{n,n}$ and $H_{n-1,n} = 0$ we obtain the well-known results

$$\alpha = \frac{1}{N}, \ E = -\frac{1}{2N^2} , \tag{3.83}$$

where $N = n + l + 1$ is the principal quantum number of the hydrogen atom [40].

3.4.2 Perturbation Theory by the Moment Method

If the moment equations (3.69) are not exactly solvable, then we apply perturbation theory. We introduce a dummy perturbation parameter λ as follows:

$$\hat{H} f_j = \sum_{m=0}^{j} H_{m,j} f_m + \lambda \sum_{m=j+1}^{\infty} H_{m,j} f_m \,, \tag{3.84}$$

so that the recurrence relation for the moments becomes

$$\sum_{m=0}^{j} (H_{m,j} - E\delta_{mj}) F_m + \lambda \sum_{m=j+1}^{\infty} H_{m,j} F_m = 0 \,. \tag{3.85}$$

Notice that we have split the recurrence relation (3.69) into a solvable part and a perturbation that vanish when $\lambda = 0$. We can therefore try approximate solutions in the form of Taylor series about $\lambda = 0$:

$$E = \sum_{i=0}^{\infty} E_i \lambda^i, \quad F_j = \sum_{i=0}^{\infty} F_{j,i} \lambda^i \,, \tag{3.86}$$

where commonly

$$E_0 = H_{n,n} \tag{3.87}$$

for a given value of n.

The normalization condition is arbitrary. Sometimes it suffices to choose $F_n = 1$, but in other cases a more elaborate normalization condition appears to be more practical, as shown in what follows. First, rewrite equation (3.85) as

$$\Delta E F_j = \sum_{m=0}^{j} \left(H_{m,j} - H_{n,n}\delta_{mj} \right) F_m + \lambda \sum_{m>j} H_{m,j} F_m \,, \tag{3.88}$$

where

$$\Delta E = E - H_{n,n} \,. \tag{3.89}$$

Second, look for a set of coefficients C_j, $j = 0, 1, \ldots, n$, such that

$$\sum_{j=0}^{n} C_j \sum_{m=0}^{j} (H_{m,j} - H_{n,n}\delta_{mj}) F_m = \sum_{m=0}^{n-1} \left[\sum_{j=m}^{n} \left(H_{m,j} - H_{n,n}\delta_{mj} \right) C_j \right] F_m = 0 \,. \tag{3.90}$$

It is convenient to choose the coefficients C_j to be solutions to the homogeneous linear system of equations

$$\sum_{j=m}^{n} \left(H_{m,j} - H_{n,n}\delta_{mj} \right) C_j = 0, \quad m = 0, 1, \ldots, n - 1 \,. \tag{3.91}$$

If the problem is nondegenerate, then there is just one linearly independent set of coefficients C_j; otherwise, there will be more than one. We will illustrate both cases later by means of appropriate examples.

Once we have the coefficients C_j, we rewrite the equation for the energy as follows

$$\Delta E \sum_{j=0}^{n} C_j F_j = \lambda \sum_{j=0}^{n} C_j \sum_{m>j} H_{m,j} F_m \ . \tag{3.92}$$

The intermediate normalization condition

$$\sum_{j=0}^{n} C_j F_j = 1 \tag{3.93}$$

leads to a useful expression for the energy in terms of the moments

$$\Delta E = \lambda \sum_{j=0}^{n} C_j \sum_{m>j} H_{m,j} F_m \ . \tag{3.94}$$

3.4.3 Nondegenerate Case

Suppose that $H_{j,j} \neq H_{n,n}$ for all $j \neq n$. Upon expanding the energy and moments in equation (3.85) in Taylor series about $\lambda = 0$ and solving for $F_{j,i}$ we obtain

$$F_{j,i} = \frac{1}{H_{n,n} - H_{j,j}} \left(\sum_{m=0}^{j-1} H_{m,j} F_{m,i} - \sum_{k=1}^{i} E_k F_{j,i-k} + \sum_{m=j+1}^{\infty} H_{m,j} F_{m,i-1} \right) \ . \tag{3.95}$$

Notice that this equation does not give us the perturbation corrections to the moment F_n, which we obtain from the arbitrary intermediate normalization $F_n = 1$ that leads to

$$F_{n,i} = \delta_{i0} \ . \tag{3.96}$$

Substitution of this normalization condition into equation (3.88) for $j = n$ yields an expression for the energy

$$\Delta E = \sum_{m=0}^{n-1} H_{m,n} F_m + \lambda \sum_{m=n+1}^{\infty} H_{m,n} F_m \ , \tag{3.97}$$

and, consequently, for its perturbation corrections

$$E_i = H_{n,n} \delta_{i0} + \sum_{m=0}^{n-1} H_{m,n} F_{m,i} + \sum_{m=n+1}^{\infty} H_{m,n} F_{m,i-1} \ . \tag{3.98}$$

Remember that $F_{j,0} = 0$ for all $j < n$ as shown above for the chosen exactly solvable models.

We easily obtain exact expressions for the perturbation corrections when $H_{i,j} = 0$ for all $|i - j|$ larger than some positive integer. As an example consider the dimensionless anharmonic oscillator

$$\hat{H} = \frac{1}{2} \left(-\frac{d^2}{dx^2} + x^2 \right) + \lambda x^4 \tag{3.99}$$

and the set of functions (3.74). The recurrence relation for the moments is

$$-\frac{(2j+s)(2j+s-1)}{2} F_{j-1} + 2(j-n) F_j - \Delta E F_j + \lambda F_{j+2} = 0 \tag{3.100}$$

and the intermediate normalization condition $F_n = 1$ leads to

$$E = 2n + s + \frac{1}{2} - \frac{(2n+s)(2n+s-1)}{2} F_{n-1} + \lambda F_{n+2} \,. \tag{3.101}$$

Expanding the energy and moments in Taylor series about $\lambda = 0$, we obtain

$$F_{j,i} = \frac{1}{2(j-n)} \left[\frac{(2j+s)(2j+s-1)}{2} F_{j-1,i} + \sum_{m=1}^{i} E_m F_{j,i-m} - F_{j+2,i-1} \right] \tag{3.102}$$

and

$$E_i = \left(2n + s + \frac{1}{2} \right) \delta_{i0} - \frac{(2n+s)(2n+s-1)}{2} F_{n-1,i} + F_{n+2,i-1} \,. \tag{3.103}$$

These two equations and the normalization condition (3.96) give us all the perturbation corrections to the energy and moments. By straightforward inspection of the equations above, one concludes that in order to obtain E_p one has to calculate the moment coefficients $F_{j,i}$ with $n - 2i \le j \le n + 2(p-i)$ when $i = 0, 1, \ldots, p-1$, and those with $n - 2p \le j \le n - 1$ when $i = p$.

Table 3.5 shows results for the first four states of the anharmonic oscillator which we calculated by means of a simple Maple program. We do not show it here because it runs more slowly and yields less general results than the set of procedures given earlier for the method of Swenson and Danforth. The only purpose of Table 3.5 is just to help the reader check his or her own calculations.

Table 3.5   Moment-Method Perturbation Theory for the Anharmonic Oscillator

$\hat{H} = -\frac{1}{2} \frac{d^2}{dx^2} + \frac{\hat{x}^2}{2} + \lambda \hat{x}^4$

Ground State

$E = \frac{1}{2} + \frac{3\lambda}{4} - \frac{21\lambda^2}{8} + \frac{333\lambda^3}{16} - \frac{30885\lambda^4}{128} + \frac{916731\lambda^5}{256} + \cdots$

$F_1 = \frac{1}{2} - \frac{3}{4}\lambda + \frac{75}{16}\lambda^2 - \frac{1527}{32}\lambda^3 + \frac{165741}{256}\lambda^4 + \cdots$

First Excited State

$E = \frac{3}{2} + \frac{15\lambda}{4} + \frac{18075\lambda^2}{256} - \frac{10982535\lambda^3}{4096} + \frac{10863836055\lambda^4}{131072} - \frac{43563085647015\lambda^5}{16777216}$
$\quad + \cdots$

$F_1 = \frac{3}{2} + \frac{1317}{64}\lambda - \frac{1399395}{2048}\lambda^2 + \frac{655649973}{32768}\lambda^3 - \frac{1275828608445}{2097152}\lambda^4 + \cdots$

Second Excited State

$E = \frac{5}{2} + \frac{93}{8}\lambda - \frac{49761}{512}\lambda^2 + \frac{12053211}{8192}\lambda^3 - \frac{15240517155}{524288}\lambda^4 + \frac{22590881117433}{33554432}\lambda^5$
$\quad + \cdots$

$F_0 = 1 - \frac{3}{8}\lambda - \frac{22239}{512}\lambda^2 + \frac{14786319}{8192}\lambda^3 - \frac{31401053397}{524288}\lambda^4 + \cdots$

Third Excited State

$E = \frac{7}{2} + \frac{579\lambda}{16} - \frac{723285\lambda^2}{1024} + \frac{138568419\lambda^3}{8192} - \frac{893001846771\lambda^4}{2097152} + \frac{214869089419155\lambda^5}{16777216}$
$\quad + \cdots$

$F_0 = 1 - \frac{53}{16}\lambda + \frac{85695}{1024}\lambda^2 - \frac{24839823}{8192}\lambda^3 + \frac{263498107665}{2097152}\lambda^4 + \cdots$

Another interesting example is the radial Hamiltonian operator for a perturbed Coulomb problem

$$\hat{H} = -\frac{1}{2}\frac{d^2}{dr^2} + \frac{l(l+1)}{2r^2} - \frac{1}{r} + \lambda r^K , \tag{3.104}$$

where $K = 1, 2, \ldots$. Choosing the set of functions (3.81) the recurrence relation for the moments is

$$-\frac{j(j+2l+1)}{2}F_{j-2} + \frac{j-n}{n+l+1}F_{j-1} - \Delta E F_j + \lambda F_{j+k} = 0 , \tag{3.105}$$

where we have substituted the value of α given by equation (3.83) and

$$\Delta E = E + \frac{1}{2(n+l+1)^2} . \tag{3.106}$$

The choice $n = 0$ selects the states with angular and principal quantum numbers $l = 0, 1, \ldots$, and $N = l + 1$, respectively, commonly denoted $1s, 2p, 3d, \ldots$ that are free from radial nodes. When $j = 0$ we obtain

$$\Delta E F_0 = \lambda F_K , \tag{3.107}$$

so that the intermediate normalization condition $F_0 = 1$ leads to

$$\Delta E = \lambda F_K . \tag{3.108}$$

We obtain an expression for F_{-1} from the general equation (3.105) with $j = 1$:

$$F_{-1} = \frac{1}{l+1}\left(\frac{1}{l+1} - \Delta E F_1 + \lambda F_{K+1}\right) . \tag{3.109}$$

The expansion of the energy and moments in Taylor series about $\lambda = 0$ leads to

$$F_{-1,i} = \frac{1}{l+1}\left(\frac{\delta_{i0}}{l+1} - \sum_{m=1}^{i} E_m F_{1,i-m} + F_{K+1,i-1}\right) \tag{3.110}$$

$$F_{0,i} = \delta_{i0} \tag{3.111}$$

$$F_{j,i} = \frac{l+1}{j+1}\left[\frac{(j+1)(j+2l+2)}{2}F_{j-1,i} + \sum_{m=1}^{i} E_m F_{j+1,i-m} - F_{j+k+1,i-1}\right] \tag{3.112}$$

The last expression follows from substituting $j+1$ for j in equation (3.105). The energy coefficients are given by

$$E_i = F_{K,i-1} . \tag{3.113}$$

In order to obtain E_p one has to calculate all the moment coefficients $F_{j,i}$ with $i = 1, 2, \ldots, p-1$, $1 \le j \le (p-i)(K+1) - 1$.

The choice $n = 1$ selects all the states with $N = l + 2$: $2s, 3p, 4d, \ldots$, and their treatment illustrates the application of equations (3.92)–(3.94). The general equation (3.105) with $n = 1$ yields

$$-\frac{F_{-1}}{l+2} - \Delta E F_0 + \lambda F_K = 0 , \tag{3.114}$$

$$-(l+1)F_{-1} - \Delta E F_1 + \lambda F_{K+1} = 0 , \tag{3.115}$$

$$-(2l+3)F_0 + \frac{F_1}{l+2} - \Delta E F_2 + \lambda F_{K+2} = 0 , \tag{3.116}$$

for $j = 0, 1$, and 2, respectively. It is difficult to explain how to develop a workable system of equations for the general case. For this reason we show how to proceed in a particular example and hope that the reader will imagine the strategy to be followed in other situations. Besides, we will discuss more illustrative examples which may give the reader additional hints later in this book. In the problem at hand we multiply equation (3.114) by $(l + 1)(l + 2)$ and subtract from equation (3.115), thus removing F_{-1} and obtaining a preliminary expression for the energy:

$$\Delta E \left[F_1 - (l + 1)(l + 2)F_0 \right] = \lambda \left[F_{K+1} - (l + 1)(l + 2)F_K \right] . \tag{3.117}$$

The intermediate normalization condition

$$F_1 - (l + 1)(l + 2)F_0 = 1 \tag{3.118}$$

leads to a simpler expression for the energy that is suitable for the application of perturbation theory:

$$\Delta E = \lambda \left[F_{K+1} - (l + 1)(l + 2)F_K \right] . \tag{3.119}$$

From equations (3.116) and (3.118) we derive the following expression for F_0:

$$F_0 = \frac{1}{l + 2} \left(\frac{1}{l + 2} - \Delta E F_2 + \lambda F_{K+2} \right) . \tag{3.120}$$

Expanding the energy and moments in Taylor series about $\lambda = 0$ we obtain all the necessary recurrence relations for the perturbation coefficients

$$F_{0,i} = \frac{1}{l + 2} \left(\frac{\delta_{i0}}{l + 2} - \sum_{m=1}^{i} E_m F_{2,i-m} + F_{K+2,i-1} \right) \tag{3.121}$$

$$F_{1,i} = \delta_{i0} + (l + 1)(l + 2)F_{0,i} , \tag{3.122}$$

$$F_{j,i} = \frac{l + 2}{j} \left[\frac{(j + 1)(j + 2l + 2)}{2} F_{j-1,i} + \sum_{m=1}^{i} E_m F_{j+1,i-m} - F_{j+k+1,i-1}, \right] \tag{3.123}$$

$$E_i = F_{K+1,i-1} - (l + 1)(l + 2)F_{K,i-1} . \tag{3.124}$$

The calculation is analogous to the one above except that in this case $2 \le j \le (p - i)(K + 1) - 1$.

Table 3.6 shows perturbation coefficients for the energy and arbitrarily selected moments when $K = 1$. The energy coefficients agree with those in Table 3.4 provided that we substitute the appropriate value of E_0 in each case: $E_0 = -1/[2(l + 1)^2]$ and $E_0 = -1/[2(l + 2)^2]$ for the states with zero and one radial node, respectively.

3.4.4 Degenerate Case

In order to illustrate the application of the moment method to a model with degenerate unperturbed states, we choose a simple nontrivial anharmonic oscillator in two dimensions with dimensionless Hamiltonian operator

$$\hat{H} = -\frac{1}{2}\nabla^2 + V(x, y), \quad V(x, y) = \frac{1}{2}(x^2 + y^2) + \lambda \left(ax^4 + by^4 + 2cx^2y^2 \right) , \tag{3.125}$$

where λ, a, b, and c are real and positive.

The potential-energy function $V(x, y)$ is a single infinite well with a minimum $V = 0$ at origin. It is invariant under the transformation $(x, y) \longleftrightarrow (-x, -y)$. It is invariant under the transformation $(x, y) \longleftrightarrow (-x, -y)$, and in the particular case that $a = b$, it is also invariant under the exchange $(x, y) \longleftrightarrow (y, x)$. The latter higher symmetry makes the

Table 3.6   Moment-Method Perturbation Theory for the Perturbed Coulomb Model

$$\hat{H} = -\frac{\nabla^2}{2} - \frac{1}{\hat{r}} + \lambda\hat{r}$$

States with No Radial Nodes

$E_1 = \frac{1}{2}(l+1)(2l+3)$

$E_2 = -\frac{1}{4}(l+1)^4(2l+3)(l+2)$

$E_3 = \frac{1}{8}(l+1)^7(2l+3)(l+2)(9+4l)$

$E_4 = -\frac{1}{32}(l+1)^{10}(2l+3)(l+2)(225l+48l^2+265)$

$E_5 = \frac{1}{32}(l+1)^{13}(2l+3)(l+2)(3128l+2562+1281l^2+176l^3)$

$E_6 = -\frac{1}{64}(l+1)^{16}(2l+3)(l+2)(93141l+59552+55031l^2+14561l^3+1456l^4)$

$F_{2,0} = \frac{1}{2}(l+1)^2(l+2)(3+2l)$

$F_{2,1} = -\frac{1}{8}(l+1)^5(3+2l)(l+2)(9+4l)$

$F_{2,2} = \frac{1}{32}(l+1)^8(3+2l)(l+2)(197+168l+36l^2)$

$F_{2,3} = -\frac{1}{128}(l+1)^{11}(3+2l)(l+2)(6375+7842l+3236l^2+448l^3)$

$F_{2,4} = \frac{1}{512}(l+1)^{14}(3+2l)(l+2)(413298l+246636l^2+65912l^3+6656l^4+261659)$

$F_{2,5} = -\frac{1}{2048}(l+1)^{17}(3+2l)(l+2)$
$\quad\quad (24183756l+18566064l^2+7188320l^3+12707073+1403696l^4+110592l^5)$

$F_{2,6} = \frac{1}{8192}(l+1)^{20}(3+2l)(l+2)(15487212722l+1434654808l^2+7155879601l^3$
$\quad\quad +1984512l^6+2027190561l^4+309252161l^5+703100137)$

States with One Radial Node

$E_1 = \frac{1}{2}(4+l)(2l+3)$

$E_2 = -\frac{1}{4}(2l+3)(l^3+11l^2+26l+22)(l+2)^2$

$E_3 = \frac{1}{8}(2l+3)(4l^5+75l^4+371l^3+890l^2+1088l+552)(l+2)^4$

$F_{0,0} = \frac{1}{(l+2)^2}$

$F_{0,1} = \frac{1}{4}(3l+2l^2+4)(2l+3)$

$F_{0,2} = -\frac{1}{8}(l+2)^2(8l^4+18l^3+27l^2+53l+54)(2l+3)$

$F_{0,3} = \frac{1}{32}(8l^7+184l^6+688l^5+1211l^4+2093l^3+4826l^2+7030l+4060)(l+2)^4$
$\quad\quad (2l+3)$

---

application of the moment method simpler as it allows the treatment of degenerate states as if they were nondegenerate [41, 42]. Here we do not restrict to such a simplification keeping our results as general as possible.

The set of functions

$$f_{ij} = x^i y^j \exp\left[-\left(x^2 + y^2\right)/2\right], \ i, j = 0, 1, \ldots \tag{3.126}$$

enables us to construct the recurrence relation for the moments

$$F_{i,j} = \langle f_{ij}|\Psi \rangle \tag{3.127}$$

necessary for the application of perturbation theory. For clarity, two subscripts label the functions (3.126) and moments (3.127) instead of only one as in the discussion of the general case in preceding subsections. The moment recurrence relation reads

$$-\frac{i(i-1)}{2}F_{i-2,j} - \frac{j(j-1)}{2}F_{i,j-2} + (i+j-N)F_{i,j}$$
$$-\Delta E F_{i,j} + \lambda\left(a F_{i+4,j} + b F_{i,j+4} + 2c F_{i+2,j+2}\right) = 0, \tag{3.128}$$

where

$$\Delta E = E - N - 1, \ N = 0, 1, \ldots . \tag{3.129}$$

Notice that the unperturbed energy is $E_0 = N + 1$. Because the subscripts of the moments in equation (3.128) are displaced by even numbers, we have four different sets of solutions denoted (e,e), (e,o), (o,e), and (o,o), where e = even and o = odd is the parity of the corresponding subscript. They certainly match the symmetry classes of the eigenfunctions.

The coefficients of the perturbation series

$$F_{i,j} = \sum_{m=0}^{\infty} F_{i,j,m}\lambda^m \tag{3.130}$$

satisfy

$$F_{i,j,m} = \frac{1}{i+j-N}\left[\frac{i(i-1)}{2}F_{i-2,j,m} + \frac{j(j-1)}{2}F_{i,j-2,m} + \sum_{k=1}^{m}E_k F_{i,j,m-k}\right.$$
$$\left. - a F_{i+4,j,m-1} - b F_{i,j+4,m-1} - 2c F_{i+2,j+2,m-1}\right], \tag{3.131}$$

which together with an appropriate expression for the energy enables us to calculate the perturbation corrections. In what follows we show how to apply the method to some of the lowest states of the anharmonic oscillator.

For the ground state we choose $N = 0$. Setting $i = j = 0$ into equation (3.128) we obtain

$$\Delta E F_{0,0} = \lambda\left(a F_{4,0} + b F_{0,4} + 2c F_{2,2}\right) . \tag{3.132}$$

The intermediate normalization condition $F_{0,0} = 1$ determines the coefficients

$$F_{0,0,m} = \delta_{0m} \tag{3.133}$$

that we cannot obtain from the general recurrence relation (3.131), and also gives us a simple expression for the energy coefficients:

$$E_m = a F_{4,0,m-1} + b F_{0,4,m-1} + 2c F_{2,2,m-1} . \tag{3.134}$$

Notice that this state belongs to the class (e,e) mentioned above. In order to obtain E_p we need the moment coefficients $F_{i,j,m}$ for all $m = 0, 1, \ldots, p-1$, $i, j = 0, 2, \ldots, 4(p-m)$.

When $N = 1$ we identify two unperturbed states with the same energy and different symmetry. Setting $i = 0$, and $j = 1$ we have

$$\Delta E F_{0,1} = \lambda \left(a F_{4,1} + b F_{0,5} + 2c F_{2,3} \right) . \tag{3.135}$$

The intermediate normalization condition $F_{0,1} = 1$ leads to

$$F_{0,1,m} = \delta_{0m} \tag{3.136}$$

and

$$E_m = a F_{4,1,m-1} + b F_{0,5,m-1} + 2c F_{2,3,m-1} . \tag{3.137}$$

In order to obtain E_p we need $F_{i,j,m}$ for all $m = 0, 1, \ldots, p-1$, $i = 0, 2, \ldots, 4(p-m)$, and $j = 1, 3, \ldots, 4(p-m) + 1$. This state belongs to the class (e,o).

The remaining state for $N = 1$ belongs to the class (o,e). Arguing as in the preceding case we obtain

$$F_{1,0,m} = \delta_{0m} \tag{3.138}$$

and

$$E_m = a F_{5,0,m-1} + b F_{1,4,m-1} + 2c F_{3,2,m-1} . \tag{3.139}$$

In order to obtain E_p we need $F_{i,j,m}$ for all $m = 0, 1, \ldots, p-1$, $i = 1, 3, \ldots, 4(p-m) + 1$, and $j = 0, 2, \ldots, 4(p-m)$.

Although the unperturbed states for $N = 1$ are degenerate we treat them as if they were nondegenerate because the perturbation does not couple them. The application of the moment method just outlined clearly discloses this independence that arises from the symmetry of the states. More precisely, the eigenfunctions $\Psi_{0,1}(x, y)$ and $\Psi_{1,0}(x, y)$ of $\hat{H}$ for the states with $N = 1$ satisfy $\Psi_{0,1}(-x, y) = \Psi_{0,1}(x, y)$, $\Psi_{0,1}(x, -y) = -\Psi_{0,1}(x, y)$, $\Psi_{1,0}(-x, y) = -\Psi_{1,0}(x, y)$, and $\Psi_{1,0}(x, -y) = \Psi_{1,0}(x, y)$ for all values of λ. Therefore the moments with subscripts (o,e) vanish when $\Psi = \Psi_{0,1}$ and those with subscripts (e,o) vanish for the other state. Notice that no explicit consideration of the eigenfunctions was necessary neither for the application of the moment method, nor for the selection of the states because the symmetry is embedded in the chosen functions $f_{i,j}$.

The states with $N = 0$ and $N = 1$ just considered do not add anything new to the one-dimensional problems discussed earlier, except for the occurrence of one more subscript in the moments. The states with $N = 2$ offer a much richer example as we will shortly see. Before proceeding, notice that the denominator in equation (3.131) vanishes when $i + j = N$ giving room for $N + 1$ degenerate unperturbed states as $i = 0, 1, \ldots, N$ and $j = N - i$ satisfy such condition. In the language of the moment method, equation (3.131) will have $N + 1$ linearly independent solutions. We have already seen that there is only one state when $N = 0$ and two states when $N = 1$, the latter belonging to different classes: (e,o) and (o,e). In general, if N is even the degenerate states belong to either (e,e) or (o,o); otherwise, they belong to either (o,e) or (e,o). When $N = 2$ there are three states; we first consider the (o,o) case.

Setting $i = j = 1$ in equation (3.128) we obtain

$$\Delta E F_{1,1} = \lambda \left(a F_{5,1} + b F_{1,5} + 2c F_{3,3} \right) , \tag{3.140}$$

which suggests the intermediate normalization condition $F_{1,1} = 1$ that leads to

$$F_{1,1,m} = \delta_{0m} \tag{3.141}$$

and

$$E_m = a F_{5,1,m-1} + b F_{1,5,m-1} + 2c F_{3,3,m-1} . \tag{3.142}$$

In order to obtain E_p we need all $F_{i,j,m}$ with $m = 0, 1, \ldots, p-1$ and $i, j = 1, 3, \ldots, 4(p-m)+1$.

The remaining two states belong to the class (e,e); consequently the perturbation couples them, and have to be explicitly treated as degenerate. When $(i, j) = (0, 0)$, $(2, 0)$, and $(0, 2)$ we have

$$-2F_{0,0} - \Delta E F_{0,0} + \lambda \left(a F_{4,0} + b F_{0,4} + 2c F_{2,2} \right) = 0 , \tag{3.143}$$
$$-F_{0,0} - \Delta E F_{2,0} + \lambda \left(a F_{6,0} + b F_{2,4} + 2c F_{4,2} \right) = 0 , \tag{3.144}$$
$$-F_{0,0} - \Delta E F_{0,2} + \lambda \left(a F_{4,2} + b F_{0,6} + 2c F_{2,4} \right) = 0 , \tag{3.145}$$

respectively. Subtracting twice equation (3.144) from equation (3.143) gives

$$\Delta E (2F_{2,0} - F_{0,0}) + \lambda \left[a \left(F_{4,0} - 2F_{6,0} \right) + b \left(F_{0,4} - 2F_{2,4} \right) + 2c \left(F_{2,2} - 2F_{4,2} \right) \right] = 0 , \tag{3.146}$$

and subtracting twice equation (3.145) from equation (3.143) yields

$$\Delta E (2F_{0,2} - F_{0,0}) + \lambda \left[a \left(F_{4,0} - 2F_{4,2} \right) + b \left(F_{0,4} - 2F_{0,6} \right) + 2c \left(F_{2,2} - 2F_{2,4} \right) \right] = 0 . \tag{3.147}$$

We arbitrarily choose the intermediate normalization condition

$$2F_{0,2} - F_{0,0} = 1 \tag{3.148}$$

that leads to

$$\Delta E = \lambda \left[a \left(2F_{4,2} - F_{4,0} \right) + b \left(2F_{0,6} - F_{0,4} \right) + 2c \left(2F_{2,4} - F_{2,2} \right) \right] . \tag{3.149}$$

Substituting equation (3.149) into equation (3.146) and dividing by λ we derive another useful equation

$$\left[a \left(2F_{4,2} - F_{4,0} \right) + b \left(2F_{0,6} - F_{0,4} \right) + 2c \left(2F_{2,4} - F_{2,2} \right) \right] (2F_{2,0} - F_{0,0})$$
$$+ a \left(F_{4,0} - 2F_{6,0} \right) + b \left(F_{0,4} - 2F_{2,4} \right) + 2c \left(F_{2,2} - 2F_{4,2} \right) = 0 . \tag{3.150}$$

We calculate the perturbation corrections to $F_{0,0}$ from equation (3.143):

$$F_{0,0,m} = \frac{1}{2} \left(a F_{4,0,m-1} + b F_{0,4,m-1} + 2c F_{2,2,m-1} - \sum_{k=1}^{m} E_k F_{0,0,m-k} \right) \tag{3.151}$$

and the perturbation corrections to $F_{0,2}$ from the intermediate normalization condition (3.148):

$$F_{0,2,m} = \frac{\delta_{m0}}{2} + \frac{1}{2} F_{0,0,m} . \tag{3.152}$$

The general recurrence relation (3.131) with $N = 2$ provides all the remaining moment coefficients $F_{i,j,m}$ except $F_{2,0,m}$ which one obtains from equation (3.150). We expand all the moments in this equation in Taylor series about $\lambda = 0$ and collect the coefficients of each power of λ. By means of equations (3.131), (3.151), and (3.152), we express every moment coefficient $F_{i,j,m}$ in terms of previously calculated ones and in terms of the only unknown $F_{2,0,m}$ which we then determine. The equation of order zero is quadratic

$$4cF_{2,0,0}^2 + 18(b - a)F_{2,0,0} - c = 0 \qquad (3.153)$$

and admits two real roots

$$F_{2,0,0} = \frac{9(a - b) \pm R}{4c}, \quad R = \sqrt{81(b - a)^2 + 4c^2} \qquad (3.154)$$

that give rise to two sets of moment coefficients $F_{i,j,m}$ which are the two independent solutions mentioned above. The occurrence of multiple solutions was already anticipated in the discussion of the general equation (3.91). The equations for perturbation orders greater than zero are linear in the unknown moment coefficient $F_{2,0,m}$. We do not show them here because they are rather complicated and do not add anything relevant to the present discussion.

We finally obtain the energy coefficients from

$$
\begin{aligned}
E_m &= a\left(2F_{4,2,m-1} - F_{4,0,m-1}\right) + b\left(2F_{0,6,m-1} - F_{0,4,m-1}\right) \\
&\quad + 2c\left(2F_{2,4,m-1} - F_{2,2,m-1}\right) .
\end{aligned}
\qquad (3.155)
$$

It is worth noticing that unlike the standard treatment of degenerate states in which the energy coefficient is a root of a secular determinant (see Section 1.2.2), here it is one of the moments that arises from a seemingly secular equation.

In the program section we show simple Maple procedures for the application of the moment method according to the equations just discussed. Table 3.7 shows sample results for the states considered above.

The moment method has recently been used to generate renormalized perturbation series for the energies of two-dimensional anharmonic oscillators [39]. The main ideas underlining that approach that yields highly accurate results may be easily understood by means of the theoretical development above, and by the discussion of the renormalized series given in Chapter 6. The moment method has also been applied to coupled Morse oscillators after expanding the potential-energy function in a Taylor series about its minimum [43]. In this case the perturbation series appear to converge for all the states considered. In Chapter 7 we will discuss the application of perturbation theory to nonpolynomial potential-energy functions by means of a simple polynomial approach.

3.4.5   Relation to Other Methods: Modified Moment Method

The moment method is a quite general strategy that reduces to other procedures under particular conditions. One such connection was already outlined above: if the complete set of vectors $\{f_j\}$ is orthonormal, then the moment method gives rise to the standard textbook approach discussed in Chapter 1. Moreover, it is not difficult to prove that the moment method yields the hypervirial theorem. In fact, if we choose the vector $f = \hat{W}\Psi$, where $\hat{W}$ is a linear operator and Ψ is an eigenfunction of $\hat{H}$, then, arguing as in Section 3.2:

$$\left\langle \left(\hat{H} - E\right) f \middle| \Psi \right\rangle = \left\langle \left[\hat{H}, \hat{W}\right] \Psi \middle| \Psi \right\rangle = -\left\langle \Psi \middle| \left[\hat{H}, \hat{W}^\dagger\right] \middle| \Psi \right\rangle = 0 . \qquad (3.156)$$

The perturbation theory by the moment method discussed above does not yield the perturbation corrections to the energy as functions of the quantum number of the unperturbed model. For this reason it is less appealing than the method of Swenson and Danforth for the treatment of separable

Table 3.7 Moment Method for the Two-Dimensional Anharmonic Oscillator

$$\hat{H} = -\frac{\nabla^2}{2} + \frac{\hat{x}^2 + \hat{y}^2}{2} + \lambda \left(a\hat{x}^4 + b\hat{y}^4 + 2c\hat{x}^2\hat{y}^2 \right) \text{ (Continued)}$$

State $N = 0$ (e,e)

$E_1 = \frac{3}{4}a + \frac{3}{4}b + \frac{1}{2}c$

$E_2 = -\frac{21}{8}a^2 - \frac{3}{2}ac - \frac{21}{8}b^2 - \frac{3}{2}bc - \frac{3}{4}c^2$

$E_3 = \frac{333}{16}a^3 + \frac{105}{8}a^2c + \frac{9}{2}abc + \frac{111}{16}ac^2 + \frac{333}{16}b^3 + \frac{105}{8}b^2c + \frac{111}{16}bc^2$
$\quad + \frac{11}{4}c^3$

$E_4 = -\frac{315}{8}a^2bc - \frac{333}{2}a^3c - \frac{3031}{32}a^2c^2 - \frac{315}{8}ab^2c - \frac{1953}{32}abc^2 - \frac{373}{8}ac^3$
$\quad - \frac{30885}{128}a^4 - \frac{333}{2}b^3c - \frac{3031}{32}b^2c^2 - \frac{30885}{128}b^4 - \frac{373}{8}bc^3 - \frac{973}{64}c^4$

$F_{2,0,0} = \frac{1}{2}$

$F_{2,0,1} = -\frac{3}{4}a - \frac{1}{4}c$

$F_{2,0,2} = \frac{75}{16}a^2 + \frac{33}{16}ac + \frac{9}{16}bc + \frac{11}{16}c^2$

$F_{2,0,3} = -\frac{81}{16}abc - \frac{1527}{32}a^3 - \frac{105}{32}c^3 - \frac{801}{32}a^2c - \frac{349}{32}ac^2 - 4b^2c - 5bc^2$

$F_{2,0,4} = \frac{165741}{256}a^4 + \frac{16421}{768}c^4 + \frac{16359}{256}a^2bc + \frac{5173}{64}abc^2 + \frac{11191}{256}b^3c$
$\quad + \frac{43859}{768}b^2c^2 + \frac{49039}{256}a^2c^2 + \frac{9669}{256}ab^2c + \frac{4433}{96}bc^3 + \frac{30995}{384}ac^3$
$\quad + \frac{98283}{256}a^3c$

State $N = 1$ (e,o)

$E_1 = \frac{3}{4}a + \frac{15}{4}b + \frac{3}{2}c$

$E_2 = -\frac{21}{8}a^2 - \frac{9}{2}ac - \frac{165}{8}b^2 - \frac{15}{2}bc - \frac{15}{4}c^2$

$E_3 = \frac{333}{16}a^3 + \frac{315}{8}a^2c + \frac{45}{2}abc + \frac{621}{16}ac^2 + \frac{3915}{16}b^3 + \frac{825}{8}b^2c + \frac{795}{16}bc^2$
$\quad + 21c^3$

$E_4 = -\frac{1575}{8}a^2bc - \frac{999}{2}a^3c - \frac{17913}{32}a^2c^2 - \frac{2475}{8}ab^2c - \frac{15525}{32}abc^2$
$\quad - \frac{3261}{8}ac^3 - \frac{30885}{128}a^4 - \frac{3915}{2}b^3c - \frac{31625}{32}b^2c^2 - \frac{520485}{128}b^4 - \frac{3845}{8}bc^3$
$\quad - \frac{10621}{64}c^4$

$F_{0,3,1} = -\frac{15}{4}b - \frac{3}{4}c$

$F_{0,3,2} = \frac{27}{16}ac + \frac{585}{16}b^2 + \frac{165}{16}bc + \frac{51}{16}c^2$

$F_{0,3,3} = -12a^2c - \frac{411}{16}ac^2 - \frac{6255}{32}b^2c - \frac{2435}{32}bc^2 - \frac{17655}{32}b^3 - \frac{727}{32}c^3 - \frac{405}{16}abc$

$F_{0,3,4} = \frac{79509}{256}a^2c^2 + \frac{1445}{4}ac^3 + \frac{33573}{256}a^3c + \frac{39665}{64}abc^2 + \frac{127845}{256}ab^2c$
$\quad + \frac{48345}{256}a^2bc + \frac{1138875}{256}b^3c + \frac{501495}{256}b^2c^2 + \frac{304115}{384}bc^3 + \frac{2719395}{256}b^4$
$\quad + \frac{160727}{768}c^4$

Table 3.7 *(Cont.)* Moment Method for the Two-Dimensional Anharmonic Oscillator

$$\hat{H} = -\frac{\nabla^2}{2} + \frac{\hat{x}^2 + \hat{y}^2}{2} + \lambda\left(a\hat{x}^4 + b\hat{y}^4 + 2c\hat{x}^2\hat{y}^2\right) \text{ (Continued)}$$

State $N = 1$ (o,e)

$E_1 = \frac{15}{4}a + \frac{3}{4}b + \frac{3}{2}c$

$E_2 = -\frac{165}{8}a^2 - \frac{15}{2}ac - \frac{21}{8}b^2 - \frac{9}{2}bc - \frac{15}{4}c^2$

$E_3 = \frac{3915}{16}a^3 + \frac{825}{8}a^2c + \frac{795}{16}ac^2 + \frac{45}{2}abc + \frac{333}{16}b^3 + \frac{315}{8}b^2c + \frac{621}{16}bc^2$

$\quad\ +21\,c^3$

$E_4 = -\frac{2475}{8}a^2bc - \frac{3915}{2}a^3c - \frac{31625}{32}a^2c^2 - \frac{1575}{8}ab^2c - \frac{15525}{32}abc^2$

$\quad\ -\frac{3845}{8}ac^3 - \frac{520485}{128}a^4 - \frac{999}{2}b^3c - \frac{17913}{32}b^2c^2 - \frac{30885}{128}b^4 - \frac{3261}{8}bc^3$

$\quad\ -\frac{10621}{64}c^4$

$F_{3,0,1} = -\frac{15}{4}a - \frac{3}{4}c$

$F_{3,0,2} = \frac{585}{16}a^2 + \frac{165}{16}ac + \frac{27}{16}bc + \frac{51}{16}c^2$

$F_{3,0,3} = -\frac{405}{16}abc - \frac{411}{16}bc^2 - \frac{17655}{32}a^3 - \frac{727}{32}c^3 - \frac{6255}{32}a^2c - \frac{2435}{32}ac^2 - 12b^2c$

$F_{3,0,4} = \frac{39665}{64}abc^2 + \frac{1138875}{256}a^3c + \frac{48345}{256}ab^2c + \frac{2719395}{256}a^4 + \frac{160727}{768}c^4$

$\quad\ + \frac{127845}{256}a^2bc + \frac{79509}{256}b^2c^2 + \frac{1445}{4}bc^3 + \frac{33573}{256}b^3c + \frac{501495}{256}a^2c^2$

$\quad\ + \frac{304115}{384}ac^3$

State $N = 2$ (o,o)

$E_1 = \frac{15}{4}a + \frac{15}{4}b + \frac{9}{2}c$

$E_2 = -\frac{165}{8}a^2 - \frac{45}{2}ac - \frac{165}{8}b^2 - \frac{45}{2}bc - \frac{63}{4}c^2$

$E_3 = \frac{3915}{16}a^3 + \frac{2475}{8}a^2c + \frac{3825}{16}ac^2 + \frac{225}{2}abc + \frac{3915}{16}b^3 + \frac{2475}{8}b^2c$

$\quad\ + \frac{3825}{16}bc^2 + \frac{477}{4}c^3$

$E_4 = -\frac{12375}{8}a^2bc - \frac{11745}{2}a^3c - \frac{164175}{32}a^2c^2 - \frac{12375}{8}ab^2c - \frac{106425}{32}abc^2$

$\quad\ -\frac{25845}{8}ac^3 - \frac{520485}{128}a^4 - \frac{11745}{2}b^3c - \frac{164175}{32}b^2c^2 - \frac{520485}{128}b^4$

$\quad\ -\frac{25845}{8}bc^3 - \frac{80205}{64}c^4$

$F_{3,1,1} = -\frac{15}{4}a - \frac{9}{4}c$

$F_{3,1,2} = \frac{585}{16}a^2 + \frac{495}{16}ac + \frac{135}{16}bc + \frac{225}{16}c^2$

$F_{3,1,3} = -\frac{17655}{32}a^3 - \frac{4341}{32}c^3 - \frac{18765}{32}a^2c - \frac{11805}{32}ac^2 - \frac{375}{4}b^2c - 165bc^2$

$\quad\ -\frac{2025}{16}abc$

$F_{3,1,4} = \frac{639225}{256}a^2bc + \frac{68725}{16}abc^2 + \frac{3416625}{256}a^3c + \frac{2591685}{256}a^2c^2 + \frac{191805}{64}bc^3$

$\quad\ + \frac{389145}{256}b^3c + \frac{749635}{256}b^2c^2 + \frac{2719395}{256}a^4 + \frac{422333}{256}c^4 + \frac{378075}{256}ab^2c$

$\quad\ + \frac{687835}{128}ac^3$

Table 3.7 *(Cont.)* Moment Method for the Two-Dimensional Anharmonic Oscillator

$$\hat{H} = -\frac{\nabla^2}{2} + \frac{\hat{x}^2 + \hat{y}^2}{2} + \lambda \left(a\hat{x}^4 + b\hat{y}^4 + 2c\hat{x}^2\hat{y}^2\right)$$

State $N = 2$ (e,e)

$$E_1 = \tfrac{21}{4}a + \tfrac{21}{4}b + \tfrac{5}{2}c + \tfrac{1}{2}R$$

$$E_2 = -\tfrac{159}{4}a^2 - 9c^2 - \tfrac{159}{4}b^2 - \tfrac{27}{2}bc - \tfrac{27}{2}ac + \left(\tfrac{2673}{8}ab^2 - \tfrac{2673}{8}b^3 - \tfrac{15}{2}ac^2\right.$$
$$\left. - 54b^2c - 12c^3 - \tfrac{15}{2}bc^2 + \tfrac{2673}{8}a^2b - \tfrac{2673}{8}a^3 + 108abc - 54a^2c\right)/R$$

$$F_{2,0,0} = \tfrac{1}{4}\frac{R}{c} + \tfrac{1}{4}\frac{9a - 9b}{c}$$

$$F_{2,0,1} = \tfrac{1}{32}\frac{324b^2 - 270a^2 + 4c^2 - 54ab - 342bc + 354ac}{c} + \tfrac{1}{32}\left(3402ab^2 - 2916b^3\right.$$
$$\left. + 24ac^2 + 3186b^2c + 1944a^2b - 2430a^3 + 8c^3 + 3186a^2c - 6372abc\right)/(cR)$$

$$F_{0,2,0} = \tfrac{1}{2}$$

$$F_{0,2,1} = \tfrac{1}{16}\frac{(3a+c)R}{c}\tfrac{1}{16}\frac{27a^2 - 27ab + 9ac - 3bc + 2c^2}{c}$$

$$F_{0,2,2} = \tfrac{1}{32}\frac{432a^2b + 243ab^2 - 48bc^2 - 675a^3 + 9a^2c - 144abc - 16c^3 + 39b^2c}{c}$$
$$+ \tfrac{1}{32}\left(-1701a^2b^2 - 2187ab^3 - 84a^2c^2 + 108b^2c^2 - 72ac^3 + 81a^3c\right.$$
$$\left. - 1215cb^3 - 24bc^3 - 6075a^4 - 1377a^2bc - 216abc^2 - 32c^4 + 9963a^3b\right.$$
$$\left. + 2511ab^2c\right)/(cR)$$

$$R = \sqrt{81(b-a)^2 + 4c^2}$$

models. However, the combination of the moment method with the method of Fernández and Castro discussed in Chapter 2 overcomes this limitation, as we illustrate in what follows by means of a simple one-dimensional problem:

$$\hat{H} = \hat{H}_0 + V_1(x), \quad \hat{H}_0 = -\frac{1}{2}\frac{d^2}{dx^2} + V_0(x). \tag{3.157}$$

Choosing a function of the form

$$F(x) = A(x)\Psi_0(x) + B(x)\Psi_0'(x), \tag{3.158}$$

where $A(x)$ and $B(x)$ are two differentiable functions, and $\Psi_0(x)$ is an eigenfunction of $\hat{H}_0$,

$$\Psi_0''(x) = 2\left[V_0(x) - E_0\right]\Psi_0(x), \tag{3.159}$$

we easily obtain

$$\left(\hat{H} - E\right)F = \left[(V_1 - \Delta E)A + 2(E_0 - V_0)B' - \frac{A''}{2} - V_0'B\right]\Psi_0$$
$$+ \left[(V_1 - \Delta E)B - A' - \frac{B''}{2}\right]\Psi_0'. \tag{3.160}$$

If

$$A' = -\frac{B''}{2} + (V_1 - \Delta E)\,B\,, \tag{3.161}$$

then the term containing Ψ_0' vanishes and

$$\left\langle \Psi_0 \left| \frac{A''}{2} + (\Delta E - V_1)\,A + 2\,(V_0 - E_0)\,B' + V'B \right| \Psi \right\rangle = 0\,, \tag{3.162}$$

which is the master equation of the modified moment method.

As a particular example consider the anharmonic oscillator

$$V_0(x) = \frac{x^2}{2},\quad V_1(x) = \lambda x^{2K},\quad K = 2, 3, \ldots\,. \tag{3.163}$$

If $B(x) = x^{2N+1}$, $N = 0, 1, \ldots$, then

$$A(x) = -\frac{2N+1}{2}x^{2N} + \frac{\lambda}{2(N+K+1)}x^{2(N+K+1)} - \frac{\Delta E}{2(N+1)}x^{2(N+1)}\,, \tag{3.164}$$

and equation (3.162) becomes

$$2(N+1)F_{N+1} - \frac{N(4N^2-1)}{2}F_{N-1} - 2(2N+1)E_0 F_N + \lambda(2N+K+1)F_{N+K}$$

$$-(2N+1)\Delta E F_N + \frac{(2N+K+2)\lambda\Delta E}{2(N+1)(N+K+1)}F_{N+K+1} - \frac{\Delta E^2}{2(N+1)}F_{N+1}$$

$$-\frac{\lambda^2}{2(N+K+1)}F_{N+2K+1} = 0\,, \tag{3.165}$$

where

$$F_N = \left\langle \Psi_0 \left| x^{2N} \right| \Psi \right\rangle\,. \tag{3.166}$$

Substituting $N-1$ for N and expanding in Taylor series about $\lambda = 0$, we obtain

$$\begin{aligned}
F_{N,i} = \frac{1}{2N}\Bigg[&\frac{(N-1)[4(N-1)^2-1]}{2}F_{N-2,i} + 2(2N-1)E_0 F_{N-1,i} \\
&- (2N+K-1)F_{N+K-1,i-1} + 2N\sum_{j=1}^{i} E_j F_{N-1,i-j} \\
&- \frac{2N+K}{2N(N+K)}\sum_{j=1}^{i-1} E_j F_{N+K,i-j-1} + \frac{1}{2N}\sum_{j=1}^{i}E_j\sum_{m=1}^{i-j} E_m F_{N,i-j-m} \\
&+ \frac{1}{2(N+K)}F_{N+2K,i-2} \Bigg]\,.
\end{aligned} \tag{3.167}$$

In order to derive an expression for the energy we choose $A = 1$ and $B = 0$, which are consistent with equation (3.161) and lead to

$$\Delta E F_0 = \lambda F_K\,. \tag{3.168}$$

Therefore, the intermediate normalization condition $F_0 = 1$ leads to

$$F_{0,i} = \delta_{0i} \qquad (3.169)$$

and

$$E_i = F_{K,i-1} . \qquad (3.170)$$

In order to obtain E_p we have to calculate $F_{N,i}$ for $i = 0, 1, \ldots, p - 1$, $N = 1, 2, \ldots, (p - i)K$.

It is not difficult to obtain the first perturbation corrections by hand. However, as the perturbation order increases, the calculation becomes tedious and the use of computer algebra is necessary. We have written a simple Maple program that we do not show here because it is quite similar to the one already given for the method of Swenson and Danforth. However, Table 3.8 shows results for the quartic anharmonic oscillator ($K = 2$) that the reader may find useful for testing. Because the perturbation corrections to the energy are exactly those displayed in Table 2.3 we only show some moment coefficients.

Table 3.8 Modified Moment Method for the Quartic Anharmonic Oscillator

$\hat{H} = -\frac{1}{2}\frac{d^2}{dx^2} + \frac{\hat{x}^2}{2} + \lambda\hat{x}^4$

$F_{1,0} = E_0$

$F_{1,1} = -\frac{3}{8} - \frac{3}{2}E_0^2$

$F_{1,2} = \frac{3831}{512}E_0 + \frac{485}{64}E_0^3 - \frac{1}{32}E_0^5$

$F_{1,3} = -\frac{28719}{2048} - \frac{93385}{768}E_0^2 - \frac{20285}{384}E_0^4 + \frac{7}{24}E_0^6$

$F_{1,4} = \frac{1442295127}{786432}E_0^3 + \frac{56302885}{131072}E_0^5 - \frac{41323}{16384}E_0^7 - \frac{259}{49152}E_0^9 + \frac{3395500527}{4194304}E_0$

$F_{2,0} = \frac{3}{8} + \frac{3}{2}E_0^2$

$F_{2,1} = -\frac{67}{16}E_0 - \frac{17}{4}E_0^3$

$F_{2,2} = \frac{1539}{256} + \frac{1707}{32}E_0^2 + \frac{375}{16}E_0^4$

$F_{2,3} = -\frac{89165}{128}E_0^3 - \frac{10689}{64}E_0^5 - \frac{305141}{1024}E_0$

$F_{2,4} = \frac{9317949}{1024}E_0^2 + \frac{587265}{64}E_0^4 + \frac{87549}{64}E_0^6 + \frac{1456569}{2048}$

$F_{3,0} = \frac{25E_0}{8} + \frac{5E_0^3}{2}$

$F_{3,1} = -\frac{945}{256} - \frac{885}{32}E_0^2 - \frac{165}{16}E_0^4$

$F_{3,2} = \frac{594853}{4096}E_0 + \frac{301195}{1024}E_0^3 + \frac{15947}{256}E_0^5 + \frac{5}{64}E_0^7$

$F_{3,3} = -\frac{491632143}{131072}E_0^2 - \frac{221146647}{65536}E_0^4 - \frac{7701}{8192}E_0^8 - \frac{3750807}{8192}E_0^6 - \frac{678272805}{2097152}$

3.5   Perturbation Theory in Operator Form

The approaches discussed above share the name of perturbation theory without wavefunction because they do not take the eigenfunctions explicitly into account. However, the eigenfunctions already appear in some way or another through expectation values or moments. We may even say that those approaches are based on unusual representations in which a recurrence relation plays the role of the Schrödinger equation, and a set of expectation values or moments is a substitute for the eigenvector. On the other hand, the perturbation theory in operator form discussed in what follows is a true perturbation theory without wavefunction because it only considers the Hamiltonian operator [44]–[47].

Perturbation theory in operator form consists of an appropriate transformation of the Hamiltonian operator

$$\hat{H} = \hat{H}_0 + \lambda \hat{H}' \tag{3.171}$$

by means of a unitary operator $\hat{U}(\lambda)$:

$$\hat{K} = \hat{U}(\lambda)\hat{H}\hat{U}(\lambda)^\dagger \tag{3.172}$$

in such a way that

$$\left[\hat{K}, \hat{H}_0\right] = 0 . \tag{3.173}$$

Consequently, the operators $\hat{K}$ and $\hat{H}_0$ share a complete set of eigenvectors. If Ψ_0 is an eigenvector of both $\hat{H}_0$ and $\hat{K}$

$$\hat{H}_0\Psi_0 = E_0\Psi_0, \ \hat{K}\Psi_0 = E\Psi_0 , \tag{3.174}$$

then $\Psi = \hat{U}^\dagger \Psi_0$ is an eigenvector of $\hat{H}$:

$$\hat{H}\Psi = \hat{H}\hat{U}^\dagger\Psi_0 = \hat{U}^\dagger\hat{U}\hat{H}\hat{U}^\dagger\Psi_0 = E\hat{U}^\dagger\Psi_0 = E\Psi . \tag{3.175}$$

Because we cannot obtain the transformation (3.172) exactly except for some simple models, we apply perturbation theory to equations (3.172) and (3.173). To this end, we expand $\hat{U}$ and $\hat{K}$ in Taylor series about $\lambda = 0$

$$\hat{U} = \sum_{j=0}^{\infty} \hat{U}_j\lambda^j, \ \hat{K} = \sum_{j=0}^{\infty} \hat{K}_j\lambda^j , \tag{3.176}$$

where

$$\hat{U}_0 = \hat{1}, \ \hat{K}_0 = \hat{H}_0 . \tag{3.177}$$

There is no unique expression for $\hat{U}$; one can choose, for example, a single exponential operator or an infinite product of exponential operators [47]. All particular representations of the operator $\hat{U}$ are equivalent, though some of them may be more practical than others. Here we write

$$\frac{d\hat{U}(\lambda)}{d\lambda} = \hat{W}(\lambda)\hat{U}(\lambda) , \tag{3.178}$$

where $\hat{W}$ is a yet unknown linear operator. Differentiating $\hat{U}\hat{U}^\dagger = \hat{1}$ with respect to λ, and taking into account that $d\hat{U}^\dagger/d\lambda = \hat{U}^\dagger\hat{W}^\dagger$, one easily proves that $\hat{W}$ is antihermitian: $\hat{W}^\dagger = -\hat{W}$; therefore

$$\frac{d\hat{U}(\lambda)^\dagger}{d\lambda} = -\hat{U}(\lambda)^\dagger\hat{W}(\lambda) . \tag{3.179}$$

It is our purpose to express the coefficients $\hat{U}_j$ in terms of the coefficients $\hat{W}_j$ of the expansion

$$\hat{W} = \sum_{j=0}^{\infty} \hat{W}_j\lambda^j . \tag{3.180}$$

It follows from equation (3.178) that both sets of coefficients are related by

$$\hat{U}_j = \frac{1}{j}\sum_{i=0}^{j-1} \hat{W}_i\hat{U}_{j-i-1} . \tag{3.181}$$

Sometimes, the use of superoperators [48] simplifies the notation and facilitates the discussion. In order to introduce them into the present perturbation approach in a natural way, consider an arbitrary operator $\hat{A}$ independent of λ. Taking into account that $d(\hat{U}A\hat{U}^\dagger)/d\lambda = [\hat{W}, \hat{U}A\hat{U}^\dagger]$ we define the superoperators $\widehat{U}$ and $\widehat{W}$ as follows:

$$\widehat{U}\hat{A} = \hat{U}A\hat{U}^\dagger, \quad \widehat{W}\hat{A} = \left[\hat{W}, \hat{A}\right] , \tag{3.182}$$

so that

$$\frac{d}{d\lambda}\widehat{U}A = \widehat{W}\widehat{U}\hat{A} . \tag{3.183}$$

Superoperators are operators that apply to a vector space of linear operators [48]. By using them, we do not obtain new results, but commonly the working equations become simpler. Since the operator $\hat{A}$ is arbitrary, it follows from equation (3.183) that $\widehat{U}(\lambda)$ satisfies

$$\frac{d\widehat{U}(\lambda)}{d\lambda} = \widehat{W}(\lambda)\widehat{U}(\lambda), \quad \widehat{U}(0) = \widehat{1} , \tag{3.184}$$

where $\widehat{1}$ is the identity superoperator that we omit from now on. It is not difficult to verify that the coefficients of the Taylor expansions

$$\widehat{U}(\lambda) = \sum_{j=0}^{\infty} \widehat{U}_j\lambda^j, \quad \widehat{W}(\lambda) = \sum_{j=0}^{\infty} \widehat{W}_j\lambda^j \tag{3.185}$$

are related by

$$\widehat{U}_j = \frac{1}{j}\sum_{i=0}^{j-1} \widehat{W}_i\widehat{U}_{j-i-1} , \tag{3.186}$$

where $\widehat{W}_i\,\hat{A} = [\hat{W}_i, \hat{A}]$ for any linear operator $\hat{A}$.

It follows from equation (3.186) and from straightforward expansion of the transformation of the Hamiltonian operator $\widehat{U}\,\hat{H} = \hat{K}$ in a Taylor series about $\lambda = 0$ that

$$\widehat{H}_0\hat{W}_j = \hat{f}_j - (j+1)\hat{K}_{j+1}, \quad \hat{f}_j = \sum_{i=0}^{j-1}\widehat{W}_i\widehat{U}_{j-i}\hat{H}_0 + (j+1)\widehat{U}_j\widehat{H}', \qquad (3.187)$$

where we have substituted $j+1$ for j and taken into account that $\widehat{W}_j\hat{H}_0 = -\widehat{H}_0\hat{W}_j$. Assuming that $\hat{f}_j$ is known, it only remains to solve the operator equation (3.187) for $\hat{K}_{j+1}$ and $\hat{W}_j$. We can formally write the solution as follows:

$$\hat{W}_j = \widehat{H}_0^{-1}\left[\hat{f}_j - (j+1)\hat{K}_{j+1}\right], \qquad (3.188)$$

provided that $(j+1)\hat{K}_j$ removes all the terms in $\hat{f}_j$ that commute with $\hat{H}_0$. In this way equation (3.188) completely determines $\hat{W}_j$ and $\hat{K}_j$.

It is particularly easy to solve equation (3.188) when $\hat{f}_j$ is a sum of eigenvectors of $\widehat{H}_0$. The reason is that if $\hat{A}$ is an eigenvector of $\widehat{H}_0$ with eigenvalue $\alpha \neq 0$ $\widehat{H}_0\hat{A} = \alpha\hat{A}$, then $\widehat{H}_0^{-1}A = \alpha^{-1}\hat{A}$. Moreover, taking into account that $\exp(t\widehat{H}_0)\hat{A} = \exp(\alpha t)\hat{A}$ we write

$$\hat{W}_j = \lim_{t\longrightarrow 0}\int^t \exp\left(s\widehat{H}_0\right)\hat{f}_j\,ds, \qquad (3.189)$$

and then

$$\hat{K}_{j+1} = \frac{1}{j+1}\left(\hat{f}_j - \widehat{H}_0\hat{W}_j\right). \qquad (3.190)$$

These equations give the desired results if we calculate the integrals as follows:

$$\int^t \exp(\alpha s)ds = \begin{cases} \exp(\alpha t)/\alpha \text{ if } \alpha \neq 0 \\ t \text{ if } \alpha = 0 \end{cases}. \qquad (3.191)$$

Notice that in this way the terms that commute with $\hat{H}_0$ ($\alpha = 0$) vanish as $t \longrightarrow 0$. The use of the exponential superoperator is practical if one plans to invert $\widehat{H}_0$ by means of an appropriate computer-algebra software.

It follows from equations (3.174) and (3.175) that the perturbation corrections to the eigenvalues and eigenvectors of $\widehat{H}$ are given by

$$\hat{K}_j\Psi_0 = E_j\Psi_0, \qquad (3.192)$$

and

$$\Psi_j = \hat{U}_j^\dagger\Psi_0, \qquad (3.193)$$

respectively, where the operators $\hat{U}_j^\dagger$ are recursively determined by the adjoint of equation (3.181). For example, the first three are

$$\hat{U}_0^\dagger = \hat{1}, \quad \hat{U}_1^\dagger = -\hat{W}_0, \quad \hat{U}_2^\dagger = \frac{1}{2}\left(\hat{W}_0^2 - \hat{W}_1\right). \qquad (3.194)$$

3.5.1 Illustrative Example: The Anharmonic Oscillator

According to our philosophy of choosing simple examples to illustrate the application of the perturbation approaches discussed in this book, in what follows we consider the dimensionless anharmonic oscillator

$$\hat{H} = \hat{H}_0 + \lambda \hat{x}^{2M}, \ \hat{H}_0 = \frac{1}{2}\left(-\hat{D}^2 + \hat{x}^2\right), \tag{3.195}$$

where $\hat{D} = \frac{d}{dx}$. The use of boson operators greatly facilitates the calculation because $\hat{a}$ and $\hat{a}^\dagger$ are eigenvectors of $\widehat{\hat{H}_0}$. Substituting

$$x = \frac{1}{\sqrt{2}}\left(\hat{a} + \hat{a}^\dagger\right), \ \hat{D} = \frac{1}{\sqrt{2}}\left(\hat{a} - \hat{a}^\dagger\right) \tag{3.196}$$

the unperturbed Hamiltonian reads

$$\hat{H}_0 = \hat{a}^\dagger \hat{a} + \frac{1}{2}. \tag{3.197}$$

Solving equation (3.188) for $\hat{W}_j$ and $\hat{K}_{j+1}$ is remarkably simple because $\hat{f}_j$ is a polynomial function

Table 3.9 Perturbation Theory in Operator Form for the Anharmonic

Oscillator $\hat{H} = \frac{1}{2}(-\frac{d^2}{dx^2} + \hat{x}^2) + \lambda \hat{x}^4$

$$\hat{W}_0 = \frac{1}{6}\left[\left(\hat{a}^\dagger\right)^4 - \hat{a}^4\right] + \frac{1}{2}\left[\left(\hat{a}^\dagger\right)^3 \hat{a} - \hat{a}^\dagger \hat{a}^3\right] + \frac{3}{4}\left[\left(\hat{a}^\dagger\right)^2 - \hat{a}^2\right]$$

$$\hat{K}_1 = \frac{3}{2}\left(\hat{a}^\dagger\right)^2 \hat{a}^2 + 3\hat{a}^\dagger \hat{a} + \frac{3}{4}$$

$$\hat{W}_1 = \frac{1}{24}\left[\hat{a}^6 - \left(\hat{a}^\dagger\right)^6\right] + \frac{3}{8}\left[\hat{a}^\dagger \hat{a}^5 - \left(\hat{a}^\dagger\right)^5 \hat{a}\right] + \frac{33}{8}\left[\left(\hat{a}^\dagger\right)^2 \hat{a}^4 - \left(\hat{a}^\dagger\right)^4 \hat{a}^2\right]$$
$$+ \frac{15}{16}\left[\hat{a}^4 - \left(\hat{a}^\dagger\right)^4\right] + \frac{33}{2}\left[\hat{a}^\dagger \hat{a}^3 - \left(\hat{a}^\dagger\right)^3 \hat{a}\right] + \frac{171}{16}\left[\hat{a}^2 - \left(\hat{a}^\dagger\right)^2\right]$$

$$\hat{K}_2 = -\frac{17}{4}\left(\hat{a}^\dagger\right)^3 \hat{a}^3 - \frac{153}{8}\left(\hat{a}^\dagger\right)^2 \hat{a}^2 - 18\hat{a}^\dagger \hat{a} - \frac{21}{8}$$

$$\hat{W}_2 = -\frac{1}{128}\hat{a}^{10} + \frac{1107}{512}\hat{a}^8 - \frac{43913}{384}\hat{a}^6 + \frac{97605}{256}\hat{a}^4 + \frac{2216529}{512}\hat{a}^2 + \frac{1}{128}\left(\hat{a}^\dagger\right)^{10}$$
$$- \frac{123}{256}\left(\hat{a}^\dagger\right)^9 \hat{a} + \frac{1099}{128}\left(\hat{a}^\dagger\right)^8 \hat{a}^2 - \frac{1107}{512}\left(\hat{a}^\dagger\right)^8 - \frac{2221}{128}\left(\hat{a}^\dagger\right)^7 \hat{a}^3 + \frac{1099}{16}\left(\hat{a}^\dagger\right)^7 \hat{a}$$
$$- \frac{25751}{64}\left(\hat{a}^\dagger\right)^6 \hat{a}^4 - \frac{46641}{256}\left(\hat{a}^\dagger\right)^6 \hat{a}^2 + \frac{43913}{384}\left(\hat{a}^\dagger\right)^6 - \frac{77253}{16}\left(\hat{a}^\dagger\right)^5 \hat{a}^3$$
$$- \frac{33081}{64}\left(\hat{a}^\dagger\right)^5 \hat{a} + \frac{25751}{64}\left(\hat{a}^\dagger\right)^4 \hat{a}^6 - \frac{525399}{32}\left(\hat{a}^\dagger\right)^4 \hat{a}^2 - \frac{97605}{256}\left(\hat{a}^\dagger\right)^4$$
$$+ \frac{2221}{128}\left(\hat{a}^\dagger\right)^3 \hat{a}^7 + \frac{77253}{16}\left(\hat{a}^\dagger\right)^3 \hat{a}^5 - \frac{69567}{4}\left(\hat{a}^\dagger\right)^3 \hat{a} - \frac{1099}{128}\left(\hat{a}^\dagger\right)^2 \hat{a}^8$$
$$+ \frac{46641}{256}\left(\hat{a}^\dagger\right)^2 \hat{a}^6 + \frac{525399}{32}\left(\hat{a}^\dagger\right)^2 \hat{a}^4 - \frac{2216529}{512}\left(\hat{a}^\dagger\right)^2 + \frac{123}{256}\hat{a}^\dagger \hat{a}^9$$
$$- \frac{1099}{16}\hat{a}^\dagger \hat{a}^7 + \frac{33081}{64}\hat{a}^\dagger \hat{a}^5 + \frac{69567}{4}\hat{a}^\dagger \hat{a}^3$$

$$\hat{K}_3 = -\frac{10689}{64}\left(\hat{a}^\dagger\right)^5 \hat{a}^5 - \frac{267225}{128}\left(\hat{a}^\dagger\right)^4 \hat{a}^4 - \frac{498865}{64}\left(\hat{a}^\dagger\right)^3 \hat{a}^3 - \frac{1283085}{128}\left(\hat{a}^\dagger\right)^2 \hat{a}^2$$
$$- 3825\hat{a}^\dagger \hat{a} - \frac{30885}{128}$$

of the boson operators and, consequently, a linear combination of eigenvectors of $\widehat{\hat{H}}_0$. Taking into account that

$$\widehat{\hat{H}}_0 \hat{a} = -\hat{a}, \ \widehat{\hat{H}}_0 \hat{a}^\dagger = \hat{a}^\dagger , \tag{3.198}$$

we easily prove that $\widehat{\hat{H}}_0 (\hat{a}^\dagger)^m \hat{a}^n = (m - n)(\hat{a}^\dagger)^m \hat{a}^n$, and $\widehat{\hat{H}}_0^{-1} (\hat{a}^\dagger)^m \hat{a}^n = (m - n)^{-1}(\hat{a}^\dagger)^m \hat{a}^n$ if $m \neq n$. Therefore, according to the discussion above, it is clear that we should choose $\hat{K}_{j+1}$ to remove all the diagonal terms $(\hat{a}^\dagger)^m \hat{a}^m$ from $\hat{f}_j$. In order to solve the equations of the operator method, one has to be careful about the order of the noncommuting operators. For that reason, keeping a given operator order facilitates the calculation. Here we adopt what is commonly called normal order in which the powers of the creation operator $\hat{a}^\dagger$ appear to the left of the powers of the annihilation operator $\hat{a}$, as in the example above.

Table 3.9 shows the first few operators $\hat{W}_j$ and $\hat{K}_{j+1}$ for the anharmonic oscillator with $M = 2$. Notice that each operator $\hat{W}_j$ is antihermitian as expected from the fact that $\hat{W}$ is antihermitian. Because one-dimensional models do not exhibit degeneracy, the eigenvectors $|n>$ of $\hat{H}_0$ are also eigenvectors of $\hat{K}$. Taking into account that [49]

$$\hat{a}|n> = \sqrt{n}|n - 1>, \ \hat{a}^\dagger|n> = \sqrt{n+1}|n + 1> , \tag{3.199}$$

we easily calculate the perturbation corrections in terms of the harmonic oscillator quantum number n according to equations (3.192) and (3.193). For example, the energy coefficients $E_{n,j}$ given by $\hat{K}_j|n> = E_{n,j}|n>$ agree with those in Table 1.1.

Perturbation theory in operator form may take many different, though equivalent, forms. For example, it is instructive to compare the present approach with an earlier one based on a particular representation of the unitary operator $\hat{U}$ [47]. The reader may easily convince himself that perturbation theory in operator form is far from being the most practical approach to treat simple quantum-mechanical models like the one-dimensional anharmonic oscillator (compare it, for example, with the method of Swenson and Danforth discussed earlier in this chapter). However, perturbation theory in operator form has certainly been the preferred approximate method for the treatment of several problems of physical interest [50]–[52].

Chapter 4

Simple Atomic and Molecular Systems

4.1 Introduction

In this chapter we apply some of the perturbation methods developed earlier in this book to simple atomic and molecular systems. Such physically motivating models are worth a separate treatment and a more detailed study. As illustrative examples we consider the Stark and Zeeman effects in hydrogen and the hydrogen molecular ion in the Born–Oppenheimer approximation [53]. The Schrödinger equation for the Stark effect in hydrogen is separable in parabolic coordinates and is suitable for illustrating the application of the method of Swenson and Danforth, discussed in Chapter 3, to a problem with separation constants other than those arising from the use of spherical coordinates. We also treat the Stark effect as a nonseparable problem in spherical coordinates so that the reader may compare both approaches. Although the latter coordinate system is not the most convenient for the Stark effect in hydrogen, the resulting equation is suitable for the application of the moment method.

No coordinate system has yet been found that renders the Schrödinger equation for the Zeeman effect in hydrogen separable. In Chapter 2 we have already treated the ground state of this problem by means of the method of Dalgarno and Stewart and logarithmic perturbation theory. Here we apply the moment method also to excited states providing interesting additional examples of the treatment of both nondegenerate and degenerate states.

The hydrogen molecular ion in the Born–Oppenheimer approximation is separable in elliptical (also called prolate spheroidal) coordinates [54]. However, we write the Schrödinger equation in spherical coordinates and apply the moment method for nonseparable problems to obtain part of the expansion of the electronic energies at large internuclear distances. This example differs from all those discussed before in that the perturbation is not a polynomial function of the coordinates. For this reason we have to expand it in a Taylor series in order to apply the moment method.

4.2 The Stark Effect in Hydrogen

4.2.1 Parabolic Coordinates

The Hamiltonian operator for the nonrelativistic isolated hydrogen atom in the coordinate representation is

$$\hat{H}_0 = -\frac{\hbar^2}{2m}\nabla^2 - \frac{e^2}{r}, \tag{4.1}$$

where m is the reduced mass, r is the distance between the nucleus and the electron, and $e > 0$ and $-e$ are, respectively, the nucleus and electron charges. Accordingly, the dipole moment of the atom is $\mathbf{d} = -e\mathbf{r}$. If the electric field is directed along the z axis $\mathbf{F} = F\mathbf{k}$, then the interaction energy is $H' = -\mathbf{d}.\mathbf{F} = eFz$. Choosing $\gamma = \hbar^2/(me^2)$ and $e^2/\gamma = me^4/\hbar^2$ as units of length and energy, respectively, and $\lambda = F\gamma^2/e = m\gamma^3 Fe/\hbar^2$ as a dimensionless perturbation parameter, then the dimensionless Hamiltonian operator $\hat{H} = \hat{H}_0 + \hat{H}'$ reads

$$\hat{H} = -\frac{1}{2}\nabla^2 - \frac{1}{r} + \lambda z \,. \tag{4.2}$$

As said before, the Schrödinger equation for this model is separable in parabolic coordinates $\xi = r - z \geq 0, \eta = r + z \geq 0, 0 \leq \phi = \tan(y/x) < 2\pi$. The inverse transformation

$$x = \sqrt{\xi\eta}\cos(\phi), \;\; y = \sqrt{\xi\eta}\sin(\phi), \;\; z = \frac{1}{2}(\xi - \eta) \tag{4.3}$$

is suitable for straightforward application of the method in Appendix A that yields the Laplacian operator

$$\nabla^2 = \frac{4}{\xi + \eta}\left(\frac{\partial}{\partial\xi}\xi\frac{\partial}{\partial\xi} + \frac{\partial}{\partial\eta}\eta\frac{\partial}{\partial\eta} + \frac{\xi + \eta}{4\xi\eta}\frac{\partial^2}{\partial\phi^2}\right) \,. \tag{4.4}$$

Factorization of the solutions of the Schrödinger equation as $\Psi_{n1,n2,m}(\xi, \eta, \phi) = F_{n_1}(\xi)G_{n_2}(\eta)$ $\exp(im\phi)/\sqrt{2\pi}, m = 0, \pm 1, \pm 2, \ldots, n_1, n_2 = 0, 1, \ldots$, leads to

$$\left(-\frac{\partial}{\partial\xi}\xi\frac{\partial}{\partial\xi} + \frac{m^2}{4\xi} - \frac{E}{2}\xi + \frac{\lambda}{4}\xi^2 - \frac{\partial}{\partial\eta}\eta\frac{\partial}{\partial\eta} + \frac{m^2}{4\eta} - \frac{E}{2}\eta - \frac{\lambda}{4}\eta^2 - 1\right) F_{n_1}(\xi)G_{n_2}(\eta) = 0 \,. \tag{4.5}$$

We split this equation into two one-dimensional parts

$$\left(-\frac{\partial}{\partial\xi}\xi\frac{\partial}{\partial\xi} + \frac{m^2}{4\xi} - \frac{E}{2}\xi + \frac{\lambda}{4}\xi^2 - A\right) F_{n_1}(\xi) = 0 \,, \tag{4.6}$$

$$\left(-\frac{\partial}{\partial\eta}\eta\frac{\partial}{\partial\eta} + \frac{m^2}{4\eta} - \frac{E}{2}\eta - \frac{\lambda}{4}\eta^2 - B\right) G_{n_2}(\eta) = 0 \,, \tag{4.7}$$

where the eigenvalues A and B are separation constants that satisfy $A + B = 1$. A pair of eigenvalues $A(E, n_1, m, \lambda)$ and $B(E, n_2, m, \lambda)$ completely determines the energy $E_{n_1,n_2,m}(\lambda)$ as a root of $A(E, n_1, m, \lambda) + B(E, n_2, m, \lambda) = 1$. Notice that both equations (4.6) and (4.7) are of the form

$$\left(-\frac{d}{du}u\frac{d}{du} + \frac{m^2}{4u} - \frac{E}{2}u + \frac{\sigma\lambda}{4}u^2 - C\right) F(u) = 0 \,, \tag{4.8}$$

where $\sigma = 1$ or $\sigma = -1$, respectively.

We are aware of two earlier applications of the method of Swenson and Danforth to this problem [55, 56]. We have tried variants of both, finally selecting an approach that in our opinion keeps the best features of each. We outline it in what follows.

First rewrite equation (4.8) in a way that resembles the radial equation for a hydrogen atom with a polynomial perturbation discussed in Chapter 3:

$$\left(-\frac{d^2}{du^2} + \frac{m^2 - 1}{4u^2} - \frac{E}{2} + \frac{\sigma\lambda}{4}u - \frac{C}{u}\right) u^{1/2}F(u) = 0 \,. \tag{4.9}$$

In this way we can apply most of the well-known results for the central-field models. For example, when $\lambda = 0$ we compare present eigenvalue equations with the radial equation for the hydrogen atom and easily obtain

$$E_0 = -\frac{A_0^2}{2k_1^2} \Leftrightarrow A_0 = k_1\sqrt{-2E_0}, \ k_1 = n_1 + \frac{|m| + 1}{2} \tag{4.10}$$

for equation (4.6), and similar expressions for equation (4.7) with B_0, k_2, and n_2. It follows from $A_0 + B_0 = 1$ that

$$E_0 = -\frac{1}{2(k_1 + k_2)^2} = -\frac{1}{2(n_1 + n_2 + |m| + 1)^2}, \tag{4.11}$$

which is the well-known energy of the isolated hydrogen atom. We can also proceed in a different way taking into account that E_0 has the same value in both equations mentioned above, and solving

$$E_0 = -\frac{A_0^2}{2k_1^2} = -\frac{(1 - A_0)^2}{2k_2^2} \tag{4.12}$$

for A_0. Only one of the two roots gives the correct result $A_0 = k_1/(k_1 + k_2)$, as the other one is unacceptable on physical grounds.

Consider the perturbed equation (4.9). Straightforward application of the hypervirial theorems as in Chapter 3 yields

$$\frac{(j + 1)E}{2}U_j + \frac{j}{4}\left(j^2 - m^2\right)U_{j-2} + \left(j + \frac{1}{2}\right)CU_{j-1}$$
$$- \frac{\sigma\lambda}{4}\left(j + \frac{3}{2}\right)U_{j+1} = 0, \tag{4.13}$$

where $U_j = <u^j>$. When $j = 0$ this equation reduces to

$$E + CU_{-1} - \frac{3\sigma\lambda}{4}U_1 = 0, \tag{4.14}$$

where we have chosen the intermediate normalization condition $U_0 = 1$. Allowing both E and C to depend on λ, the Hellmann–Feynman theorem takes the form

$$\frac{\partial E}{\partial \lambda} = -2\frac{\partial C}{\partial \lambda}U_{-1} + \frac{\sigma}{2}U_1. \tag{4.15}$$

Substitute the Taylor series

$$E = \sum_{i=0}^{\infty} E_i\lambda^i, \ C = \sum_{i=0}^{\infty} C_i\lambda^i, \ U_j = \sum_{i=0}^{\infty} U_{j,i}\lambda^i \tag{4.16}$$

into the equations given above to derive

$$
U_{j,p} = \frac{2}{(j+1)E_0}\left[\frac{j}{4}\left(m^2-j^2\right)U_{j-2,p} - \left(j+\frac{1}{2}\right)\sum_{i=0}^{p}C_iU_{j-1,p-i}\right.
$$
$$
\left. -\frac{j+1}{2}\sum_{i=1}^{p}E_iU_{j,p-i} + \frac{\sigma}{4}\left(j+\frac{3}{2}\right)U_{j+1,p-1}\right], \tag{4.17}
$$

$$
U_{0,p} = \delta_{0p}, \tag{4.18}
$$

$$
U_{-1,p} = \frac{1}{C_0}\left(-E_p + \frac{3\sigma}{4}U_{1,p-1} - \sum_{i=1}^{p}C_iU_{-1,p-i}\right), \tag{4.19}
$$

$$
E_p = \frac{1}{p}\left(\frac{\sigma}{2}U_{1,p-1} - 2\sum_{i=1}^{p}iC_iU_{-1,p-i}\right). \tag{4.20}
$$

The calculation of the energy coefficients is similar to that in Chapter 3 for the perturbed Coulomb problem. It is not difficult to verify that E_j is linear in C_j and nonlinear in the coefficients C_i with $i < j$. Substitute

$$
E_0 = -\frac{A_0^2}{2k_1^2}, \quad C_j = A_j, \quad A_0 = \frac{k_1}{k_1+k_2}, \quad \sigma = 1, \tag{4.21}
$$

and

$$
E_0 = -\frac{(1-A_0)^2}{2k_2^2}, \quad C_j = \delta_{j0} - A_j, \quad \sigma = -1 \tag{4.22}
$$

to obtain two sets of energy coefficients E_j^I and E_j^{II}, respectively. Solve $E_j^I = E_j^{II}$, $j = 1, 2, \ldots$, for A_j, and then substitute the result back into either E_j^I or E_j^{II}. The calculation is straightforward because each equation $E_j^I - E_j^{II} = 0$ for $j > 0$ is linear in the only unknown A_j. In this way we obtain both A_j and E_j in terms of the quantum numbers n_1, n_2, and m. In the program section we show a set of simple Maple procedures for the calculation just described.

Table 4.1 shows the first coefficients A_j and E_j in terms of the quantum numbers. For comparison purposes we express the results in terms of

$$
n = k_1 + k_2 = n_1 + n_2 + |m| + 1, \quad q = k_1 - k_2 = n_1 - n_2, \tag{4.23}
$$

which lead to simpler expressions. As far as we know, the most extensive calculation of analytical perturbation corrections to the Stark effect in hydrogen has been carried out by a Maple program running an algorithm based on algebraic methods [57].

4.2.2 Spherical Coordinates

Before discussing the particular case of the Stark effect in spherical coordinates, it is convenient to consider the application of the moment method to a hydrogen atom with a more general perturbation $\hat{H}'$:

$$
\hat{H} = -\frac{1}{2}\nabla^2 - \frac{1}{r} + \lambda\hat{H}'(r,\theta,\phi). \tag{4.24}
$$

Table 4.1 Method of Swenson and Danforth for the Stark Effect in Hydrogen in Parabolic Coordinates

$$A_1 = -\tfrac{1}{8} n^2 \left(m^2 - 1 - 3 n^2 + 3 q^2\right)$$

$$A_2 = -\tfrac{1}{16} n^5 q \left(-n^2 + q^2 - 6 + 6 m^2\right)$$

$$A_3 = -\tfrac{1}{256} n^8 \left(-171 n^4 - 622 n^2 + 186 n^2 q^2 + 82 m^2 n^2 + 30 m^2 + 390 q^2 + 150 m^2 q^2 \right.$$
$$\left. -55 + 25 m^4 - 15 q^4\right)$$

$$A_4 = -\tfrac{1}{512} n^{11} q \left(419 n^4 + 1035 m^2 n^2 + 15 n^2 - 380 n^2 q^2 - 39 q^4 + 300 m^4 + 285 m^2 q^2 \right.$$
$$\left. +2400 m^2 - 2700 - 1335 q^2\right)$$

$$A_5 = -\tfrac{1}{4096} n^{14} \left(-27024 n^6 + 17135 m^2 n^4 + 26892 q^2 n^4 - 333515 n^4 - 425654 n^2 \right.$$
$$+305550 n^2 q^2 + 240 q^4 n^2 + 112800 m^2 n^2 + 3110 m^4 n^2 + 5970 m^2 n^2 q^2 - 108 q^6$$
$$-29663 + 3885 q^4 + 4320 m^2 q^2 + 6690 m^4 q^2 + 975 m^2 q^4 + 7467 m^4 + 21561 m^2$$
$$\left. +298734 q^2 + 635 m^6\right)$$

$$A_6 = -\tfrac{1}{8192} n^{17} q \left(223309 n^6 + 2746578 n^4 + 146382 m^2 n^4 - 223639 q^2 n^4 + 1270579 n^2 \right.$$
$$+12420 m^2 n^2 q^2 + 63 q^4 n^2 + 25275 m^4 n^2 + 1847130 m^2 n^2 - 2904900 n^2 q^2$$
$$-2344614 + 39774 m^4 + 2294490 m^2 + 23445 m^4 q^2 + 978 q^4 - 3190819 q^2$$
$$\left. +24390 m^2 q^2 + 10350 m^6 + 267 q^6 - 1458 m^2 q^4\right)$$

$$E_1 = \frac{3}{2} n q$$

$$E_2 = \frac{1}{16} \left(-17 n^2 + 3 q^2 - 19 + 9 m^2\right) n^4$$

$$E_3 = \frac{3}{32} \left(23 n^2 + 11 m^2 + 39 - q^2\right) q n^7$$

$$E_4 = \frac{1}{1024} \left(-5487 n^4 - 1806 n^2 q^2 - 35182 n^2 + 3402 m^2 n^2 + 8622 m^2 - 16211 - 5754 q^2 \right.$$
$$\left. +549 m^4 + 1134 m^2 q^2 - 147 q^4\right) n^{10}$$

$$E_5 = \frac{3}{1024} \left(10563 n^4 + 90708 n^2 + 772 m^2 n^2 + 98 n^2 q^2 + 59293 - 21 q^4 + 220 m^2 q^2 \right.$$
$$\left. +830 m^2 + 780 q^2 + 725 m^4\right) q n^{13}$$

$$E_6 = \frac{1}{8192} \left(-547262 n^6 - 685152 q^2 n^4 - 9630693 n^4 + 429903 m^2 n^4 - 22691096 n^2 \right.$$
$$+25470 m^2 n^2 q^2 + 4786200 m^2 n^2 - 7787370 n^2 q^2 - 390 q^4 n^2 + 16200 m^4 n^2$$
$$-7335413 + 62100 m^2 q^2 - 765 m^2 q^4 - 1185 q^4 - 7028718 q^2 + 372 q^6$$
$$\left. +16845 m^4 + 36450 m^4 q^2 + 6951 m^6 + 4591617 m^2\right) n^{16}$$

By means of the method developed in Appendix A we easily obtain the Laplacian in spherical coordinates

$$\nabla^2 = \frac{1}{r^2}\frac{\partial}{\partial r}r^2\frac{\partial}{\partial r} + \frac{1}{r^2\sin(\theta)}\frac{\partial}{\partial\theta}\sin(\theta)\frac{\partial}{\partial\theta} + \frac{1}{r^2\sin(\theta)^2}\frac{\partial^2}{\partial\phi^2}\ . \tag{4.25}$$

In order to build an appropriate recurrence relation for the moments of the eigenfunctions of $\hat{H}$ we choose the set of functions

$$f_{i,j,k,m} = \sin(\theta)^i\cos(\theta)^j r^k \exp(-\alpha r + \iota m\phi)\ , \tag{4.26}$$

where $i, j, k = 0, 1, \ldots$ and $m = 0, \pm 1, \ldots$. To avoid confusion ι denotes the imaginary number. It is not difficult to verify that

$$\begin{aligned}\left(\hat{H} - E\right)f_{i,j,k,m} &= \frac{(i+j)(i+j+1) - k(k+1)}{2}f_{i,j,k-2,m} \\ &+ [\alpha(k+1) - 1]f_{i,j,k-1,m} + \frac{m^2 - i^2}{2}f_{i-2,j,k-2,m} \\ &- \frac{j(j-1)}{2}f_{i,j-2,k-2,m} - \Delta E f_{i,j,k,m} + \lambda\hat{H}' f_{i,j,k,m}\ ,\end{aligned} \tag{4.27}$$

where $\Delta E = E + \alpha^2/2$. In order to obtain this equation we have systematically rewritten expressions of the form $\sin(\theta)^{i-2}\cos(\theta)^{j+2}$ as $[\sin(\theta)^{i-2} - \sin(\theta)^i]\cos(\theta)^j$.

The application of the moment method to this problem is straightforward if we can write $\hat{H}'$ as a polynomial function of r and trigonometric functions of θ and ϕ. For simplicity, here we assume that $\hat{H}'$ does not depend on ϕ, so that $\hat{L}_z$ is a constant of the motion and m is a good quantum number. This fact is reflected in that the subscript m does not change in the recurrence relation (4.27) and can therefore be omitted.

We choose

$$\alpha = \frac{1}{n},\ n = 1, 2, \ldots \tag{4.28}$$

that makes the second term on the right-hand side of equation (4.27) vanish when $k = n - 1$ simplifying the problem. Therefore,

$$\Delta E = E - E_0 = E + \frac{1}{2n^2} \tag{4.29}$$

is the energy shift with respect to the energy of the isolated hydrogen atom $E_0 = -1/(2n^2)$.

The recurrence relation (4.27) for the Stark effect becomes

$$\begin{aligned}(\hat{H} - E)f_{i,j,k,m} &= \frac{(i+j)(i+j+1) - k(k+1)}{2}f_{i,j,k-2,m} \\ &+ [\alpha(k+1) - 1]f_{i,j,k-1,m} + \frac{m^2 - i^2}{2}f_{i-2,j,k-2,m} \\ &- \frac{j(j-1)}{2}f_{i,j-2,k-2,m} - \Delta E f_{i,j,k,m} + \lambda f_{i,j+1,k+1,m}\ .\end{aligned} \tag{4.30}$$

Notice that the subscript i changes in only one term that vanishes when $i = |m|$, and in that case the moments of the eigenfunction Ψ

$$F_{j,k} = \langle f_{i,j,k,m}|\Psi\rangle,\ i = |m| \tag{4.31}$$

satisfy the recurrence relation

$$\frac{(i+j)(i+j+1)-k(k+1)}{2}F_{j,k-2} + \frac{k-n+1}{n}F_{j,k-1} - \frac{j(j-1)}{2}F_{j-2,k-2}$$
$$- \Delta E F_{j,k} + \lambda F_{j+1,k+1} = 0 . \tag{4.32}$$

Present moment method does not allow the simultaneous treatment of all the Stark states because the Schrödinger equation is not separable in spherical coordinates. However, we can treat classes of states determined by the relation between $|m|$ and n [58]. With this purpose in mind we define $k = |m| + 1 + t, t = -1, 0, 1, \ldots$ and

$$G_{j,t} = F_{j,k-1} \tag{4.33}$$

so that the recurrence relation (4.32) becomes

$$\frac{(i+j)(i+j+1)-k(k+1)}{2}G_{j,t-1} + \frac{k-n+1}{n}G_{j,t} - \frac{j(j-1)}{2}G_{j-2,t-1}$$
$$- \Delta E G_{j,t+1} + \lambda G_{j+1,t+2} = 0 . \tag{4.34}$$

Expanding the energy and the new moments in Taylor series about $\lambda = 0$

$$E = \sum_{p=0}^{\infty} E_p \lambda^p, \ G_{j,t} = \sum_{p=0}^{\infty} G_{j,t,p} \lambda^p \tag{4.35}$$

we obtain the master recurrence relation

$$G_{j,t,p} = \frac{n}{k-n+1}\left[\frac{k(k+1)-(i+j)(i+j+1)}{2}G_{j,t-1,p} + \frac{j(j-1)}{2}G_{j-2,t-1,p} \right.$$
$$\left. + \sum_{q=1}^{p} E_q G_{j,t+1,p-q} - G_{j+1,t+2,p-1} \right] \tag{4.36}$$

valid for all the states discussed below.

We first consider states with $|m| = n - 1$ ($k = n + t$). When $j = 0$ and $t = -1$ equation (4.34) reduces to $-\Delta E G_{0,0} + \lambda G_{1,1} = 0$; therefore, if we choose the arbitrary normalization condition $G_{0,0} = 1$, then we obtain a suitable expression for the energy: $\Delta E = \lambda G_{1,1}$. We easily calculate all the perturbation corrections to the energy and moments by means of equation (4.36) supplemented with

$$G_{0,0,q} = \delta_{0q}, \ E_q = G_{1,1,q-1} \tag{4.37}$$

that come from the normalization condition and from the energy equation, respectively. The calculation of E_p requires the moment coefficients $G_{j,t,q}$ with $q = 0, 1, \ldots, p-1, j = 0, 1, \ldots, p-q$, $t = 0, 1, \ldots, 2(p-q) - 1$.

We next consider the states with $|m| = n-2$. Setting $j = 1$ and $t = 0$ we obtain $\Delta E G_{1,1} = \lambda G_{2,2}$ that suggests the intermediate normalization condition $G_{1,1} = 1$ leading to the simple energy equation $\Delta E = \lambda G_{2,2}$. When $(j = 0, t = -1)$ and $(j = 0, t = 0)$ we obtain two equations

$$\frac{1}{n}G_{0,-1} + \Delta E G_{0,0} - \lambda = 0, \ (n-1)G_{0,-1} + \Delta E G_{0,1} - \lambda G_{1,2} = 0 \tag{4.38}$$

which lead to

$$\left[n(n-1)G_{0,0} - G_{0,1}\right]\Delta E + \left[G_{1,2} - n(n-1)\right]\lambda = 0 \tag{4.39}$$

after elimination of the moment $G_{0,-1}$ between them. Substitution of the expression for the energy into equation (4.39) yields the secular equation

$$\left[n(n-1)G_{0,0} - G_{0,1}\right]G_{2,2} + G_{1,2} - n(n-1) = 0 \,. \tag{4.40}$$

Another useful equation arises when $j = 1$ and $t = 1$:

$$G_{1,0} = \frac{1}{n}\left(\frac{1}{n} - \Delta E G_{1,2} + \lambda G_{2,3}\right)\,. \tag{4.41}$$

The equations just derived enable the calculation of all the perturbation corrections to the energy and moments for the states with $|m| = n - 2$. Expanding ΔE and the moments $G_{j,t}$ in Taylor series about $\lambda = 0$ we obtain

$$G_{1,1,q} = \delta_{q0} \,, \quad E_q = G_{2,2,q-1} \,, \tag{4.42}$$

$$\sum_{q=0}^{p}\left[n(n-1)G_{0,0,q} - G_{0,1,q}\right]G_{2,2,p-q} + G_{1,2,p} - n(n-1)\delta_{0p} = 0 \,, \tag{4.43}$$

and the master equation (4.36) with $i = |m| = n - 2$. It is not difficult to verify that the perturbation equations (4.36) and (4.42) leave undetermined only the moment coefficients $G_{0,0,p}$, $p = 0, 1, \ldots$. We obtain them from equation (4.43) which is quadratic in $G_{0,0,0}$ when $p = 0$, and linear in $G_{0,0,p}$ for all $p > 0$. For $p = 0$ we have

$$G_{0,0,0} = \pm\frac{1}{n^2} \,. \tag{4.44}$$

Each sign corresponds to one of the two Stark states arising from degenerate unperturbed hydrogenic states. In order to obtain E_p we have to calculate $G_{j,t,q}$ for all $q = 0, 1, \ldots, p-1, j = 0, 1, \ldots, p - q + 1$ and $t = 1, 2, \ldots, 2(p - q)$.

Table 4.2 shows energy coefficients for the two cases just discussed, where $\sigma = \pm 1$ selects each of the two Stark states arising from degenerate unperturbed states with $|m| = n - 2$. We do not show the simple Maple procedures used to obtain the results in Table 4.2, and it is left to the reader to write them following the lines of other programs in the program section. The energy coefficients in Table 4.2 agree with those obtained by means of an earlier application of the moment method [58] Moreover, if we set $(q = 0, |m| = n - 1)$, and $(q = \sigma, |m| = n - 2)$, the energy coefficients of Table 4.1 reduce to those in Table 4.2.

It is well known that the projection of the angular momentum along the field direction is a constant of the motion (that is to say $\hat{L}_z$ commutes with $\hat{H}$), and, consequently, $m = 0, \pm 1, \pm 2, \ldots$ is a good quantum number. The moment method gives us another quantum number $\sigma = \pm 1$ to label some pairs of Stark states. In parabolic coordinates we clearly have three quantum numbers: m, n_1, and n_2, or, alternatively, m, n, and q. When comparing the results of Tables 4.1 and 4.2 we saw that $q = \sigma$ in the second case studied by means of the moment method. In spherical coordinates it is customary to label the Stark states by means of the quantum numbers of the isolated hydrogen atom: $n = 1, 2, \ldots, l = 0, 1, \ldots, n - 1$, and $m = 0, \pm 1, \pm 2, \ldots, \pm l$ which may be suitable at low field strengths. In the first case discussed above $|m| = n - 1 = l$ and the Stark states correspond to the hydrogenic $1s$, $2p_{\pm 1}$, $3d_{\pm 2}$, etc. In the second case $|m| = n - 2$, and the two possible values of the angular momentum quantum number $l = n - 2, n - 1$ show that the perturbation couples the pairs of hydrogenic states $(2s, 2p_0)$, $(3p_{\pm 1}, 3d_{\pm 1})$, etc.

Table 4.2 Moment Method for the Stark Effect in Hydrogen in Spherical Coordinates

States with $|m| = n - 1$

$E_{2j+1} = 0$

$E_2 = -\frac{1}{8} n^4 (n+1)(4n+5)$

$E_4 = -\frac{1}{128} n^{10} (n+1) \left(192 n^3 + 933 n^2 + 1550 n + 880\right)$

$E_6 = -\frac{1}{1024} n^{16} (n+1)$
$\qquad \left(415522 n^3 + 109013 n^4 + 340000 + 821540 n + 814928 n^2 + 11776 n^5\right)$

$E_8 = -\frac{1}{32768} n^{22} (n+1) \left(1104000000 + 2933036518 n^3 + 1313502002 n^4\right.$
$\qquad \left. + 3189097200 n + 4047270620 n^2 + 4063232 n^7 + 363981946 n^5 + 57826285 n^6\right)$

$E_{10} = -\frac{1}{262144} n^{28} (n+1) \left(8246600607 n^8 + 419168256 n^9 + 6314922783568 n^3\right.$
$\qquad + 3680066142092 n^4 + 4868385352960 n + 7167165224192 n^2$
$\qquad \left. + 74557383526 n^7 + 1474270752706 n^5 + 406670914358 n^6 + 1502988800000\right)$

States with $|m| = n - 2$

$\sigma^2 = 1$

$E_1 = \frac{3}{2} n \sigma$

$E_2 = -\frac{1}{4} n^4 (n+5)(2n-1)$

$E_3 = \frac{3}{16} n^7 \left(41 - 22 n + 17 n^2\right) \sigma$

$E_4 = -\frac{1}{64} n^{10} \left(3537 n + 28 n^2 + 1125 n^3 + 96 n^4 - 1606\right)$

$E_5 = \frac{3}{256} n^{13} \left(-6850 n + 28086 n^2 - 2222 n^3 + 3015 n^4 + 18963\right) \sigma$

$E_6 = -\frac{1}{512} n^{16} \left(5888 n^6 + 120789 n^5 + 182838 n^4 + 1331475 n^3 + 1353240 n + 210794 n^2\right.$
$\qquad \left. - 346528\right)$

$E_7 = \frac{3}{2048} n^{19} \left(7828405 + 355761 n^6 + 293230 n^5 + 8203515 n^4 + 4806230 n^3\right.$
$\qquad \left. + 23997287 n^2 + 5413380 n\right) \sigma$

$E_8 = -\frac{1}{16384} n^{22} \left(-20876640 + 3755981880 n + 3505527310 n^2 + 61889517 n^7\right.$
$\qquad + 6653206413 n^3 + 2396964444 n^4 + 1677325062 n^5 + 224958382 n^6$
$\qquad \left. + 2031616 n^8\right)$

$E_9 = \frac{3}{65536} n^{25} \sigma \left(24292851427 + 59251864516 n + 126499280380 n^2 + 600211882 n^7\right.$
$\qquad + 86572048058 n^3 + 74879641210 n^4 + 18968318968 n^5 + 9152464636 n^6$
$\qquad \left. + 190178763 n^8\right)$

$E_{10} = -\frac{1}{131072} n^{28} \left(209584128 n^{10} + 8665768863 n^9 + 54208739210 n^8\right.$
$\qquad + 1444091466972 n^6 + 4992252806403 n^5 + 8043191827530 n^4$
$\qquad + 12467418860376 n^3 + 477827155350 n^7 + 9247913436768 n^2$
$\qquad \left. + 5395565208960 n + 867032805376\right)$

4.3   The Zeeman Effect in Hydrogen

We have briefly discussed the Zeeman effect in hydrogen in Chapter 2 to illustrate the application of the method of Dalgarno and Stewart and logarithmic perturbation theory to a nonseparable problem. It was shown that the relevant part of the dimensionless Hamiltonian operator reads

$$\hat{H} = -\frac{1}{2}\nabla^2 - \frac{1}{r} + \lambda r^2 \sin^2\theta \ . \tag{4.45}$$

Therefore, arguing as in the preceding section we obtain the recurrence relation

$$\left(\hat{H} - E\right) f_{i,j,k,m} = \frac{(i+j)(i+j+1) - k(k+1)}{2} f_{i,j,k-2,m}$$
$$+ [\alpha(k+1) - 1] f_{i,j,k-1,m} + \frac{m^2 - i^2}{2} f_{i-2,j,k-2,m}$$
$$- \frac{j(j-1)}{2} f_{i,j-2,k-2,m} - \Delta E f_{i,j,k,m} + \lambda f_{i+2,j,k+2,m} \ , \tag{4.46}$$

where we notice that the subscript j does not change if $j(j-1) = 0$. We thus obtain two disjoint sets of states, one for each value of j, $j = 0$ or $j = 1$, which plays the role of a quantum number. With either choice the moments

$$F_{i,k} = \left\langle f_{i,j,k,m} | \Psi \right\rangle \tag{4.47}$$

satisfy the recurrence relation

$$\frac{(i+j)(i+j+1) - k(k+1)}{2} F_{i,k-2} + \frac{k-n+1}{n} F_{i,k-1} + \frac{m^2 - i^2}{2} F_{i-2,k-2}$$
$$- \Delta E F_{i,k} + \lambda F_{i+2,k+2} = 0 \ . \tag{4.48}$$

In order to derive a moment recurrence relation that applies not only to individual states but also to whole classes of them we define $i = |m| + 2s, k = |m| + j + 1 + t, s = 0, 1, \ldots, t = -1, 0, 1, \ldots,$ and

$$G_{s,t} = F_{i,k-1} \ . \tag{4.49}$$

Consequently, equation (4.48) becomes

$$\frac{(i+j)(i+j+1) - k(k+1)}{2} G_{s,t-1} + \frac{k-n+1}{n} G_{s,t} + \frac{m^2 - i^2}{2} G_{s-1,t-1}$$
$$- \Delta E G_{s,t+1} + \lambda G_{s+1,t+3} = 0 \ . \tag{4.50}$$

Expanding the new moments and the energy in Taylor series about $\lambda = 0$ we obtain a master equation

$$G_{s,t,p} = \frac{n}{k-n+1} \left[\frac{k(k+1) - (i+j)(i+j+1)}{2} G_{s,t-1,p} + \frac{i^2 - m^2}{2} G_{s-1,t-1,p} \right.$$
$$\left. + \sum_{q=1}^{p} E_q G_{s,t+1,p-q} - G_{s+1,t+3,p-1} \right] \tag{4.51}$$

that applies to all the cases studied here.

We first consider states with $j = 0$. The simplest case is $|m| = n - 1$ because three terms of equation (4.50) vanish when $s = 0$ and $t = -1$ giving a single expression for the energy $\Delta E G_{0,0} = \lambda G_{1,2}$ that suggests the intermediate normalization condition $G_{0,0} = 1$. We thus obtain two additional expressions,

$$G_{0,0,q} = \delta_{q0}, \quad E_q = G_{1,2,q-1}, \tag{4.52}$$

to supplement the master equation (4.51). In order to obtain E_p we calculate all the moment coefficients $G_{s,t,q}$ with $q = 0, 1, \ldots, p - 1$, $s = 0, 1, \ldots, p - q$, and $t = 0, 1, \ldots, 3(p - q) - 1$.

When $|m| = n - 2$ ($j = 0$) we cannot make three terms of the recurrence relation (4.50) vanish simultaneously. Choosing $(s, t) = (0, -1)$ and $(s, t) = (0, 0)$ we obtain $n^{-1} G_{0,-1} + \Delta E G_{0,0} - \lambda G_{1,2} = 0$ and $(n-1)G_{0,-1} + \Delta E G_{0,1} - \lambda G_{1,3} = 0$, respectively. Removing $G_{0,-1}$ from them we obtain a useful expression for the energy: $[n(n-1)G_{0,0} - G_{0,1}]\Delta E - [n(n-1)G_{1,2} - G_{1,3}]\lambda = 0$. The normalization condition $n(n - 1)G_{0,0} - G_{0,1} = 1$ leads to a simpler formula $\Delta E = \lambda[n (n-1)G_{1,2} - G_{1,3}]$. When $s = 0$ and $t = 1$ equation (4.50) becomes $n G_{0,0} + n^{-1} + \Delta E G_{0,2} - \lambda G_{1,4} = 0$ from which we obtain $G_{0,0}$. Summarizing, we have

$$G_{0,0} = \frac{1}{n}\left(-\frac{1}{n} - \Delta E G_{0,2} + \lambda G_{1,4}\right), \tag{4.53}$$

$$G_{0,1} = n(n-1)G_{0,0} - 1, \tag{4.54}$$

and

$$\Delta E = \lambda\left[n(n-1)G_{1,2} - G_{1,3}\right]. \tag{4.55}$$

Expanding the energy and moments in Taylor series about $\lambda = 0$ we obtain

$$G_{0,0,p} = \frac{1}{n}\left(-\frac{\delta_{p0}}{n} - \sum_{q=1}^{p} E_q G_{0,2,p-q} + G_{1,4,p-1}\right), \tag{4.56}$$

$$G_{0,1,p} = n(n-1)G_{0,0,p} - \delta_{p0}, \tag{4.57}$$

and

$$E_p = n(n-1)G_{1,2,p-1} - G_{1,3,p-1} \tag{4.58}$$

in addition to the master equation (4.51). In order to obtain the energy coefficient E_p we need $G_{s,t,q}$ with $q = 0, 1, \ldots, p - 1$, $s = 0, 1, \ldots, p - q$, and $t = 1, 2, \ldots, 3(p - q)$.

The states just discussed can be treated as nondegenerate because the perturbation does not couple them. A different situation takes place when $|m| = n - 3$. From the general recurrence relation (4.50) for the moments we obtain the following equations

$$\frac{2}{n}G_{0,-1} + \Delta E G_{0,0} - \lambda G_{1,2} = 0 \tag{4.59}$$

$$(2n - 3)G_{0,0} + \Delta E G_{0,2} - \lambda G_{1,4} = 0 \tag{4.60}$$

$$\Delta E G_{1,2} + (2n - 4)G_{0,0} - \lambda G_{2,4} = 0 \tag{4.61}$$

$$(n - 2)G_{0,-1} + \frac{1}{n}G_{0,0} + \Delta E G_{0,1} - \lambda G_{1,3} = 0 \tag{4.62}$$

when (s, t) is respectively equal to $(0, -1)$, $(0, 1)$, $(1, 1)$, and $(0, 0)$. It is left to the reader to derive the following working expressions

$$G_{0,0} = \frac{1}{2n-3}\left(-\Delta E G_{0,2} + \lambda G_{1,4}\right) , \tag{4.63}$$

$$G_{1,2} = \frac{2n-4}{2n-3}G_{0,2} - 1 , \tag{4.64}$$

$$\Delta E = \lambda\left(\frac{2n-4}{2n-3}G_{1,4} - G_{2,4}\right) , \tag{4.65}$$

and

$$\frac{1}{2n-3}\left[\frac{G_{0,2}}{n(2n-3)} + \frac{n(n-2)}{2}G_{0,0} - G_{0,1}\right]\left[(2n-4)G_{1,4} - (2n-3)G_{2,4}\right]$$
$$- \frac{n(n-2)}{2}G_{1,2} - \frac{G_{1,4}}{n(2n-3)} + G_{1,3} = 0 , \tag{4.66}$$

where equation (4.64) is simply an arbitrary normalization condition. Expanding those equations in Taylor series about $\lambda = 0$ and using the master equation (4.51) with the appropriate values of $|m|$ and j we calculate all the energy and moment coefficients. In order to obtain E_p we need $G_{s,t,q}$ with $q = 0, 1, \ldots, p-1$, $s = 0, 1, \ldots, p-q+1$, and $t = 2, 3, \ldots, 3(p-q)+1$. The coefficient of order q of the expansion of equation (4.66) is linear in the moment coefficient $G_{0,1,q}$ when $q > 0$ and quadratic in $G_{0,1,0}$ when $q = 0$. The latter case yields the secular equation for the two degenerate states coupled by the perturbation. The two roots are

$$G_{0,1,0} = \frac{(3 - 2n)[8n^2 - 24n + 13 \pm (2n-3)\sqrt{16n^2 - 48n + 41}]}{20n^2(n-1)(n-2)} . \tag{4.67}$$

We next consider the states with $j = 1$. The simplest case is $|m| = n - 2$ (the reader may verify that one cannot obtain suitable working equations when $|m| = n - 1$). Choosing the arbitrary normalization condition $G_{0,0} = 1$ we obtain the energy expression $\Delta E = \lambda G_{1,2}$; consequently,

$$G_{0,0,p} = \delta_{p0} , \quad E_p = G_{1,2,p-1} . \tag{4.68}$$

In order to obtain E_p we have to calculate $G_{s,t,q}$ for $q = 0, 1, \ldots, p-1$, $s = 0, 1, \ldots, p-q$, and $t = 0, 1, \ldots, 3(p-q) - 1$.

The case $|m| = n - 3$ exhibits no additional difficulty and is therefore left to the reader; however, for the sake of completeness we show the main equations in what follows. We obtain all the energy and moment coefficients from the master equation (4.51) and the additional expressions

$$G_{0,0,p} = \frac{1}{n}\left(-\frac{\delta_{p0}}{n} - \sum_{q=1}^{p} E_q G_{0,2,p-q} + G_{1,4,p-1}\right) , \tag{4.69}$$

$$G_{0,1,p} = n(n-1)G_{0,0,p} - \delta_{0p} , \tag{4.70}$$

and

$$E_p = n(n-1)G_{1,2,p-1} - G_{1,3,p-1} , \tag{4.71}$$

where equation (4.70) is an arbitrary normalization condition. To obtain E_p we have to calculate the moment coefficients $G_{s,t,q}$ with $q = 0, 1, \ldots, p-1$, $s = 0, 1, \ldots, p-q$, and $t = 1, 2, \ldots, 3(p-q)$.

Table 4.3 Moment Method for the Zeeman Effect in Hydrogen *(Continued)*

States with $j = 0$

$|m| = n - 1$

$E_1 = n^3 (n + 1)$

$E_2 = -\frac{1}{6} n^7 (n + 1) \left(12 n^2 + 27 n + 14\right)$

$E_3 = \frac{1}{18} n^{11} (n + 1) \left(1089 n^3 + 21 n^4 + 2048 n^2 + 528 + 1700 n\right)$

$E_4 = -\frac{1}{1080} n^{15} (n + 1) \left(926640 + 6072790 n^3 + 3225070 n^4 + 926235 n^5 + 112320 n^6 \right.$

$\left. + 6524514 n^2 + 3789828 n\right)$

$E_5 = \frac{1}{16200} n^{19} (n + 1) \left(17625600 n^8 + 8634176720 n^3 + 6675896034 n^4 + 3380668050 n^5 \right.$

$\left. + 1095766700 n^6 + 207771075 n^7 + 7126778904 n^2 + 3419028000 n \right.$

$\left. + 725587200\right)$

$E_6 = -\frac{1}{3402000} n^{23} (n + 1) \left(4902104138750 n^8 + 678352208625 n^9 + 43436736000 n^{10} \right.$

$\left. + 214608912662148 n^3 + 203462151787266 n^4 + 135522396155608 n^5 \right.$

$\left. + 64417497299806 n^6 + 21613830539000 n^7 + 151582655926080 n^2 \right.$

$\left. + 64403176052160 n + 12418815628800\right)$

$|m| = n - 2$

$E_1 = n^2 (n + 5) (n - 1)$

$E_2 = -\frac{1}{3} n^6 (n - 1) \left(6 n^3 + 75 n^2 - 19 n + 168\right)$

$E_3 = \frac{4}{9} n^{10} (n - 1) \left(27 n^5 + 585 n^4 + 26 n^3 + 3649 n^2 - 1239 n + 2772\right)$

$E_4 = -\frac{1}{540} n^{14} (n - 1) \left(56160 n^7 + 1801575 n^6 + 1698625 n^5 + 24246755 n^4 - 718497 n^3 \right.$

$\left. + 48450030 n^2 - 13165128 n + 18230400\right)$

$E_5 = \frac{1}{4050} n^{18} (n - 1) \left(4406400 n^9 + 191700675 n^8 + 437243425 n^7 + 4821233400 n^6 \right.$

$\left. + 3564450084 n^5 + 20437328903 n^4 + 2324786331 n^3 + 20035153782 n^2 \right.$

$\left. - 3630669480 n + 4191004800\right)$

$E_6 = -\frac{1}{1701000} n^{22} (n - 1) \left(21718368000 n^{11} + 1211753064375 n^{10} + 4826789599625 n^9 \right.$

$\left. + 52080472580500 n^8 + 94399639455562 n^7 + 424804836198743 n^6 \right.$

$\left. + 372530794254741 n^5 + 921146642791662 n^4 + 272163391589592 n^3 \right.$

$\left. + 511320228053040 n^2 - 38324443075200 n + 57774342528000\right)$

Table 4.3 *(Cont.)* Moment Method for the Zeeman Effect in Hydrogen *(Continued)*

$$|m| = n - 3$$

$$E_1 = R\,n^2 + n^2\left(n^2 + 3n - 7\right)$$

$$E_2 = -\frac{1}{2}\frac{\left(160\,n^4 - 592\,n^3 + 1316\,n^2 - 2159\,n + 1578\right)n^6\,R}{16\,n^2 - 48\,n + 41}$$

$$\qquad -\frac{1}{2}\frac{n^6\left(64\,n^6 + 432\,n^5 - 3388\,n^4 + 9823\,n^3 - 17665\,n^2 + 19583\,n - 9758\right)}{16\,n^2 - 48\,n + 41}$$

$$E_3 = \frac{1}{6}\left(79104\,n^8 - 515968\,n^7 + 2011136\,n^6 - 6258120\,n^5 + 14406296\,n^4 - 22747119\,n^3\right.$$
$$\left.+24278234\,n^2 - 16659692\,n + 5678856\right)n^{10}\,R/\left(16\,n^2 - 48\,n + 41\right)^2$$
$$+\frac{1}{6}n^{10}\left(-36834072 + 129285956\,n + 18432\,n^{10} - 176972205\,n^4 + 236171613\,n^3\right.$$
$$-38200632\,n^6 - 219094386\,n^2 + 95404713\,n^5 + 11659552\,n^7 + 223488\,n^9$$
$$\left.-2480640\,n^8\right)/\left(16\,n^2 - 48\,n + 41\right)^2$$

$$E_4 = -\frac{1}{360}n^{14}\left(596916429824\,n^8 - 4600155967690\,n^5 + 4256844241306\,n^2\right.$$
$$+2849807540976\,n^6 + 404761711680 - 6095473972271\,n^3 + 6079342696994\,n^4$$
$$+41872783360\,n^{10} - 8039997440\,n^{11} + 965099520\,n^{12} - 1870253070144\,n$$
$$\left.-180350726656\,n^9 - 1470111015264\,n^7\right)R/\left(16\,n^2 - 48\,n + 41\right)^3$$
$$-\frac{1}{360}n^{14}\left(13417446708816\,n - 36823941120\,n^{12} + 222705301760\,n^{11}\right.$$
$$-995770429696\,n^{10} + 2873733120\,n^{13} + 153354240\,n^{14}$$
$$-62461978186622\,n^4 + 55210737998015\,n^3 - 36138167068989\,n^6$$
$$-34210809912178\,n^2 + 53134403287234\,n^5 + 20398070759667\,n^7$$
$$\left.+3488722075408\,n^9 - 9476088954080\,n^8 - 2573851988160\right)/\left(16\,n^2 - 48\,n + 41\right)^3$$

$$R = \pm\sqrt{16\,n^2 - 48\,n + 41}$$

States with $j = 1$

$$|m| = n - 2$$

$$E_1 = n^2\left(n - 1\right)\left(n + 1\right)$$

$$E_2 = -n^7\left(n - 1\right)\left(2n + 3\right)\left(n + 1\right)$$

$$E_3 = \frac{2}{3}n^{11}\left(n - 1\right)\left(n + 1\right)\left(18\,n^3 + 63\,n^2 + 62\,n + 10\right)$$

$$E_4 = -\frac{1}{180}n^{15}\left(n - 1\right)\left(n + 1\right)$$
$$\qquad\left(18720\,n^5 + 110565\,n^4 + 245390\,n^3 + 247915\,n^2 + 114162\,n + 26184\right)$$

$$E_5 = \frac{1}{1350}n^{19}\left(n - 1\right)\left(n + 1\right)\left(1468800\,n^7 + 12700800\,n^6 + 46304000\,n^5 + 92453775\,n^4\right.$$
$$\left.+110837196\,n^3 + 82846915\,n^2 + 37353426\,n + 7488360\right)$$

$$E_6 = -\frac{1}{567000}n^{23}\left(n - 1\right)\left(n + 1\right)\left(7239456000\,n^9 + 84514397625\,n^8 + 442123228750\,n^7\right.$$
$$+1370327540750\,n^6 + 2804697695128\,n^5 + 3986041260369\,n^4$$
$$\left.+3975789877658\,n^3 + 2681185159704\,n^2 + 1093513436880\,n + 203616662400\right)$$

Table 4.3 *(Cont.)* Moment Method for the Zeeman Effect in Hydrogen

$$|m| = n - 3$$

$$E_1 = n^2 (n + 5) (n - 2)$$

$$E_2 = -\frac{1}{2} n^6 (n - 2) \left(4 n^3 + 47 n^2 - 31 n + 182\right)$$

$$E_3 = \frac{1}{6} n^{10} (n - 2) \left(72 n^5 + 1449 n^4 - 1835 n^3 + 12372 n^2 - 12712 n + 20856\right)$$

$$E_4 = -\frac{1}{360} n^{14} (n - 2) \left(37440 n^7 + 1113435 n^6 - 1269295 n^5 + 17598810 n^4 \right.$$
$$\left. - 29044904 n^3 + 64577580 n^2 - 62479176 n + 59883840\right)$$

$$E_5 = \frac{1}{5400} n^{18} (n - 2) \left(5875200 n^9 + 237147075 n^8 - 127220525 n^7 + 6124217100 n^6 \right.$$
$$- 10608481964 n^5 + 38257425072 n^4 - 59035205096 n^3 + 83364260448 n^2$$
$$\left. - 71175114720 n + 48166185600\right)$$

$$E_6 = -\frac{1}{1134000} n^{22} (n - 2) \left(14478912000 n^{11} + 750746768625 n^{10} + 343872883375 n^9 \right.$$
$$+ 29610399572250 n^8 - 40041106465976 n^7 + 276011011998456 n^6$$
$$- 475002238895108 n^5 + 1011743637657984 n^4 - 1384404310926576 n^3$$
$$\left. + 1481297714534400 n^2 - 1112814105840000 n + 581205028320000\right)$$

Table 4.3 shows the first energy coefficients for all the states considered above in terms of the principal quantum number n and $R = \pm\sqrt{16 n^2 - 48 n + 41}$. Analytical expressions of greater order are much longer and, most probably, of no use for the reader.

Having sufficient computer memory one easily calculates more analytic energy coefficients than those shown in Table 4.3 by means of simple Maple procedures. In the program section we show only the most difficult case ($j = 0$, $|m| = n - 3$); the reader may easily derive other cases by straightforward modification of the main procedure given there. The calculation is considerably faster and requires less computer memory if one sets the value of the principal quantum number $n = 1, 2, \ldots$ for a given particular state.

Many authors have already calculated energy coefficients for the Zeeman effect in hydrogen and surprisingly their results exhibit a good deal of disagreement as noticed in an earlier application of the moment method [59]. We believe that the energy coefficients displayed in Table 4.3, which agree with those in reference [59], are correct.

With respect to the calculation of energy coefficients the moment method is much simpler and easier to apply than the method of Dalgarno and Stewart and logarithmic perturbation theory discussed in Chapter 2. The advantage of the moment method is particularly noticeable in the treatment of excited states. Another powerful approach, which exhibits the additional advantage of producing a more useful representation of the eigenfunctions, is the expansion of the perturbed state in a basis set of unperturbed states, aided by an algebraic approach to calculate the necessary matrix elements systematically [12]. The algebraic approach is preferable if one is interested in the calculation of system properties other than the energy; otherwise, the moment method leads to simpler programs [12].

Finally, we discuss the classification and labelling of states within the moment method. Notice that we have made no explicit use of the well-known properties of the Zeeman states during the calculation. The model Hamiltonian is invariant under the substitutions $\theta \to -\theta$ and $\theta \to \theta + \pi$, and the eigenfunctions are either even or odd with respect to them. The functions chosen to construct the moments satisfy

$$f_{i,j,k,m}(r, -\theta, \phi) = (-1)^i f_{i,j,k,m}(r, \theta, \phi) = (-1)^{|m|} f_{i,j,k,m}(r, \theta, \phi) , \tag{4.72}$$

and

$$f_{i,j,k,m}(r, \theta + \pi, \phi) = (-1)^{i+j} f_{i,j,k,m}(r, \theta, \phi) = (-1)^{|m|+j} f_{i,j,k,m}(r, \theta, \phi) . \tag{4.73}$$

The symmetry of the functions $f_{i,j,k,m}(r, \theta, \phi)$ (determined by the values of j and $|m|$) has to match the symmetry of the chosen Zeeman state in order to obtain a nontrivial recurrence relation for the moments.

The unperturbed eigenfunctions are radial factors times the spherical harmonics $Y_{l,m}(\theta, \phi)$ that satisfy $Y_{l,m}(-\theta, \phi) = (-1)^{|m|} Y_{l,m}(\theta, \phi)$ and $Y_{l,m}(\theta + \pi, \phi) = (-1)^l Y_{l,m}(\theta, \phi)$ [40]. Therefore, when $\lambda = 0$ we expect that $l = |m| + j + 2u$, $u = 0, 1, \ldots$. For the class of states with $j = 0$ and $|m| = n - 1$ we have $l = |m|$ and; therefore, the Zeeman states arise from the hydrogenic ones $1s$, $2p_{\pm 1}, 3d_{\pm 2}, \ldots$. When $j = 0$ and $|m| = n - 2$ we conclude that $l = |m|$ and the Zeeman states come from the hydrogenic ones $2s$, $3p_{\pm 1}, 4d_{\pm 2}, \ldots$. The choice $j = 0$ and $|m| = n - 3$ gives room for two possibilities, $l = |m|$ and $l = |m| + 2$, and we have pairs of hydrogenic states $(3s, 3d_0)$, $(4p_{\pm 1}, 4f_{\pm 1}), \ldots$ coupled by the perturbation. When $j = 1$ and $|m| = n - 2$ the unperturbed states are $2p_0, 3d_{\pm 1}, 4f_{\pm 2}, \ldots$. Finally, for $j = 1$ and $|m| = n - 3$ we have $3p_0, 4d_{\pm 1}, 5f_{\pm 2}, \ldots$.

It is not difficult to apply the moment method to more general perturbations than those discussed here. An example already studied is the hydrogen atom in parallel electric and magnetic fields [60].

4.4   The Hydrogen Molecular Ion

The hydrogen molecular ion is the simplest diatomic molecule having only one electron. Here we consider some electronic states under the Born–Oppenheimer approximation that separates the electronic and nuclear motions [53]. The dimensionless model Hamiltonian in this frozen-nuclei approach reads

$$\hat{H} = -\frac{1}{2}\nabla^2 - \frac{1}{r_A} - \frac{1}{r_B} , \tag{4.74}$$

where r_A and r_B are the distances between the electron and protons A and B, respectively [61].

The Schrödinger equation for this model is separable in elliptical coordinates [61]; however, here we choose spherical coordinates and apply the moment method for nonseparable problems because it is particularly simple and straightforward for the naive perturbation approach developed below. To this end we place the molecule along the z axis with proton A at the coordinate origin and proton B at a distance R in the positive direction. Therefore, if $\mathbf{r}$ denotes the position of the electron with respect to the coordinate origin, then $r_A = r = |\mathbf{r}|$ and $r_B = \sqrt{R^2 - 2R\cos(\theta) + r^2}$, where θ is the angle between $\mathbf{r}$ and the z axis. We rewrite $1/r_B = C(\beta)/R$, where

$$C(\beta) = \frac{1}{\sqrt{1 - 2\beta\cos(\theta) + \beta^2}} , \tag{4.75}$$

$\beta = r/R$, and expand $C(\beta)$ as follows:

$$C(\beta) = \sum_{j=0}^{\infty} C_j(\cos(\theta))\beta^j . \tag{4.76}$$

The functions $C_j(\cos(\theta))$ are the well-known Legendre polynomials [62]. Taking into account that $[1 + \beta^2 - 2\beta\cos(\theta)]C'(\beta) = [\cos(\theta) - \beta]C(\beta)$ one easily verifies that the functions C_j satisfy the recurrence relation

$$C_{j+1} = \frac{1}{j+1}\left[(2j+1)\cos(\theta)C_j - jC_{j-1}\right] \tag{4.77}$$

and are therefore polynomials of the form

$$C_j = \sum_{i=0}^{j} C_{j,i}\cos(\theta)^i . \tag{4.78}$$

Moreover, the coefficients $C_{j,i}$ satisfy

$$C_{j+1,i} = \frac{1}{j+1}\left[(2j+1)C_{j,i-1} - jC_{j-1,i}\right] , \tag{4.79}$$

where $j = 0, 1, \ldots$, $i = 0, 1, \ldots, j$, $C_{0,0} = 1$, and $C_{j,i} = 0$ if a subscript is negative. In the end we have

$$\frac{1}{r_B} = \sum_{u=0}^{\infty} \lambda^{u+1} r^u \sum_{v=0}^{u} C_{u,v}\cos(\theta)^v , \tag{4.80}$$

where $\lambda = 1/R$ is the perturbation parameter.

We apply the moment method as in the Stark effect discussed above. The recurrence relation for the moments reads

$$\frac{(i+j)(i+j+1) - k(k+1)}{2}F_{j,k-2} + \frac{k-n+1}{n}F_{j,k-1} - \frac{j(j-1)}{2}F_{j-2,k-2}$$
$$-\Delta E F_{j,k} - \sum_{u=0}^{\infty} \lambda^{u+1} \sum_{v=0}^{u} C_{u,v}F_{j+v,k+u} = 0 , \tag{4.81}$$

where $i = |m|$. In order to treat all the states with $|m| = n - 1$ simultaneously we define $k = n + t$ and $G_{j,t} = F_{j,k-1}$. Arguing as in the case of the Stark effect we derive the master equation

$$G_{j,t,p} = \frac{n}{k-n+1}\left[\frac{k(k+1) - (i+j)(i+j+1)}{2}G_{j,t-1,p} + \frac{j(j-1)}{2}G_{j-2,t-1,p}\right.$$
$$\left. + \sum_{q=1}^{p} E_q G_{j,t+1,p-q} + \sum_{u=0}^{p-1}\sum_{v=0}^{u} C_{u,v}G_{j+v,t+u+1,p-u-1}\right] , \tag{4.82}$$

the normalization condition

$$G_{0,0,p} = \delta_{0p} , \tag{4.83}$$

and the energy expression

$$E_p = -\sum_{u=0}^{p-1}\sum_{v=0}^{u} C_{u,v}G_{v,u,p-u-1} . \tag{4.84}$$

Table 4.4   Moment Method for the Hydrogen Molecular Ion

States with $|m| = n - 1$

$E_1 = -1$

$E_2 = 0$

$E_3 = \frac{1}{2} n^2 (n-1)(n+1)$

$E_4 = -\frac{1}{8} n^4 (n+1)(4n+5)$

$E_5 = -\frac{3}{8} n^4 (n-1)(n-2)(n+2)(n+1)$

$E_6 = \frac{1}{4} n^6 (n+1)\left(4n^3 + 14n^2 - 5n - 28\right)$

$E_7 = \frac{1}{16} n^6 (n+1)\left(13n^5 - 49n^4 - 269n^3 - 121n^2 + 180n - 180\right)$

$E_8 = -\frac{1}{128} n^8 (n+1)\left(384n^5 + 2181n^4 + 910n^3 - 7930n^2 - 1362n + 13572\right)$

$E_9 = -\frac{1}{128} n^8 (n+1)\left(323n^7 - 1907n^6 - 15895n^5 - 14189n^4 + 44232n^3 + 44172n^2\right.$
$\left. -20160n + 20160\right)$

$E_{10} = \frac{1}{64} n^{10} (n+1)\left(704n^7 + 5668n^6 + 4744n^5 - 45013n^4 - 77061n^3 + 52830n^2\right.$
$\left. +38694n - 150084\right)$

	$n = 1$	$n = 2$
E_1	-1	-1
E_2	0	0
E_3	0	6
E_4	$\frac{-9}{4}$	-78
E_5	0	0
E_6	$\frac{-15}{2}$	2400
E_7	$\frac{-213}{4}$	-33888
E_8	$\frac{-7755}{64}$	201552
E_9	$\frac{-1773}{2}$	1835904
E_{10}	$\frac{-84759}{16}$	-28483200

To obtain E_p we need $G_{j,t,q}$ for $q = 0, 1, \ldots, p-1$, $j = 0, 1, \ldots, p-q-1$, and $t = 0, 1, \ldots,$ $p-q-1$. Table 4.4 shows analytical energy coefficients in terms of n and particular results for $n = 1$ and $n = 2$ suitable for comparison with expressions available in the literature. One easily obtains more perturbation corrections by means of the Maple procedures shown in the program section.

There is a vast literature on the asymptotic expansion of the electronic energies of the hydrogen molecular ion at large internuclear distances. Here we select References [63] through [65] where the reader may find other relevant papers on the subject. Our results agree with those obtained earlier by other authors after appropriate corrections. For example, the coefficient of order p of reference [65] is $(2/n)^p$ times ours.

We could have chosen elliptical coordinates to separate the Schrödinger equation and apply, for example, the method of Swenson and Danforth as in the Stark effect discussed above, thus obtaining more general results valid for all states. However, we have preferred spherical coordinates and the moment method because we think it is simpler to develop the working equations. The reader may verify that the application of the method of Swenson and Danforth is straightforward although rather more tedious.

According to the perturbation method just discussed, the asymptotic expansion for the electronic energy of the ground state is

$$E(R) = -\frac{1}{2} - \frac{1}{R} - \frac{9}{4R^4} - \frac{15}{2R^6} - \cdots .$$
(4.85)

However, a more careful analysis reveals the occurrence of exponential and logarithmic terms [64]–[66] originated in the double-well nature of the problem [66]. Figure 4.1 (produced by Maple *plot3d* command) clearly shows that the potential-energy function of the hydrogen molecular ion is a double well. In order to plot the function in three dimensions we set $y = 0$. It is not surprising that the exponential terms do not appear in the naive perturbation approach developed above because functions of the form $\exp(-R/n) = \exp[-1/(n\lambda)]$ and all its derivatives vanish as $\lambda \downarrow 0$.

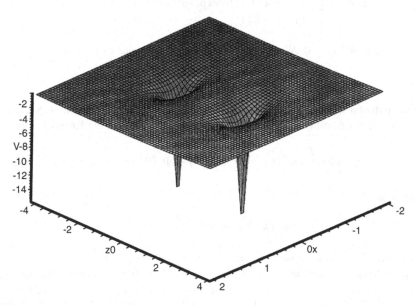

FIGURE 4.1

Dimensionless potential-energy function for the hydrogen molecular ion in the Born–Oppenheimer approximation ($y = 0$, $R = 3$).

4.5   The Delta Molecular Ion

A rigorous discussion of the perturbation expansion for the hydrogen molecular ion that accounts for the exponential and logarithmic terms at large internuclear distances is beyond the scope of this book. We do not even consider the perturbation treatment of simpler double wells in one dimension. However, a pedagogical approach to the perturbation expansion of a one-dimensional, one-electron molecule at large internuclear distances is possible for a simple, exactly solvable model. In the delta molecular ion one simulates the Coulomb interaction by means of the much simpler delta interaction. The dimensionless Schrödinger equation reads

$$\Psi''(x) = 2[V(x) - E]\Psi(x), \quad V(x) = -\delta(x) - \delta(x - R), \tag{4.86}$$

where $\delta(x)$ is the Kronecker delta function and R is the internuclear separation. The only properties of this function that we need here are $\delta(x) = 0$ if $x \neq 0$, and

$$\int_{-\epsilon}^{\epsilon} \Psi(x)\delta(x)\,dx = \Psi(0), \quad \epsilon > 0. \tag{4.87}$$

Notwithstanding the delta potential is a extremely short range interaction, it is commonly chosen as a one-dimensional model for the Coulomb interaction [67].

Although the bound states of this model are well known, [67] in what follows we briefly show how to obtain them for completeness. First of all notice that a square-integrable solution of the Schrödinger equation (4.86) is of the form

$$\Psi(x) = \begin{cases} A_1 \exp(kx) & \text{if} \quad x < 0 \\ A_2 \exp(kx) + B_2 \exp(-kx) & \text{if} \quad 0 < x < R \\ A_3 \exp(-kx) & \text{if} \quad x > R \end{cases}, \tag{4.88}$$

where $k = \sqrt{-2E}$ and $E < 0$. We require that it be continuous at $x = 0$ and $x = R$:

$$\Psi\left(0^-\right) = \Psi\left(0^+\right), \quad \Psi\left(R^-\right) = \Psi\left(R^+\right). \tag{4.89}$$

Because the potential-energy function is singular at $x = 0$ and $x = R$ the first derivative $\Psi'(x)$ is not continuous at those points as follows from the property (4.87) of the Kronecker delta function:

$$\lim_{\epsilon \to 0} \int_{-\epsilon}^{\epsilon} \Psi''(x)\,dx = \Psi'\left(0^+\right) - \Psi'\left(0^-\right) = -2\Psi(0). \tag{4.90}$$

Analogously, at $x = R$ we have

$$\Psi'\left(R^+\right) - \Psi'\left(R^-\right) = -2\Psi(R). \tag{4.91}$$

It is not difficult to obtain a suitable expression for the energy from equations (4.88)–(4.91) if we proceed orderly. First, obtain A_2 and B_2 in terms of A_1 from the two equations that give the boundary conditions at $x = 0$. Second, rewrite the two equations giving the boundary conditions at $x = R$ conveniently and divide one by the other in order to remove A_3. Finally, substitute the values of A_2 and B_2 obtained previously, remove A_1, and derive an equation solely in terms of k and R. One easily rewrites the resulting equation as follows:

$$(k - 1)^2 = \exp(-2kR). \tag{4.92}$$

The two roots of this equation

$$k_\pm = 1 \pm \exp(-kR) \tag{4.93}$$

give us the only two bound-state energies of the delta molecular ion in terms of R:

$$E_\pm = -\frac{k_\pm^2}{2} \, . \tag{4.94}$$

Notice that $0 < k_- < 1 < k_+$ leads to $E_+ < -1/2 < E_- < 0$ and tells us that E_+ is the ground-state energy. Here we have not taken into account the parity of the eigenfunctions as in a previous pedagogical treatment of this model [67]. Our simpler expression for the energy is suitable for the application of perturbation theory to obtain an expansion at large internuclear distances.

When $R \to \infty$ we obtain the approximation of order zero $k_{\pm 0} = 1$. We then write $k_\pm \approx k_{\pm 0} + k_{\pm 1}$, where $k_{\pm 1}$ is a small first-order correction to $k_{\pm 0}$, so that at first order we have $1 + k_{\pm 1} \approx 1 \pm \exp[-(1 + k_{\pm 1})R] \approx 1 \pm \exp(-R)$ from which it follows that $k_{\pm 1} = \pm \exp(-R)$. In order to make this procedure more systematic we substitute the expansion

$$k(\sigma) = \sum_{j=0}^{\infty} k_j \sigma^j \tag{4.95}$$

into

$$k(\sigma) = 1 + \sigma \exp(-kR) \, , \tag{4.96}$$

and then solve for the coefficients k_j term by term, and finally substitute the actual value of the perturbation parameter $\sigma = \pm 1$. A straightforward calculation (greatly facilitated by Maple) yields

$$k_\pm = 1 \pm \exp(-R) - R \exp(-2R) \pm \frac{3R^2}{2} \exp(-3R) - \frac{8R^3}{3} \exp(-4R) + \cdots , \tag{4.97}$$

from which it follows that

$$
\begin{aligned}
E_\pm = \ & -\frac{1}{2} \mp \exp(-R) + \left(R - \frac{1}{2}\right) \exp(-2R) \pm R \left(1 - \frac{3R}{2}\right) \exp(-3R) \\
& + R^2 \left(\frac{8R}{3} - 2\right) \exp(-4R) + \cdots .
\end{aligned}
\tag{4.98}
$$

At large internuclear distances the energies of the two bound states are almost degenerate, their difference being

$$\Delta E = E_- - E_+ = 2 \exp(-R) + \cdots . \tag{4.99}$$

This trivial model clearly shows that exponential terms take place in the expansion of the energies of diatomic molecules at large internuclear distances. In this oversimplified example there is no expansion in powers of $\lambda = 1/R$, and if we tried it we would find that all the coefficients vanish. On the other hand, the perturbation expansion for the hydrogen molecular ion exhibits a series in powers of $1/R$ in addition to exponential and logarithmic terms, [64]–[66] and we were able to obtain the former by means of the moment method in the preceding section. Exponentially small energy gaps like that in equation (4.99) are typical of double well potential-energy functions [66].

Chapter 5

The Schrödinger Equation on Bounded Domains

5.1 Introduction

In this chapter we focus our attention on the Schrödinger equation with boundary conditions for finite values of the coordinates. In particular, we first consider a particle in a box with impenetrable walls that give rise to Dirichlet boundary conditions. Models with such features appear in many branches of physics and chemistry, among which we mention the rate of escape of stars from clusters, [68] the theory of solids, [69, 70] molecular interactions, [71] electrons in crystals within electric fields, [72, 73] quantum wells, [74] and magnetic properties of metals [75]. This is just a sample of the vast literature on the subject [32, 76, 77]. Here we restrict ourselves to simple nontrivial one-particle systems in one and three dimensions.

In this chapter we also consider periodic boundary conditions. It is impossible to enumerate all the physical problems that require the solution of the Schrödinger equation with such boundary conditions. We only mention well-known models for the study of molecular rotation spectra that are based on a variety of perturbed rigid rotors [78, 79]. For concreteness we concentrate on a particular model and apply perturbation theory in the cases of weak and strong interaction.

5.2 One-Dimensional Box Models

We first consider a particle of mass m in a box with impenetrable walls at $x = L_1$ and $x = L_2$, under the effect of a potential-energy function $V(x)$ in $L_1 < x < L_2$. The state vector $\Psi(x)$ vanishes outside the box where the potential-energy function is infinite and continuity requires that $\Psi(L_1) = \Psi(L_2) = 0$. In the coordinate representation the Hamiltonian operator reads

$$\hat{H} = -\frac{\hbar^2}{2m}\frac{d^2}{dx^2} + V(x) \tag{5.1}$$

for all $L_1 < x < L_2$. By means of the change of variables

$$x = Lq + L_1, \quad \Phi(q) = \sqrt{L}\Psi(Lq + L_1), \tag{5.2}$$

where $L = L_2 - L_1$ is the box length, we rewrite the Schrödinger equation for $\Phi(q)$ as

$$\left[-\frac{1}{2}\frac{d^2}{dq^2} + \lambda v(q)\right]\Phi(q) = \epsilon\Phi(q), \tag{5.3}$$

where

$$\lambda v(q) = \frac{mL^2}{\hbar^2} V(Lq + L_1), \quad \epsilon = \frac{mL^2 E}{\hbar^2}. \tag{5.4}$$

The solution $\Phi(q)$ satisfies the simpler boundary conditions $\Phi(0) = \Phi(1) = 0$.

In order to apply perturbation theory we choose

$$\hat{H}_0 = -\frac{1}{2}\frac{d^2}{dq^2}, \quad \hat{H}' = v(q) \tag{5.5}$$

and expand

$$\Phi = \sum_{j=0}^{\infty} \Phi_j \lambda^j, \quad \epsilon = \sum_{j=0}^{\infty} \epsilon_j \lambda^j, \tag{5.6}$$

where we omit explicit reference to the quantum number n. The solutions to the unperturbed equation

$$-\frac{1}{2}\frac{d^2}{dq^2}\Phi_0(q) = \epsilon_0 \Phi_0(q) \tag{5.7}$$

are

$$\Phi_0(q) = \sqrt{2}\sin(n\pi q), \quad \epsilon_0 = \frac{n^2\pi^2}{2}, \quad n = 1, 2, \ldots. \tag{5.8}$$

For simplicity we require that $\Phi_j(0) = \Phi_j(1) = 0$ for all j to be sure that the approximate solution $\Phi(q)$ satisfies the appropriate boundary conditions at any perturbation order.

5.2.1 Straightforward Integration

In the particular case of the particle in the box just considered, the perturbation equations developed in Chapter 1 take the form

$$\left(\frac{d^2}{dq^2} + n^2\pi^2\right)\Phi_j = f_j, \quad j = 0, 1, 2, \ldots, \tag{5.9}$$

where $f_0 = 0$, and

$$f_j = 2v\Phi_{j-1} - 2\sum_{k=1}^{j}\epsilon_k\Phi_{j-k}, \quad j > 0. \tag{5.10}$$

The method developed in Appendix B proves suitable for solving the set of inhomogeneous ordinary differential equations of second order with constant coefficients (5.9). A real solution of equation (5.9) that satisfies the boundary condition at $q = 0$ is

$$\Phi_j(q) = C_j \sin(n\pi q) + \frac{1}{n\pi}\int_0^q \sin\left[n\pi(q - q')\right] f_j(q')\, dq', \tag{5.11}$$

where C_j is an arbitrary integration constant. The boundary condition at the other end point $\Phi_j(1) = 0$ determines the energy coefficient ϵ_j. The resulting equation is equivalent to the general expression (1.10) for the energy coefficient of order j that in the present case takes the form

$$\epsilon_j = \int_0^1 \Phi_0(q)v(q)\Phi_{j-1}(q)dq - \sum_{k=1}^{j-1}\epsilon_k\int_0^1 \Phi_0(q)\Phi_{j-k}(q)\, dq. \tag{5.12}$$

To derive this equation from $\Phi(1) = 0$ simply rewrite $\sin[n\pi(1 - q')]$ as $-\cos(n\pi)\sin(n\pi q')$. Finally, an appropriate normalization condition determines the remaining integration constant C_j; here we choose equation (1.11):

$$\sum_{i=0}^{j} \int_0^1 \Phi_i(q)\Phi_{j-i}(q) = \delta_{j0} . \tag{5.13}$$

It is possible to derive exact analytical perturbation corrections from equation (5.11) for many potential-energy functions $v(q)$. Maple greatly facilitates the systematic calculation of the integrals in equations (5.11) and (5.13), and we show a simple set of procedures for that purpose in the program section. For example, a particularly simple class of perturbations is given by polynomial potential-energy functions $V(x) = V_0 + V_1 x + V_2 x^2 + \cdots + V_M x^M$; we consider some illustrative examples in what follows.

The simplest case is the linear interaction $V(x) = V_1 x$ that has proved suitable for the investigation of the effect of electric fields on electrons in crystals [72, 73] and quantum wells [74]. Without loss of generality we choose $L_1 = 0$ and set

$$\lambda = \frac{mV_1 L^3}{\hbar^2}, \quad v(q) = q . \tag{5.14}$$

Notice that the dimensionless perturbation parameter λ is given by the ratio of the maximum value of the potential energy $V_1 L$ to a kind of characteristic kinetic energy $\hbar^2/(mL^2)$. Moreover, it can also be written as $\lambda = (L/L_0)^3$, where $L_0 = [\hbar^2/(mV_1)]^{1/3}$ is a characteristic length for the particle in the field.

Table 5.1 shows the first perturbation corrections to the energy and eigenfunction of the nth state in terms of $\omega = n\pi$. One easily obtains many more by means of the Maple procedures in the program section.

5.2.2 The Method of Swenson and Danforth

The application of the method of Swenson and Danforth to the particle in a box requires a careful discussion. It is well known that all the zeros of a solution $\Phi(q)$ of the Schrödinger equation at finite values of the coordinate are simple, because otherwise Φ vanishes everywhere. This conclusion follows from the fact that if both $\Phi(q)$ and $\Phi'(q)$ vanish at q_0, then all the derivatives of $\Phi(q)$ vanish at that point. The method of Swenson and Danforth is based on the hypervirial theorem $< \Phi \mid [\hat{H}, \hat{W}] \mid \Phi >= 0$ as discussed in Chapter 3. This expression is valid provided that $\hat{W}\Phi$ belongs to the state space, which in the present case is given by differentiable functions that vanish at $q = 0$ and $q = 1$. Because in the method of Swenson and Danforth we choose $\hat{W} = f(\hat{q})\hat{D} + g(\hat{q})$, then $f(q)$ has to vanish at $q = 0$ and $q = 1$. If $f(q)$ does not satisfy those boundary conditions, we can still apply the method of Swenson and Danforth provided that we modify the hypervirial relations [32]. Here we choose $f(q)$ to satisfy the boundary conditions in order to keep the standard form of the hypervirial theorem. All earlier applications of the approach were based on the modified hypervirial theorem [32].

For polynomial potential-energy functions we choose $f(q) = q^j - q$, $j = 2, 3, \ldots$ so that the hypervirial relations developed in Chapter 3 become

$$2j\epsilon\left\langle q^{j-1}\right\rangle + \frac{j(j-1)(j-2)}{4}\left\langle q^{j-3}\right\rangle - \lambda\left\langle q^j v'\right\rangle - 2j\lambda\left\langle q^{j-1}v\right\rangle$$
$$- 2\epsilon + \lambda\left\langle qv'\right\rangle + 2\lambda < v >= 0 . \tag{5.15}$$

Table 5.1 Straightforward Integration of the Perturbation Equations for a Particle in a Box with a Linear Interaction $v(q) = q$

$\omega = n\pi$

$\varepsilon_1 = \frac{1}{2}$

$\Phi_1 = \frac{\sqrt{2}\,(2q-1)\,\sin(\omega q)}{4\omega^2} - \frac{\sqrt{2}\,q\,(q-1)\,\cos(\omega q)}{2\omega}$

$\varepsilon_2 = \frac{\omega^2 - 15}{24\,\omega^4}$

$\Phi_2 = -\sqrt{2}\,\left(\omega^4 - 20\,\omega^2 - 405 + 150\,q\,\omega^2 - 150\,\omega^2\,q^2 + 30\,\omega^4\,q^4 - 60\,q^3\,\omega^4 + 30\,q^2\,\omega^4\right)$

$\frac{\sin(\omega q)}{240\,\omega^6} - \frac{5\sqrt{2}\,(q-1)\,q\,(2q-1)\,\cos(\omega q)}{24\,\omega^3}$

$\varepsilon_3 = 0$

$\Phi_3 = -\frac{\sqrt{2}}{960}$

$\left(70\,\omega^4\,q^4 - 140\,q^3\,\omega^4 - 450\,\omega^2\,q^2 + 70\,q^2\,\omega^4 + 450\,q\,\omega^2 + 195 - 75\,\omega^2 + \omega^4\right)$

$(2q-1)\,\sin(\omega q)/\omega^8 + \frac{\sqrt{2}}{480}$

$\left(10\,\omega^4\,q^4 - 20\,q^3\,\omega^4 + 10\,q^2\,\omega^4 - 250\,\omega^2\,q^2 + 250\,q\,\omega^2 - 50\,\omega^2 + 195 + \omega^4\right)$

$(q-1)\,q\,\cos(\omega q)/\omega^7$

$\varepsilon_4 = \dfrac{-210\,\omega^2 + 1980 + \omega^4}{288\,\omega^{10}}$

$\Phi_4 = \frac{\sqrt{2}}{241920}\,\left(468720\,\omega^4\,q^4 + 1408050\,q\,\omega^2 + 616770\,q^2\,\omega^4 - 937440\,q^3\,\omega^4\right.$

$+900900\,\omega^2 - 13059900 - 1408050\,\omega^2\,q^2 + 630\,\omega^8\,q^8 - 2520\,\omega^8\,q^7 + 3780\,\omega^8\,q^6$

$-2520\,q^5\,\omega^8 - 47460\,\omega^6\,q^6 - 252\,\omega^8\,q^3 + 126\,\omega^8\,q^2 - 153300\,\omega^6\,q^4$

$+69300\,\omega^6\,q^3 + 142380\,q^5\,\omega^6 + \omega^8 - 148050\,q\,\omega^4 + 630\,\omega^6\,q - 11550\,\omega^6\,q^2$

$\left. +756\,\omega^8\,q^4 + 13020\,\omega^4 - 130\,\omega^6\right)\,\sin(\omega q)/\omega^{12} + \frac{\sqrt{2}}{1152}$

$\left(18\,\omega^4\,q^4 - 36\,q^3\,\omega^4 + 18\,q^2\,\omega^4 - 438\,\omega^2\,q^2 + 438\,q\,\omega^2 + 2235 - 89\,\omega^2 + \omega^4\right)$

$(q-1)\,q\,(2q-1)\,\cos(\omega q)/\omega^9$

As an illustrative example we consider a second-order polynomial

$$v(q) = \alpha q + \beta q^2 \tag{5.16}$$

that accounts for most of the applications of physical interest of such oversimplified models. Arguing as in Chapter 3 we easily derive the following equations

$$Q_{0,p} = \delta_{0p}, \tag{5.17}$$

$$Q_{j,p} = \frac{1}{2(j+1)\epsilon_0}\left[\frac{j(1-j^2)}{4}Q_{j-2,p} - 2(j+1)\sum_{i=1}^{p}\epsilon_i Q_{j,p-i}\right.$$

$$\left. + (2j+3)\alpha Q_{j+1,p-1} + 2(j+2)\beta Q_{j+2,p-1} + 2\epsilon_p - 3\alpha Q_{1,p-1}\right.$$

$$-4\beta Q_{2,p-1}\Bigg], \quad j = 1, 2, \ldots, \tag{5.18}$$

$$\epsilon_p = \frac{1}{p}\left(\alpha Q_{1,p-1} + \beta Q_{2,p-1}\right), \tag{5.19}$$

where $Q_{j,p}$, $p = 0, 1, \ldots$ are the perturbation corrections to the expectation value $Q_j = <q^j>$, equation (5.17) is the normalization condition, and equation (5.19) follows from the Hellmann–Feynman theorem. In order to obtain ϵ_p we need $Q_{j,s}$, $s = 0, 1, \ldots, p-1$, $j = 1, 2, \ldots, 2(p-s)$.

For the harmonic oscillator $V(x) = kx^2/2$ we have

$$\lambda = \frac{mkL^4}{\hbar^2}, \quad v(q) = \frac{1}{2}(q - q_0)^2, \quad q_0 = -\frac{L_1}{L}. \tag{5.20}$$

Notice that in this case we can also write the dimensionless perturbation parameter λ either as the ratio of a potential kL^2 to a kinetic $\hbar^2/(mL^2)$ energy or as $\lambda = (L/L_0)^4$, where $L_0 = [\hbar^2/(mk)]^{1/4}$ is the characteristic length of the oscillator. The harmonic oscillator is symmetrical when $L_2 = -L_1 > 0$, in which case $q_0 = 1/2$. Equations (5.17)–(5.19) apply to the general case provided that we add $q_0^2/2$ to the perturbation correction of first order.

The calculation of exact analytical perturbation corrections by means of equations (5.17)–(5.19) is straightforward even by hand. However, if one is interested in relatively great perturbation orders, the use of computer algebra becomes necessary. Table 5.2 shows some results in terms of $\omega = n\pi$ and q_0 obtained by means of a simple Maple program similar to those discussed earlier in Chapter 3. They are valid for both symmetrical and nonsymmetrical harmonic oscillators and also for the inverted ones ($k < 0 \Rightarrow \lambda < 0$).

In order to test the perturbation coefficients just obtained and estimate the rate of convergence of the perturbation series, we compare the partial sums

$$S_N = \sum_{j=0}^{N} \epsilon_j \lambda^j \tag{5.21}$$

with exact energies of excited states of the harmonic oscillator with the usual boundary conditions $\Phi(\pm\infty) = 0$. For example, from its second excited state

$$\Phi(q) = \left(2q^2 - 1\right)\exp\left(-q^2/2\right) \tag{5.22}$$

we obtain the dimensionless energy $\epsilon = 5/2$ of the ground state of a harmonic oscillator $v(q) = q^2/2$ in a symmetrical box with walls at $L_2 = -L_1 = 1/\sqrt{2}$. The third excited state

$$\Phi(q) = q\left(2q^2 - 3\right)\exp\left(-q^2/2\right) \tag{5.23}$$

gives us the energy $\epsilon = 7/2$ of the ground state of a harmonic oscillator in a box with walls at $L_1 = 0$ and $L_2 = \sqrt{3/2}$ as well as of the first excited state of a harmonic oscillator in a box with walls $L_2 = -L_1 = \sqrt{3/2}$. $\log(|\epsilon - S_N|)$ provides a reasonable measure of the rate of convergence of the partial sums (5.21)). Figure 5.1 shows that the perturbation series for the first two cases just mentioned converge rapidly. The numerical results for the third case exactly agree with those for the second one, and do not add anything new to the present discussion. We appreciate that the rate of convergence is slightly greater for the symmetric oscillator.

Table 5.2   Method of Swenson and Danforth for the Dimensionless Bounded Harmonic

Oscillator $\hat{H} = -\dfrac{1}{2}\dfrac{d^2}{dq^2} + \dfrac{(q-q_0)^2}{2}$

$\omega = n\pi$

$\varepsilon_0 = \dfrac{\omega^2}{2}$

$\varepsilon_1 = -\dfrac{q_0}{2} - \dfrac{1}{4\omega^2} + \dfrac{1}{6} + \dfrac{q_0^2}{2}$

$\varepsilon_2 = \dfrac{-\dfrac{1}{24}q_0 + \dfrac{1}{24}q_0^2 + \dfrac{1}{90}}{\omega^2} + \dfrac{\dfrac{5}{8}q_0 - \dfrac{5}{8}q_0^2 - \dfrac{5}{24}}{\omega^4} + \dfrac{7}{16}\dfrac{1}{\omega^6}$

$\varepsilon_3 = \dfrac{\top 240\,q_0^2 + \dfrac{1}{945} - \dfrac{1}{240}q_0}{\omega^4} + \dfrac{\dfrac{5}{8}q_0 - \dfrac{5}{8}q_0^2 - \dfrac{1}{6}}{\omega^6} + \dfrac{\dfrac{93}{16}q_0^2 - \dfrac{93}{16}q_0 + \dfrac{31}{16}}{\omega^8} - \dfrac{121}{32}\dfrac{1}{\omega^{10}}$

$\varepsilon_4 = \dfrac{\dfrac{1}{288}q_0^4 - \dfrac{1}{144}q_0^3 + \dfrac{1}{2835} + \dfrac{139}{24192}q_0^2 - \dfrac{55}{24192}q_0}{\omega^6}$

$\quad + \dfrac{\dfrac{623}{960}q_0 - \dfrac{16}{135} + \dfrac{35}{24}q_0^3 - \dfrac{35}{48}q_0^4 - \dfrac{441}{320}q_0^2}{\omega^8}$

$\quad + \dfrac{-\dfrac{55}{4}q_0^3 + \dfrac{55}{8}q_0^4 + \dfrac{683}{160} + \dfrac{3149}{128}q_0^2 - \dfrac{2269}{128}q_0}{\omega^{10}} + \dfrac{\dfrac{14573}{128}q_0 - \dfrac{14573}{128}q_0^2 - \dfrac{14573}{384}}{\omega^{12}}$

$\quad + \dfrac{17771}{256\,\omega^{14}}$

$\varepsilon_5 = \dfrac{-\dfrac{1}{360}q_0^3 - \dfrac{397}{518400}q_0 + \dfrac{7}{66825} + \dfrac{1117}{518400}q_0^2 + \dfrac{1}{720}q_0^4}{\omega^8}$

$\quad + \dfrac{-\dfrac{17291}{11520}q_0^2 + \dfrac{15}{8}q_0^3 + \dfrac{6491}{11520}q_0 - \dfrac{223}{2700} - \dfrac{15}{16}q_0^4}{\omega^{10}}$

$\quad + \dfrac{\dfrac{101783}{15120} + \dfrac{1617}{32}q_0^4 - \dfrac{150293}{3840}q_0 - \dfrac{1617}{16}q_0^3 + \dfrac{344333}{3840}q_0^2}{\omega^{12}}$

$\quad + \dfrac{\dfrac{135121}{192}q_0 - \dfrac{213601}{192}q_0^2 - \dfrac{1635}{4}q_0^4 + \dfrac{1635}{2}q_0^3 - \dfrac{115501}{720}}{\omega^{14}}$

$\quad + \dfrac{\dfrac{938927}{768} + \dfrac{938927}{256}q_0^2 - \dfrac{938927}{256}q_0}{\omega^{16}} - \dfrac{1094647}{512\,\omega^{18}}$

$Q_{1,0} = \dfrac{1}{2}$

$Q_{1,1} = \dfrac{-\dfrac{1}{12}q_0 + \dfrac{1}{24}}{\omega^2} + \dfrac{-\dfrac{5}{8} + \dfrac{5}{4}q_0}{\omega^4}$

$Q_{1,2} = \dfrac{-\dfrac{1}{120}q_0 + \dfrac{1}{240}}{\omega^4} + \dfrac{-\dfrac{5}{8} + \dfrac{5}{4}q_0}{\omega^6} + \dfrac{-\dfrac{93}{8}q_0 + \dfrac{93}{16}}{\omega^8}$

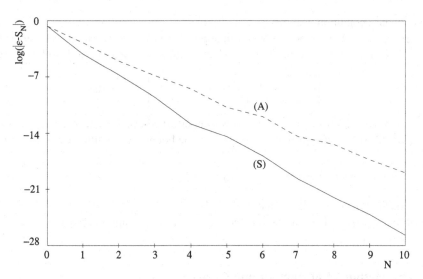

FIGURE 5.1
Rate of convergence of the perturbation series for the ground state of the symmetric (S) ($L_2 = -L_1 = 1/\sqrt{2}$) and asymmetric (A) ($L_1 = 0$, $L_2 = \sqrt{3/2}$) harmonic oscillator in a one-dimensional box.

5.3 Spherical-Box Models

A particle of mass m in a spherical box of radius R with an isotropic interaction potential proves also to be an interesting model with physical applications. The stationary states are solutions to the Schrödinger equation

$$\left[-\frac{\hbar^2}{2m} \nabla^2 + V(r) \right] \Psi(\mathbf{r}) = E \Psi(\mathbf{r}) \tag{5.24}$$

with the Dirichlet boundary condition $\Psi(\mathbf{r}) = 0$ on the box surface $r = |\mathbf{r}| = R$. The Schrödinger equation (5.24) is separable in spherical coordinates because both the potential-energy function and the box are spherically symmetric. One easily verifies that the radial factor of $\Psi(\mathbf{r}) = \chi(r) Y_{l,m}(\theta, \phi)$ satisfies the eigenvalue equation

$$\left[-\frac{\hbar^2}{2m} \left(\frac{d^2}{dr^2} + \frac{2}{r} \frac{d}{dr} \right) + \frac{\hbar^2 l(l+1)}{2mr^2} + V(r) \right] \chi(r) = E \chi(r) \tag{5.25}$$

with the boundary condition $\chi(R) = 0$. By means of the change of variables $r = Rq, \varphi(q) = R^{3/2} \chi(Rq)$ we derive the dimensionless equation

$$\left[-\frac{1}{2} \left(\frac{d^2}{dq^2} + \frac{2}{q} \frac{d}{dq} \right) + \frac{l(l+1)}{2q^2} + \lambda v(q) \right] \varphi(q) = \epsilon \varphi(q) \,, \tag{5.26}$$

where

$$\lambda v(q) = \frac{mR^2}{\hbar^2} V(Rq), \ \epsilon = \frac{mR^2 E}{\hbar^2} \,, \tag{5.27}$$

and the boundary condition becomes $\varphi(1) = 0$.

The particle in a spherical box provides an appropriate unperturbed model. In terms of the new variable $z = \sqrt{2\epsilon_0} q$ the eigenvalue equation (5.26) with $\lambda = 0$ becomes a Bessel equation [80]

$$\left[z^2 \frac{d^2}{dz^2} + 2z \frac{d}{dz} + z^2 - l(l+1) \right] \Xi(z) = 0 , \qquad (5.28)$$

where $\Xi(z) \propto \varphi(z/\sqrt{2\epsilon_0})$, and ϵ_0 denotes the unperturbed dimensionless energy. Since $\Xi(z) \propto j_l(z) = \sqrt{[\pi/(2z)]} J_{l+1/2}(z)$ [80] we conclude that the boundary condition $\varphi(1) = 0$ leads to

$$\epsilon_0 = \frac{j_{v,n}^2}{2} , \qquad (5.29)$$

where $j_{v,n}$ is the nth zero of $J_v(z)$, and $v = l + 1/2$.

In what follows we apply perturbation theory by means of two approaches discussed in Chapters 2 and 3.

5.3.1   The Method of Fernández and Castro

The method of Fernández and Castro developed in Chapter 2 is particularly simple when the auxiliary functions $A(q)$ and $B(q)$ are polynomials. For this reason it is convenient to rewrite equation (5.26) as

$$\left[\frac{d^2}{dq^2} + \frac{2(l+1)}{q} \frac{d}{dq} - 2\lambda v(q) + 2\epsilon \right] \vartheta(q) = 0 , \qquad (5.30)$$

where $\vartheta(q) = q^{-l} \varphi(q)$. We then write

$$\vartheta(q) = A(q)\vartheta_0(q) + B(q)\vartheta_0'(q) , \qquad (5.31)$$

where $\vartheta_0(q)$ is a solution to the unperturbed equation

$$\left[\frac{d^2}{dq^2} + \frac{2(l+1)}{q} \frac{d}{dq} + 2\epsilon_0 \right] \vartheta_0(q) = 0 \qquad (5.32)$$

with the boundary condition $\vartheta_0(1) = 0$. The auxiliary functions satisfy the equations

$$\begin{aligned}
qA'' + 2(l+1)A' - 4\epsilon_0 qB' + 2q(\Delta\epsilon - \lambda v)A &= 0 , \\
q^2 B'' - 2(l+1)qB' + 2q^2 A' + 2(l+1)B + 2q^2(\Delta\epsilon - \lambda v)B &= 0 ,
\end{aligned} \qquad (5.33)$$

where $\Delta\epsilon = \epsilon - \epsilon_0$ is the energy shift caused by the perturbation. We then expand the auxiliary functions and the dimensionless energy in Taylor series about $\lambda = 0$

$$A(q) = \sum_{j=0}^{\infty} A_j(q)\lambda^j, \ B(q) = \sum_{j=0}^{\infty} B_j(q)\lambda^j, \ \epsilon = \sum_{j=0}^{\infty} \epsilon_j \lambda^j , \qquad (5.34)$$

where $A_0 = 1$ and $B_0 = 0$. Straightforward substitution of the series (5.34) into equations (5.33) leads to the perturbation equations for the coefficients A_j, B_j, and ϵ_j

$$qA_j'' + 2(l+1)A_j' - 4\epsilon_0 qB_j' + 2q \sum_{i=1}^{j} \epsilon_i A_{j-i} - 2qvA_{j-1} = 0 ,$$

$$q^2 B_j'' - 2(l+1)qB_j' + 2q^2 A_j' + 2(l+1)B_j + 2q^2 \sum_{i=1}^{j} \epsilon_i B_{j-i} - 2q^2 vB_{j-1} = 0 . \ (5.35)$$

As an illustrative example we consider the isotropic harmonic oscillator with potential-energy function $V(r) = kr^2/2$ that leads to

$$\lambda = \frac{mkR^4}{\hbar^2}, \quad v(q) = \frac{q^2}{2} . \tag{5.36}$$

By inspection of the perturbation equations with the harmonic potential we conclude that

$$A_j(q) = \sum_{k=1}^{n_{a,j}} a_{j,k} q^{2k}, \quad B_j(q) = \sum_{k=0}^{n_{b,j}} b_{j,k} q^{2k+1} , \tag{5.37}$$

where $n_{a,1} = n_{b,1} = 1$ and $n_{a,j} = n_{b,j-1} + 2, n_{b,j} = n_{a,j-1} + 1$ for all $j > 1$. The energy is given either by the equations in Chapter 1 or by the boundary condition $B(1) = 0$ that leads to

$$b_{j,n_{b,j}} = - \sum_{k=0}^{n_{b,j}-1} b_{j,k} , \tag{5.38}$$

which enables one to be removed of the unknown coefficients. Both the calculation procedure and the corresponding Maple program are similar to those already discussed in Chapter 2. Table 5.3 shows analytical expressions in terms of ϵ_0 and l. After appropriate modification, present results agree with those obtained earlier by straightforward application of this approach to the isotropic harmonic oscillator in a spherical box in a space of D dimensions [81].

5.3.2 The Method of Swenson and Danforth

In order to simplify the application of the method of Swenson and Danforth it is convenient to transform the eigenvalue equation (5.26) into

$$\left[-\frac{1}{2}\frac{d^2}{dq^2} + \frac{l(l+1)}{2q^2} + \lambda v(q) \right] \Phi(q) = \epsilon \Phi(q) , \tag{5.39}$$

where the solution $\Phi(q) = q\varphi(q)$ satisfies boundary conditions similar to those for the one-dimensional box: $\Phi(0) = \Phi(1) = 0$. In this way we can apply all the perturbation equations developed earlier by simply adding the centrifugal term $l(l+1)/(2q^2)$ to the potential-energy function $\lambda v(q)$. Arguing as before we easily derive the hypervirial relation

$$2j\epsilon \left\langle q^{j-1} \right\rangle + (j-1)\frac{j(j-2) - 4l(l+1)}{4} \left\langle q^{j-3} \right\rangle - \lambda \left\langle q^j v' \right\rangle$$
$$- 2j\lambda \left\langle q^{j-1} v \right\rangle - 2\epsilon + \lambda \left\langle q v' \right\rangle + 2\lambda < v >= 0 . \tag{5.40}$$

As an illustrative example we consider the isotropic harmonic oscillator with dimensionless potential-energy function $v(q) = q^2/2$. Expanding the expectation values $Q_j = < q^{2j} >$ and the energy ϵ in Taylor series about $\lambda = 0$ in the usual way, we derive the following working equations:

$$Q_{0,p} = \delta_{0p} , \tag{5.41}$$

$$Q_{j,p} = \frac{1}{2(2j+1)\epsilon_0} \left\{ \frac{j[4l(l+1) - 4j^2 + 1]}{2} Q_{j-1,p} - 2(2j+1) \sum_{i=1}^{p} \epsilon_i Q_{j,p-i} \right.$$

$$\left. +2(j+1)Q_{j+1,p-1} + 2\epsilon_p - 2Q_{1,p-1} \right\}, \quad j = 1, 2, \ldots, \tag{5.42}$$

$$\epsilon_p = \frac{1}{2p} Q_{1,p-1} . \tag{5.43}$$

Table 5.3   Method of Fernández and Castro for the Isotropic Harmonic Oscillator in a Spherical Box

$A_0 = 1$

$B_0 = 0$

$\varepsilon_0 = \frac{j_{\nu,n}^2}{2}$

$A_1(q) = \frac{(1-2l)\,q^2}{24\,\varepsilon_0}$

$B_1(q) = -\frac{q\,(q-1)\,(q+1)}{12\,\varepsilon_0}$

$\varepsilon_1 = \frac{(-3+4\,\varepsilon_0+4l\,(l+1))}{24\,\varepsilon_0}$

$A_2(q) = \left(-40\,q^6\,\varepsilon_0^2 + \left(-76\,\varepsilon_0\,l + 80\,\varepsilon_0^2 + 20\,l^2\,\varepsilon_0 + 129\,\varepsilon_0\right)\,q^4\right.$
$\left. + \left(-40\,\varepsilon_0^2 + 80\,\varepsilon_0\,l - 24\,l^3 - 12\,l^2 + 222\,l - 40\,\varepsilon_0 - 105\right)\,q^2\right) / \left(5760\,\varepsilon_0^3\right)$

$B_2(q) = -\frac{q\,(q-1)\,(q+1)\,(48\,q^2\,\varepsilon_0 - 20\,\varepsilon_0\,l + 12\,l^2 + 12\,l - 105 - 22\,\varepsilon_0)}{2880\,\varepsilon_0^3}$

$\varepsilon_2 = \frac{(32\,\varepsilon_0^2 + 315 - 300\,\varepsilon_0 + (-112\,\varepsilon_0 - 456)\,l\,(l+1) + 48\,l^2\,(l+1)^2)}{5760\,\varepsilon_0^3}$

$A_3(q) = \left(\left(-4984\,\varepsilon_0^3 + 560\,\varepsilon_0^3\,l\right)\,q^8 + \left(-13922\,\varepsilon_0^2\,l + 756\,l^2\,\varepsilon_0^2 + 11424\,\varepsilon_0^3\right.\right.$
$\left. +45207\,\varepsilon_0^2 - 280\,l^3\,\varepsilon_0^2\right)\,q^6 + \left(-7896\,\varepsilon_0^3 + 1008\,\varepsilon_0\,l^4 + 16800\,\varepsilon_0^2\,l\right.$
$+96408\,\varepsilon_0\,l - 1680\,\varepsilon_0^3\,l - 4128\,l^3\,\varepsilon_0 + 672\,l^2\,\varepsilon_0^2 - 138537\,\varepsilon_0 - 5496\,l^2\,\varepsilon_0$
$\left. -56952\,\varepsilon_0^2\right)\,q^4 + \left(-259050\,l - 2080\,l^5 - 2016\,l^2\,\varepsilon_0^2 + 31400\,l^2 - 74592\,\varepsilon_0\,l\right.$
$+60720\,l^3 - 7392\,\varepsilon_0^2\,l + 8064\,l^3\,\varepsilon_0 + 20328\,\varepsilon_0^2 + 114345 + 1120\,\varepsilon_0^3\,l$
$\left.\left. +35280\,\varepsilon_0 + 1456\,\varepsilon_0^3 + 4032\,l^2\,\varepsilon_0 - 3120\,l^4\right)\,q^2\right) / \left(2903040\,\varepsilon_0^5\right)$

$B_3(q) = -q\,(q-1)\,(q+1)\,\left(-280\,q^6\,\varepsilon_0^3 + 7479\,q^4\,\varepsilon_0^2 + 560\,q^4\,\varepsilon_0^3 + 140\,q^4\,\varepsilon_0^2\,l\right.$
$+140\,q^4\,\varepsilon_0^2\,l^2 - 2296\,q^2\,\varepsilon_0^2\,l + 3072\,q^2\,\varepsilon_0\,l - 280\,q^2\,\varepsilon_0^3 - 280\,q^2\,l^2\,\varepsilon_0^2$
$+3072\,q^2\,\varepsilon_0\,l^2 - 5310\,q^2\,\varepsilon_0^2 - 48384\,q^2\,\varepsilon_0 + 1284\,\varepsilon_0^2 + 1568\,\varepsilon_0^2\,l$
$+560\,l^2\,\varepsilon_0^2 + 6348\,\varepsilon_0\,l - 3480\,l^2\,\varepsilon_0 + 126\,\varepsilon_0 - 303601\,l + 1040\,l^4 - 29320\,l^2$
$\left. +2080\,l^3 - 1008\,l^3\,\varepsilon_0 + 114345\right) / \left(1451520\,\varepsilon_0^5\right)$

$\varepsilon_3 = \left(768\,\varepsilon_0^3 - 60480\,\varepsilon_0^2 + 351540\,\varepsilon_0 - 343035 + \left(31776\,\varepsilon_0 + 6912\,\varepsilon_0^2\right.\right.$
$\left. + 548460\right)\,l\,(l+1) + 4160\,l^3\,(l+1)^3 +$
$\left(-124560 - 11968\,\varepsilon_0\right)\,l^2\,(l+1)^2) / \left(2903040\,\varepsilon_0^5\right)$

In order to obtain ϵ_p we need $Q_{j,s}$ for all $s = 0, 1, \ldots, p - 1$, $j = 1, 2, \ldots, p - s$.

Table 5.4 shows sample analytical expressions of perturbation corrections in terms of the unperturbed energy ϵ_0. The Maple program that provided such results is similar to those for earlier applications of the method of Swenson and Danforth. The dimensionless energy coefficients ϵ_j in Tables 5.3 and 5.4 appear to be different simply because their terms are arranged differently.

The application of the method of Swenson and Danforth to a particle in a spherical box is straightforward when the potential-energy function is a polynomial with only even powers of the radial coordinate. The reader may easily verify that problems arise when there are also odd powers. When $l = 0$ the spherical models are similar to those in one dimension, and we can therefore apply the approach already described above. On the other hand, when $l > 0$ one cannot calculate all the perturbation coefficients which result to be functions of the unknown $Q_{-1,p}$ that remain undetermined. In what follows we show how to overcome such difficulty in the case of the Coulomb interaction, and consider a hydrogen-like atom in a spherical box of radius R as an illustrative example. More precisely, the potential-energy function inside the box is $V(r) = -Ze^2/r$, where $-e < 0$ and $Ze > 0$ are the electronic and nuclear charges, respectively.

The perturbation parameter and dimensionless potential-energy function read

$$\lambda = \frac{mRZe^2}{\hbar^2}, \; v(q) = -\frac{1}{q}, \tag{5.44}$$

respectively, where λ/Z is the radius of the box in units of $\hbar^2/(me^2)$. For concreteness, from now on we choose $Z = 1$ (hydrogen atom) without loss of generality.

In this case the method of Swenson and Danforth does not apply, not even when $l = 0$, because the hypervirial recurrence relations do not provide the perturbation coefficients $Q_{-1,p}$. We can however overcome this problem by means of a Liouville transformation [82] of the Schrödinger equation for the hydrogen-like atom into the Schrödinger equation for a harmonic oscillator.

The change of independent and dependent variables according to $s = q^{1/2}$ and $F(s) = \Phi(s^2)/\sqrt{s}$, respectively, transforms equation (5.39) into

$$\left[-\frac{1}{2} \frac{d^2}{ds^2} + \frac{L(L+1)}{2s^2} + \frac{\beta s^2}{2} \right] F(s) = eF(s), \tag{5.45}$$

where

$$L = 2l + \frac{1}{2}, \; \beta = -8\epsilon, \; e = 4\lambda. \tag{5.46}$$

Since the boundary conditions are $F(0) = F(1) = 0$, the perturbation corrections to the energy in Tables 5.3 and 5.4 give us the coefficients of the series

$$e(\beta) = \sum_{j=0}^{\infty} e_j \beta^j = \sum_{j=0}^{\infty} e_j(-8\epsilon)^j = 4\lambda. \tag{5.47}$$

Given λ we obtain the energy as a root of equation (5.47) after substituting a partial sum for the series. We expect the perturbation series (5.47) to be more accurate the smaller the value of $|\epsilon|$. In particular, the box radii for which $\epsilon = 0$ are given by just the first term

$$\lambda_{l,n} = \frac{j_{\nu,n}^2}{8}, \; \nu = L + \frac{1}{2} = 2l + 1. \tag{5.48}$$

In order to test the equations just derived, we consider states of the unbounded hydrogen atom with radial nodes that provide exact results for the spherical-box model. For example, the 2s state

Table 5.4 Method of Swenson and Danforth for the Isotropic Harmonic Oscillator in a Spherical Box

$$\varepsilon_0 = \frac{j_{v,n}^2}{2}$$

$$\varepsilon_1 = \frac{1}{6} + \frac{\frac{1}{6}l(l+1) - \frac{1}{8}}{\varepsilon_0}$$

$$\varepsilon_2 = \frac{1}{180\,\varepsilon_0} + \frac{-\frac{7}{360}l(l+1) - \frac{5}{96}}{\varepsilon_0^2} + \frac{\frac{1}{120}l^2(l+1)^2 - \frac{19}{240}l(l+1) + \frac{7}{128}}{\varepsilon_0^3}$$

$$\varepsilon_3 = \frac{1}{3780\,\varepsilon_0^2} + \frac{-\frac{1}{48} + \frac{1}{420}l(l+1)}{\varepsilon_0^3} + \frac{\frac{31}{256} + \frac{331}{30240}l(l+1) - \frac{187}{45360}l^2(l+1)^2}{\varepsilon_0^4}$$
$$+ \frac{\frac{13}{9072}l^3(l+1)^3 + \frac{3047}{16128}l(l+1) - \frac{173}{4032}l^2(l+1)^2 - \frac{121}{1024}}{\varepsilon_0^5}$$

$$\varepsilon_4 = \frac{11}{22680\,\varepsilon_0^3} + \frac{-\frac{4l(l+1)}{14175} - \frac{1}{135}}{\varepsilon_0^4} + \frac{\frac{683}{5120} + \frac{1129\,l^2(l+1)^2}{907200} - \frac{2083\,l(l+1)}{201600}}{\varepsilon_0^5}$$
$$+ \frac{\frac{120671\,l(l+1)}{1935360} + \frac{57719\,l^2(l+1)^2}{2419200} - \frac{14573}{24576} - \frac{6967\,l^3(l+1)^3}{5443200}}{\varepsilon_0^6}$$
$$+ \frac{\frac{281\,l^4(l+1)^4}{777600} - \frac{85301\,l(l+1)}{92160} + \frac{17771}{32768} + \frac{198443\,l^2(l+1)^2}{691200} - \frac{5993\,l^3(l+1)^3}{259200}}{\varepsilon_0^7}$$

$$Q_{1,1} = \frac{1}{45\,\varepsilon_0} + \frac{-\frac{7}{90}l(l+1) - \frac{5}{24}}{\varepsilon_0^2} + \frac{\frac{1}{30}l^2(l+1)^2 - \frac{19}{60}l(l+1) + \frac{7}{32}}{\varepsilon_0^3}$$

$$Q_{1,2} = \frac{1}{630\,\varepsilon_0^2} + \frac{-\frac{1}{8} + \frac{1}{70}l(l+1)}{\varepsilon_0^3} + \frac{\frac{93}{128} + \frac{331}{5040}l(l+1) - \frac{187}{7560}l^2(l+1)^2}{\varepsilon_0^4}$$
$$+ \frac{\frac{13}{1512}l^3(l+1)^3 + \frac{3047}{2688}l(l+1) - \frac{173}{672}l^2(l+1)^2 - \frac{363}{512}}{\varepsilon_0^5}$$

$$Q_{1,3} = \frac{1}{2835\,\varepsilon_0^3} + \frac{-\frac{32}{14175}l(l+1) - \frac{8}{135}}{\varepsilon_0^4} + \frac{\frac{683}{640} + \frac{1129}{113400}l^2(l+1)^2 - \frac{2083}{25200}l(l+1)}{\varepsilon_0^5}$$
$$+ \frac{\frac{120671}{241920}l(l+1) + \frac{57719}{302400}l^2(l+1)^2 - \frac{14573}{3072} - \frac{6967}{680400}l^3(l+1)^3}{\varepsilon_0^6}$$
$$+ \frac{\frac{281}{97200}l^4(l+1)^4 - \frac{85301}{11520}l(l+1) + \frac{17771}{4096} + \frac{198443}{86400}l^2(l+1)^2 - \frac{5993}{32400}l^3(l+1)^3}{\varepsilon_0^7}$$

of the dimensionless unbounded hydrogen atom

$$\Phi(q) = q(2 - q)\exp(-q/2) \tag{5.49}$$

is a solution of the hydrogen atom in a spherical box of radius $\lambda = 2$ with exact energy $\epsilon = -1/8$. Analogously, the $3s$ state

$$\Phi(q) = q\left(27 - 18q + 2q^2\right)\exp(-q/3) \tag{5.50}$$

gives the energy $\epsilon = -1/18$ of a hydrogen atom in a box of radius $\lambda = (9 - 3\sqrt{3})/2$. For the calculation we need the first zero of the Bessel function $J_1(z)$, which one either obtains from the standard bibliography [83] or easily calculates by means of Maple. In the latter case we first expand $Bessel J(1, z)$ in a Taylor series to a sufficiently great degree, then convert it into a polynomial, and finally obtain the desired zero by means of Maple root finder $fsolve$.

The convergence of the partial sums to the exact energy values indicated above confirms that the perturbation coefficients in Table 5.4 for the harmonic oscillator and equation (5.47) are both correct. The ground-state energy given in Table 5.5 for some values of λ improves results obtained previously by means of perturbation theory and agrees with energy values coming from other methods. The third column in Table 5.5 lists energy values obtained from a Taylor expansion of the eigenfunction of equation (5.39) about $q = 0$

$$\Phi(q) = \sum_{j=0}^{\infty} c_j(\epsilon, \lambda)q^{j+l+1} \tag{5.51}$$

that is forced to satisfy the boundary condition at $q = 1$

$$\sum_{j=0}^{N} c_j(\epsilon, \lambda) = 0 \tag{5.52}$$

for sufficiently great values of N. Such nonperturbative results are a good test for the perturbative ones.

5.4   Perturbed Rigid Rotors

The rigid rotor has proved to be a suitable approximate model for the study of purely rotational molecular spectra [78, 79]. Fitting the energy levels of a polar rigid rotor in a classical external field to appropriate microwave molecular spectra enables one to estimate molecular structure constants, such as moments of inertia, [78, 79, 84] dipole moments, [78, 79, 84] and polarizability anisotropies [84, 85]. Such applications require sufficiently accurate energy levels in terms of the quantum numbers and molecular structure constants. Rayleigh–Schrödinger perturbation theory is particularly useful for this purpose because it produces analytical expressions for the rotational energies. For this reason there has been great interest in the calculation of energy coefficients of both the weak-field expansions [81, 85], [86]–[89] and the strong-field expansions [89]–[92].

There is a vast literature about the practical applications of the perturbed rigid rotor, and the calculation of its energies and states by perturbation theory. We will not try to be exhaustive and will just mention those works that are relevant to the present discussion. The reader may look up additional references in those cited here.

Table 5.5 Ground-State Energy of the Hydrogen Atom in a Spherical Box for Several Values of λ

λ	Perturbation Theory	Power Series
0.80	4.543380181	4.543380181
1.00	2.373990866	2.373990866
1.20	1.269315015	1.269315015
1.40	0.6471051144	0.6471051144
1.60	0.2713123126	0.2713123126
1.80	0.03255625279	0.03255625279
2.00	−0.1250000000	−0.1250000000
2.20	−0.2320333715	−0.2320333715
2.40	−0.3063980071	−0.3063980071
2.60	−0.3589782940	−0.3589782940
2.80	−0.3966665967	−0.3966665967
3.00	−0.4239672878	−0.4239672877
3.20	−0.4439029787	−0.4439029780
3.40	−0.458547621	−0.4585476156
3.60	−0.46935119	−0.4693511607
3.80	−0.4773435	−0.4773433863
4.00	−0.483266	−0.4832652506

For the sake of concreteness we restrict the discussion below to a linear rigid rotor under a perturbation $V(\theta)$ that depends only on the polar angle θ. The unperturbed Hamiltonian operator is $\hat{H}_0 = \hat{J}^2/(2I)$, where $\hat{J}^2$ is the angular momentum operator and I is the moment of inertia of the rotating rigid body. The dimensionless Schrödinger equation in the coordinate representation for the stationary states reads

$$\left[-\frac{1}{\sin(\theta)} \frac{\partial}{\partial \theta} \sin(\theta) \frac{\partial}{\partial \theta} - \frac{1}{\sin(\theta)^2} \frac{\partial^2}{\partial \phi^2} + \lambda v(\theta) \right] \Psi(\theta, \phi) = \epsilon \Psi(\theta, \phi) , \tag{5.53}$$

where $\epsilon = 2IE/\hbar^2$, $\lambda v(\theta) = 2IV(\theta)/\hbar^2$, and λ is a perturbation parameter. When $\lambda = 0$ the unperturbed energy is $\epsilon_0 = J(J+1)$, where $J = 0, 1, \ldots$ is the rotational quantum number [78, 79]. The eigenvalue equation (5.53) is separable; writing $\Psi(\theta, \phi) = \Theta(\theta) \exp(im\phi), m = 0, \pm 1, \pm 2, \ldots$ we are left with an eigenvalue equation for $\Theta(\theta)$:

$$\left[-\frac{1}{\sin(\theta)} \frac{d}{d\theta} \sin(\theta) \frac{d}{d\theta} + \frac{m^2}{\sin(\theta)^2} + \lambda v(\theta) \right] \Theta(\theta) = \epsilon \Theta(\theta) . \tag{5.54}$$

5.4.1 Weak-Field Expansion by the Method of Fernández and Castro

We first transform the Sturm–Liouville equation (5.54) into another one which is more suitable for the application of the method of Fernández and Castro [81]. In terms of the new variable $x = \cos(\theta)$, equation (5.54) becomes

$$\left(1 - x^2\right) \chi''(x) - 2x\chi'(x) + \left(\epsilon - \frac{m^2}{1 - x^2} - \lambda v \right) \chi(x) = 0 , \tag{5.55}$$

where $\chi(\cos(\theta)) \propto \Theta(\theta)$. In order to obtain simple polynomial solutions to the pair of coupled equations occurring in the method of Fernández and Castro, it is necessary to remove the singular points at $x = \pm 1$ from equation (5.55). Close to the singular points the solution of equation (5.55) behaves asymptotically as $\chi(x) \sim (1 - x^2)^{M/2}$, where $M = |m|$; consequently, writing $\chi(x) = (1 - x^2)^{M/2} \Phi(x)$ we obtain a more convenient Sturm–Liouville equation for the new function $\Phi(x)$:

$$\left(x^2 - 1\right) \Phi''(x) + 2(M + 1)x\Phi'(x) + [M(M + 1) + \lambda v - \epsilon]\Phi(x) = 0 . \tag{5.56}$$

If we now apply the method of Fernández and Castro with $P(x) = (x^2 - 1)$, $Q(x) = 2(M + 1)x$, $R(x) = M(M + 1) + \lambda v - \epsilon$, and $R_0(x) = M(M + 1) - \epsilon_0$, we obtain the following system of perturbation equations:

$$\left(x^2 - 1\right) A_j''(x) \quad + \quad 2(M + 1)x A_j'(x) - 2R_0(x^2 - 1)B_j'(x) - 2R_0 x B_j(x)$$

$$+ \quad v A_{j-1}(x) - \sum_{k=1}^{j} \epsilon_k A_{j-k}(x) = 0 , \tag{5.57}$$

$$\left(x^2 - 1\right) B_j''(x) \quad - \quad 2(M - 1)x B_j'(x) - 2M B_j(x) + 2A_j'(x)$$

$$+ \quad v B_{j-1}(x) - \sum_{k=1}^{j} \epsilon_k B_{j-k}(x) = 0 . \tag{5.58}$$

In what follows we consider a polar diatomic molecule in an electric field as a particular interaction. The Stark effect in a polar rigid rotor is of relevance to the study of molecular structure [78, 79] and has consequently been discussed by several authors, [85]–[88], [90]–[92] among many others. If $\mathbf{d}$ is the molecular dipole moment, and the electric field $\mathbf{F}$ is chosen to be along the z axis, then the classical interaction between them is $-dF\cos(\theta)$, where $d = |\mathbf{d}|$ and $F = |\mathbf{F}|$. Its dimensionless form reads $\lambda v(\theta) = -\lambda\cos(\theta) = -\lambda x$, where $\lambda = 2IdF/\hbar^2$.

By simple inspection of equations (5.57) and (5.58) it is not difficult to convince oneself that the solutions for the Stark effect are polynomial functions of the form

$$A_j(x) = \sum_{k=0}^{j} a_{jk}x^k, \quad B_j(x) = \sum_{k=0}^{j-1} b_{jk}x^k . \tag{5.59}$$

Before proceeding with the discussion of results it is worth noticing that the present application of the method of Fernández and Castro to the perturbed rigid rotor is simpler than a previous one [81] in which the solution of the perturbed Sturm–Liouville equation was written $\Phi(x) = A(x)\Phi_0(x) + B(x)\Phi_0'(x)$. The more convenient form $\Phi(x) = A(x)\Phi_0(x) + B(x)P(x)\Phi_0'(x)$ leads to simpler perturbation equations and to a simpler polynomial function $B_j(x)$. Moreover, since the latter expression of the perturbed function $\Phi(x)$ satisfies the boundary conditions $\Phi(\pm 1) = 0$, then the exact perturbation corrections to the energy ϵ_j arise by simply requiring that $A_j(x)$ and $B_j(x)$ be polynomial functions. The present implementation of the method of Fernández and Castro follows the outline in Chapter 2 and the corresponding references listed there that are supposed to be an improvement on an earlier treatment of the Stark effect in the rigid rotor [81].

The perturbation corrections to the energy and eigenfunction depend on M and $\epsilon_0 = J(J + 1)$, where $J = M, M+1, \ldots$. The energy coefficients shown in Table 5.6 agree with those derived earlier by other authors [81, 88]. For simplicity we have arbitrarily chosen the undetermined coefficients $a_{j0} = 0$ for all $j > 0$, thus neglecting terms proportional to Φ_0 which we may add later in order to normalize the perturbed eigenfunction to unity. The Maple program that produced the results in Table 5.6 is similar to those applications of the method of Fernández and Castro discussed earlier in this book.

Table 5.6   Method of Fernández and Castro for a Polar Linear Rigid Rotor in an Electric Field

$A_1 = \frac{M\,x}{2\varepsilon_0}$

$B_1 = \frac{1}{2\varepsilon_0}$

$\varepsilon_1 = 0$

$A_2 = \frac{(-3\,M + 2\,\varepsilon_0)\,x^2}{4\,\varepsilon_0\,(-3 + 4\,\varepsilon_0)}$

$B_2 = -\frac{3x}{4\,\varepsilon_0\,(-3 + 4\,\varepsilon_0)}$

$\varepsilon_2 = \frac{-3\,M^2 + \varepsilon_0}{2\,\varepsilon_0\,(-3 + 4\,\varepsilon_0)}$

$A_3 = \frac{(-3\,\varepsilon_0\,M^3 - 4\,\varepsilon_0{}^2\,M^3 - 18\,M^3 - 9\,\varepsilon_0{}^2\,M^2 + 18\,M^2\,\varepsilon_0 + 4\,\varepsilon_0{}^3\,M - 8\,\varepsilon_0{}^2\,M + 5\,\varepsilon_0{}^3)\,x}{24\,\varepsilon_0{}^3\,(-3 + 4\,\varepsilon_0)\,(\varepsilon_0 - 2)}$

$\qquad + \frac{(2\,M\,\varepsilon_0 + 6\,M - 5\,\varepsilon_0)\,x^3}{24\,\varepsilon_0\,(-3 + 4\,\varepsilon_0)\,(\varepsilon_0 - 2)}$

$B_3 = \frac{-3\,M^2\,\varepsilon_0 - 4\,\varepsilon_0{}^2\,M^2 - 18\,M^2 + 4\,\varepsilon_0{}^3 - 8\,\varepsilon_0{}^2}{24\,\varepsilon_0{}^3\,(-3 + 4\,\varepsilon_0)\,(\varepsilon_0 - 2)} + \frac{(\varepsilon_0 + 3)\,x^2}{12\,\varepsilon_0\,(-3 + 4\,\varepsilon_0)\,(\varepsilon_0 - 2)}$

$\varepsilon_3 = 0$

$\varepsilon_4 = \frac{(612\,\varepsilon_0{}^2 + 513\,\varepsilon_0 - 405)\,M^4 + (-504\,\varepsilon_0{}^3 + 90\,\varepsilon_0{}^2)\,M^2 + 20\,\varepsilon_0{}^4 + 33\,\varepsilon_0{}^3}{8\,(-3 + 4\,\varepsilon_0)^3\,(4\,\varepsilon_0 - 15)\,\varepsilon_0{}^3}$

5.4.2   Weak-Field Expansion by the Method of Swenson and Danforth

It is not difficult to apply the method of Swenson and Danforth to the Sturm–Liouville equation (5.54). However, in what follows we first transform it into a Schrödinger-like eigenvalue equation in order to make direct use of the results of Chapter 3. The function $\Phi(\theta) = \sin(\theta)^{1/2}\Theta(\theta)$ satisfies

$$\left[-\frac{d^2}{d\theta^2} + \frac{\alpha}{\sin(\theta)^2} + \lambda v(\theta) - \mathcal{E} \right] \Phi(\theta) = 0 , \qquad (5.60)$$

where

$$\alpha = m^2 - \frac{1}{4}, \quad \mathcal{E} = \epsilon + \frac{1}{4} . \qquad (5.61)$$

If the potential-energy function $v(\theta)$ is even, then the eigenfunctions are either even or odd, and we have to choose odd functions $f(\theta)$ in the method of Swenson and Danforth (refer to Chapter 3 for more details) in order to obtain a nontrivial recurrence relation. In what follows we show that the set of functions $f(\theta) = \sin(\theta)^i \cos(\theta)^j$, $i = 1, 3, \ldots, j = 0, 1, \ldots$, is suitable for the application of perturbation theory.

According to equation (3.16) a term of the form

$$2\sin(\theta)^{-2} f' + f\left[\sin(\theta)^{-2} \right]' = 2(i - 1)\sin(\theta)^{i-3}\cos(\theta)^{j+1} - 2j\sin(\theta)^{i-1}\cos(\theta)^{j-1} \qquad (5.62)$$

will appear in the recurrence relation, and we realize that if $f(\theta) = \sin(\theta)\cos(\theta)^j$, then no undesired negative power of $\sin(\theta)$ will occur.

When $v(\theta) = -\cos(\theta)$ the recurrence relation reads

$$-\frac{(j+1)^3}{2}C_{j+1} + j\left(j^2 + 1\right)C_{j-1} - \frac{j(j-1)(j-2)}{2}C_{j-3} + 2(j+1)\mathcal{E}C_{j+1}$$
$$- 2j\mathcal{E}C_{j-1} + 2\alpha j \dot{C}_{j-1} - \lambda[(2j+1)C_j - (2j+3)C_{j+2}] = 0, \qquad (5.63)$$

where

$$C_j = \left\langle \cos(\theta)^j \right\rangle = \int_0^\pi \Phi(\theta)^2 \cos(\theta)^j \, d\theta, \qquad (5.64)$$

and $C_0 = 1$ if the eigenfunction is normalized to unity. To these equations we add the Hellmann–Feynman theorem $d\mathcal{E}/d\lambda = -C_1$.

From straightforward application of perturbation theory

$$\mathcal{E} = \sum_{k=0}^{\infty} \mathcal{E}_k \lambda^k, \quad C_j = \sum_{k=0}^{\infty} C_{j,k} \lambda^k \qquad (5.65)$$

we easily obtain

$$
\begin{aligned}
C_{j+1,p} = {} & \frac{2}{(j+1)[4\mathcal{E}_0 - (j+1)^2]} \Bigg[j\left(2\mathcal{E}_0 - 2\alpha - 1 - j^2\right) C_{j-1,p} \\
& + \frac{j(j-1)(j-2)}{2} C_{j-3,p} + 2j \sum_{s=1}^{p} \mathcal{E}_s C_{j-1,p-s} - 2(j+1) \sum_{s=1}^{p} \mathcal{E}_s C_{j+1,p-s} \\
& + (2j+1)C_{j,p-1} - (2j+3)C_{j+2,p-1} \Bigg]
\end{aligned}
\qquad (5.66)
$$

and

$$\mathcal{E}_p = -\frac{1}{p} C_{1,p-1}. \qquad (5.67)$$

In order to obtain $\mathcal{E}_p$ we have to calculate $C_{j+1,q}$ for all $q = 0, 1, \ldots, p-1$ and $j = 0, 1, \ldots, p-q-1$, taking into account the normalization condition $C_{0,p} = \delta_{0p}$. The perturbation coefficients for the energy and expectation values are given in terms of M and

$$\mathcal{E}_0 = J(J+1) + \frac{1}{4} = \left(J + \frac{1}{2}\right)^2. \qquad (5.68)$$

Taking into account that the eigenvalue equation (5.60) is invariant under the transformation $(\lambda, \theta) \to (-\lambda, \theta + \pi)$, it is not difficult to prove that $\epsilon_{2j+1} = 0$, $j = 0, 1, \ldots$, and $C_{j,p} = 0$, $j+p = 1, 3, \ldots$. We briefly discuss this point in Appendix C.

Present straightforward application of the method of Swenson and Danforth to the perturbed linear rotor is different from an earlier adaptation of that method for Sturm–Liouville equations [93]–[95]. The advantage of the approach given here is that the same eigenvalue equation (5.60) is suitable for both weak and strong fields as we will shortly see.

The method of Swenson and Danforth is faster and requires less computer memory than the method of Fernández and Castro. However, the former provides expectation values of chosen trigonometric functions instead of the eigenfunctions, whereas the latter yields the eigenfunctions explicitly and is therefore more suitable for the calculation of properties other than the energy.

Table 5.7 shows some moment and energy coefficients. One easily obtains many more by means of Maple procedures similar to those given earlier for other applications of the method of Swenson and Danforth. The energy coefficients agree with those in Table 5.6.

Table 5.7   Method of Swenson and Danforth for the Polar Linear Rigid Rotor in an Electric Field. Notice that $\varepsilon_{2p+1} = 0$, $C_{j,\,p} = 0$ for all $p = 0, 1, \ldots$, and $j + p = 1, 3, \ldots$

$$\varepsilon_0 = J\,(J+1)$$

$$\varepsilon_2 = \frac{\varepsilon_0 - 3\,M^2}{2\,\varepsilon_0\,(4\,\varepsilon_0 - 3)}$$

$$\varepsilon_4 = \frac{(612\,\varepsilon_0{}^2 + 513\,\varepsilon_0 - 405)\,M^4 + (-504\,\varepsilon_0{}^3 + 90\,\varepsilon_0{}^2)\,M^2 + 20\,\varepsilon_0{}^4 + 33\,\varepsilon_0{}^3}{8\,\varepsilon_0{}^3\,(4\,\varepsilon_0 - 3)^3\,(4\,\varepsilon_0 - 15)}$$

$$\begin{aligned}
\varepsilon_6 = \Big(&\big(1505952\,\varepsilon_0{}^3 - 306099\,\varepsilon_0{}^2 + 255150 - 147840\,\varepsilon_0{}^5 - 249840\,\varepsilon_0{}^4 - 650025\,\varepsilon_0\big)\,M^6 \\
&+ \big(-170100\,\varepsilon_0{}^2 + 177216\,\varepsilon_0{}^6 - 1057644\,\varepsilon_0{}^4 + 773550\,\varepsilon_0{}^3 - 63360\,\varepsilon_0{}^5\big)\,M^4 \\
&+ \big(-42240\,\varepsilon_0{}^7 + 54360\,\varepsilon_0{}^5 + 5481\,\varepsilon_0{}^4 + 72336\,\varepsilon_0{}^6\big)\,M^2 + 2848\,\varepsilon_0{}^7 + 576\,\varepsilon_0{}^8 \\
&- 5180\,\varepsilon_0{}^6 - 5640\,\varepsilon_0{}^5\Big) \big/ \Big(16\,\varepsilon_0{}^5\,(4\,\varepsilon_0 - 3)^5\,(\varepsilon_0 - 2)\,(4\,\varepsilon_0 - 15)\,(4\,\varepsilon_0 - 35)\Big)
\end{aligned}$$

$$\begin{aligned}
\varepsilon_8 = \Big(&\big(162753806250 - 620272839375\,\varepsilon_0 + 99436055625\,\varepsilon_0{}^2 + 2155215016800\,\varepsilon_0{}^3 \\
&- 2376614966715\,\varepsilon_0{}^4 - 6957377280\,\varepsilon_0{}^8 - 198105678720\,\varepsilon_0{}^7 \\
&- 370733751900\,\varepsilon_0{}^5 + 892218823200\,\varepsilon_0{}^6 + 3122058240\,\varepsilon_0{}^9\big)\,M^8 \\
&+ \big(-159137055000\,\varepsilon_0{}^2 + 26535269376\,\varepsilon_0{}^9 - 1214163167616\,\varepsilon_0{}^7 \\
&+ 171499341312\,\varepsilon_0{}^8 + 1721754031248\,\varepsilon_0{}^6 + 924763108500\,\varepsilon_0{}^3 \\
&- 1750322803500\,\varepsilon_0{}^4 + 570123095940\,\varepsilon_0{}^5 - 4896141312\,\varepsilon_0{}^{10}\big)\,M^6 \\
&+ \big(-719148451464\,\varepsilon_0{}^7 + 337056493248\,\varepsilon_0{}^8 - 17097435648\,\varepsilon_0{}^{10} + 17135363700\,\varepsilon_0{}^4 \\
&- 13004234496\,\varepsilon_0{}^9 - 133846229790\,\varepsilon_0{}^5 + 412609935582\,\varepsilon_0{}^6 + 2108282880\,\varepsilon_0{}^{11}\big) \\
&M^4 + \big(-226652160\,\varepsilon_0{}^{12} + 1541835000\,\varepsilon_0{}^6 + 14690097360\,\varepsilon_0{}^8 + 14470172100\,\varepsilon_0{}^7 \\
&- 9888439680\,\varepsilon_0{}^9 + 2100188160\,\varepsilon_0{}^{11} - 3219770880\,\varepsilon_0{}^{10}\big)\,M^2 + 8235264\,\varepsilon_0{}^{12} \\
&+ 228561696\,\varepsilon_0{}^{10} - 990203130\,\varepsilon_0{}^7 - 1212854283\,\varepsilon_0{}^8 + 930530052\,\varepsilon_0{}^9 \\
&+ 1504256\,\varepsilon_0{}^{13} - 155906432\,\varepsilon_0{}^{11}\Big) \\
&\Big/ \Big(128\,\varepsilon_0{}^7\,(4\,\varepsilon_0 - 3)^7\,(\varepsilon_0 - 2)\,(4\,\varepsilon_0 - 15)^3\,(4\,\varepsilon_0 - 35)\,(4\,\varepsilon_0 - 63)\Big)
\end{aligned}$$

$$C_{1,\,1} = \frac{-\varepsilon_0 + 3\,M^2}{\varepsilon_0\,(4\,\varepsilon_0 - 3)}$$

$$C_{1,\,3} = \frac{(-612\,\varepsilon_0{}^2 - 513\,\varepsilon_0 + 405)\,M^4 + (504\,\varepsilon_0{}^3 - 90\,\varepsilon_0{}^2)\,M^2 - 20\,\varepsilon_0{}^4 - 33\,\varepsilon_0{}^3}{2\,\varepsilon_0{}^3\,(4\,\varepsilon_0 - 3)^3\,(4\,\varepsilon_0 - 15)}$$

$$C_{2,\,0} = \frac{2\,\varepsilon_0 - 1 - 2\,M^2}{4\,\varepsilon_0 - 3}$$

$$C_{2,\,2} = \frac{(252\,\varepsilon_0 - 45)\,M^4 + (-200\,\varepsilon_0{}^2 + 174\,\varepsilon_0 - 90)\,M^2 + 12\,\varepsilon_0{}^3 - 13\,\varepsilon_0{}^2 + 15\,\varepsilon_0}{(4\,\varepsilon_0 - 3)^3\,\varepsilon_0\,(4\,\varepsilon_0 - 15)}$$

5.4.3 Strong-Field Expansion

The perturbation expansion in powers of λ is suitable for sufficiently weak interactions, and its radius of convergence will be discussed in Chapter 6. In addition to the λ-power series it is also possible to derive an expansion for large values of λ that we discuss in what follows. The stronger the interaction, the deeper the potential well $v(\theta)$ as shown in Figure 5.2 for the dimensionless Stark potential $v(\theta) = -\lambda \cos(\theta)$. When the interaction is sufficiently strong the system oscillates about the minimum of $v(\theta)$ which we can approximate by its Taylor expansion in a way that resembles the approach of small oscillations in classical mechanics. Although we discuss such polynomial approximation in more detail in Chapter 7, we believe it appropriate to round off the present study of perturbed rigid rotors with the treatment of strong interactions.

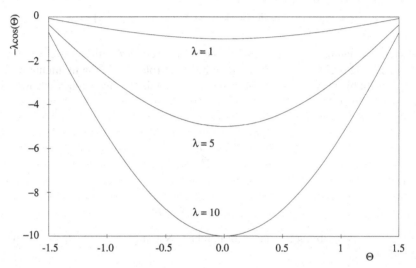

FIGURE 5.2
Dimensionless potential-energy function $v(\theta) = -\lambda \cos(\theta)$ for increasing values of λ.

For concreteness we consider a parity-invariant potential-energy function with a minimum at $\theta = 0$; that is to say, $v(-\theta) = v(\theta)$, $v'(0) = 0$, and $v''(0) > 0$. We expand it in a Taylor series about the minimum

$$v(\theta) = \sum_{j=0}^{\infty} v_j \theta^{2j}. \tag{5.69}$$

It is convenient to define the function

$$F(\theta) = \frac{\theta^2}{\sin(\theta)^2} = \sum_{j=0}^{\infty} F_j \theta^{2j}. \tag{5.70}$$

By comparing the Taylor series of both sides of $\sin(\theta)^2 F(\theta)/\theta^2 = 1$ it is not difficult to prove that the coefficients F_j satisfy the recurrence relation

$$F_j = \delta_{j0} + \sum_{i=1}^{j} (-1)^{i+1} \frac{2^{2i+1}}{(2i+2)!} F_{j-i} \tag{5.71}$$

which is useful to obtain them by means of computer algebra.

Rewriting the eigenvalue equation (5.60) in terms of the new variable

$$q = \frac{\theta}{\sqrt{\xi}}, \; \xi = \frac{1}{\sqrt{\lambda v_1}} \,, \tag{5.72}$$

we obtain

$$\left[-\frac{d^2}{dq^2} + q^2 + \frac{\alpha}{q^2} + \sum_{j=1}^{\infty} (a_j q^{2j+2} + b_j q^{2j-2}) \xi^j - e \right] \Phi = 0 \,, \tag{5.73}$$

where

$$e = \xi \left(\epsilon + \frac{1}{4} \right) - \xi \alpha F_1 - \lambda \xi v_0, \; a_j = \frac{v_{j+1}}{v_1}, \; b_j = \left(1 - \delta_{j1} \right) \alpha F_j \,. \tag{5.74}$$

The Schrödinger equation (5.73) describes a harmonic oscillator with a power series perturbation and accounts for the oscillation of the strongly hindered rotor about the minimum of the potential well. On solving it by means of perturbation theory we obtain the eigenvalues as power series of the perturbation parameter ξ:

$$e = \sum_{j=0}^{\infty} e_j \xi^j \,, \tag{5.75}$$

where [30]

$$e_0 = 2(2n + M + 1) \,, \tag{5.76}$$

and $n = 0, 1, \ldots$ is a vibrational quantum number. Solving for ϵ we obtain the eigenvalues of equation (5.54) as

$$\epsilon = v_0 \lambda - \frac{1}{4} + \alpha F_1 + \sum_{j=0}^{\infty} e_j (v_1 \lambda)^{(1-j)/2} \,. \tag{5.77}$$

It only remains to calculate the energy coefficients e_j by means of any of the methods described in previous chapters. For simplicity, here we choose the method of Swenson and Danforth. Straightforward application of the hypervirial and Hellmann–Feynman theorems, and expansion of the energy e and the expectation values $Q_N = < q^N >$ in Taylor series about $\xi = 0$, yield

$$Q_{N+1,p} = \frac{1}{4(N+1)} \left\{ N \left(4N^2 - 4m^2 \right) Q_{N-1,p} + 2(2N+1) \sum_{k=0}^{p} e_k Q_{N,p-k} \right. \tag{5.78}$$

$$\left. - \sum_{j=1}^{p} \left[2(2N+j+2) a_j Q_{N+j+1,p-j} + 2\alpha(2N+j)(1-\delta_{j1}) F_j Q_{N+j-1,p-j} \right] \right\} \,,$$

$$e_p = \frac{1}{p} \sum_{j=1}^{p} j \left[a_j Q_{j+1,p-j} + \alpha \left(1 - \delta_{j1} \right) F_j Q_{j-1,p-j} \right] \,. \tag{5.79}$$

In order to obtain e_p we need all the moment coefficients $Q_{N+1,s}$, $s = 0, 1, \ldots, p-1$, and $N = 0, 1, \ldots, p-s$, where $Q_{0,s} = \delta_{0s}$.

By means of the equations just derived and a Maple program similar to those for earlier applications of the method of Swenson and Danforth, one easily obtains as many perturbation corrections as

Table 5.8 Energy of a Polar Rigid Rotor in a Strong Electric Field

$$K = 2n + M + 1 = 2J - M + 1$$

$$\varepsilon = -\lambda + \frac{3M^2}{8} - \frac{3}{8} + K\sqrt{2}\sqrt{\lambda} - \frac{1}{8}K^2 + \frac{K(-3+9M^2-K^2)\sqrt{2}}{128\sqrt{\lambda}}$$

$$- \frac{34K^2 - 102M^2K^2 + 5K^4 + 9 - 42M^2 + 33M^4}{2048\lambda}$$

$$- \frac{K(-1722M^2 + 813M^4 - 1230M^2K^2 + 405 + 410K^2 + 33K^4)\sqrt{2}}{65536\lambda^{3/2}}$$

$$+ \frac{9}{262144}\left(-54 + 1350M^2K^2 - 495M^4K^2 + 420M^2K^4 - 327K^2 - 314M^4 - 140K^4\right.$$

$$\left. + 286M^2 + 82M^6 - 7K^6\right)/\lambda^2 + \cdots$$

desired. Table 5.8 shows an analytical expression for the energy through order four in terms of the quantum numbers J, M, and $K = 2n + M + 1$; the latter introduced with the only purpose of comparing present results with earlier ones [91].

In order to match the weak- and strong-field series for a given state, one should just take into account that $J = M + n$. Therefore, writing $K = 2J - M + 1$ we obtain the strong-field expansion solely in terms of the rotational quantum numbers J and $M = |m|$ [90].

In Chapter 6 we show that the radius of convergence of the weak-field perturbation series is nonzero and increases with the quantum number J. On the other hand, the strong-field perturbation series is divergent but still useful for sufficiently great values of λ provided that we truncate it properly. In what follows we show that both series match smoothly at an intermediate point. As an illustrative example we consider the ground state ($M = 0$, $J = 0$), and arbitrarily choose maximum perturbation orders $p = 20$ and $p = 5$ for the weak-field (WF) and strong-field (SF) series, respectively.

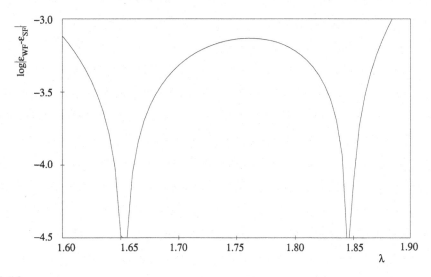

FIGURE 5.3

Logarithmic absolute difference between the weak-field ϵ_{WF} and strong-field ϵ_{SF} series for the dimensionless ground-state energy of a polar rigid rotor in an electric field of intermediate strength.

Figure 5.3 shows $\log |\epsilon_{WF} - \epsilon_{SF}|$ for an intermediate region $1.6 < \lambda < 1.9$ where we clearly see that both series match smoothly and even cross twice. If, for example, we use the weak-field series for $\lambda < 1.65$ and the strong-field expansion for $\lambda > 1.65$, then we expect to obtain reasonable results for all values of λ with the greatest error somewhere around the matching point. At $\lambda = 1.65$ the "exact" (accurately calculated by a numerical method), weak-field, and strong-field dimensionless energies are, respectively, $\epsilon_{exact} = -0.3975020830$, $\epsilon_{WF} = -0.3972789054$, and $\epsilon_{SF} = -0.397311692$. The percent error is 0.056 and 0.048 for the weak-field and strong-field series, respectively, and must be smaller than these values for $\lambda < 1.65$ and $\lambda > 1.65$ where the former and latter series, respectively, are expected to improve. We have obtained the "exact" results by means of an accurate summation of the weak-field series in a way described in Chapter 6.

Chapter 6

Convergence of the Perturbation Series

6.1 Introduction

In preceding chapters we developed several methods for solving the perturbation equations for the time-independent Schrödinger equation and showed that it is sometimes possible to obtain as many perturbation coefficients as desired for the energies and eigenfunctions. Except for some trivial cases perturbation expansions are infinite power series like

$$\sum_{j=0}^{\infty} E_j \lambda^j \,, \tag{6.1}$$

and we are faced with the problem of finding a number $E(\lambda)$ which may be properly called the sum of that series for a given value of λ. A rigorous discussion of such a subject is beyond the scope of this book. However, in this chapter we briefly comment on some of the results and conclusions derived by other authors keeping present contribution as simple as possible. Moreover, we mainly concentrate on the practical aspect of obtaining accurate results from the perturbation series.

If the rate of convergence of the perturbation series is sufficiently great, we may obtain accurate results without difficulty by summing all available terms. If, on the other hand, the series is divergent or slowly convergent we may need an appropriate summation algorithm to obtain acceptable results. For concreteness and simplicity we focus on the perturbation series for the energy.

6.2 Convergence Properties of Power Series

The investigation of the convergence properties of the series

$$\sum_{j=0}^{\infty} a_j \tag{6.2}$$

is based on the examination of its partial sums $S_N = a_0 + a_1 + \cdots a_N$ to find out whether or not they tend to a finite limit as $N \to \infty$.

There are several well-known convergence tests that one may apply to a given series [96]. For example, consider an infinite series (6.2) of positive terms such that

$$\lim_{j \to \infty} \frac{a_{j+1}}{a_j} = L \,. \tag{6.3}$$

The ratio test states that if $L < 1$ the series converges, if $L > 1$ the series diverges, and if $L = 1$ the test is inconclusive. Accordingly, the power series (6.1) converges for all values of λ satisfying [96]

$$|\lambda| < \lim_{j \to \infty} \frac{|E_j|}{|E_{j+1}|} . \tag{6.4}$$

If we can prove that the power series (6.1) converges for all values of λ such that $|\lambda| < R$, then we say that R is the radius of convergence of the series. For example, the geometric series $1 + \lambda + \lambda^2 + \cdots$ converges to $E(\lambda) = 1/(1 - \lambda)$ for all $|\lambda| < 1$. In this case the radius of convergence is determined by a singular point at $\lambda = 1$ where $|E(\lambda)|$ becomes infinite. Although the function $E(\lambda)$ is well defined for all $|\lambda| \neq 1$, the power series is meaningful only for $|\lambda| < 1$.

In general, we say that $\lambda = \lambda_p$ is a pole of the function $E(\lambda)$ of order n if $A = \lim_{\lambda \to \lambda_p} (\lambda - \lambda_p)^n E(\lambda)$ is a finite nonzero number for some positive integer n. We find that the function $E(\lambda)$ behaves approximately as $A(\lambda - \lambda_p)^{-n}$ in the neighborhood of λ_p. The radius of convergence of the geometric series considered above is determined by a pole of order $n = 1$ at $\lambda_p = 1$.

The power series

$$\sum_{j=0}^{\infty} (-1)^j \begin{pmatrix} 1/2 \\ j \end{pmatrix} \lambda^j = 1 - \frac{\lambda}{2} - \frac{\lambda^2}{8} - \frac{\lambda^3}{16} - \cdots , \tag{6.5}$$

where $\begin{pmatrix} a \\ j \end{pmatrix}$ denotes the combinatorial numbers, converges to $\sqrt{1 - \lambda}$ for all $|\lambda| < 1$. The radius of convergence is in this case determined by a square-root branch point at $\lambda = 1$. To understand this kind of singular point consider the simple function $f(z) = z^{1/2}$ in the complex z plane. In polar representation $z = re^{i\theta}$, where $r = |z|$. If we circle once around $z = 0$ we arrive at the same point $z' = re^{i(\theta + 2\pi)} = z$ but the value of the function $f(z)$ differs from the initial one: $f(z') = -f(z)$. In other words, $f(z)$ is not single-valued. We have to circle twice around $z = 0$ in order that $f(z)$ returns its initial value. Consider the more general case $f(z) = z^{1/n}$, where n is a positive integer. Solving $w^n = z = re^{i\theta}$ for w we obtain $w_k = r^{1/n} e^{i\theta/n + 2ik\pi/n}$, where $k = 0, 1, \ldots, n-1$, and the solution w exhibits n branches. Starting at the kth branch and circling once around $z = 0$ we arrive at the $(k+1)$th branch: $w_k(\theta + 2\pi) = r^{1/n} e^{i\theta/n + 2i(k+1)\pi/n} = w_{k+1}(\theta)$. We say that the function $z^{1/n}$ exhibits a branch point of order n at $z = 0$.

6.2.1   Straightforward Calculation of Singular Points from Power Series

We say that a function $f(z)$ is analytic (also regular or holomorphic) in a given region of the complex plane if it is differentiable and single-valued there [97]. The simplest nonanalytic functions with algebraic singularities are of the form

$$f(z) = A (z - z_0)^a , \tag{6.6}$$

where a is not a positive integer. It is not difficult to obtain the coefficients of the Taylor series $f(z) = f_0 + f_1 z + \cdots$ explicitly; however, it is more convenient for the discussion below to derive a recurrence relation for them from the differential equation

$$(z - z_0) f'(z) = af(z) \tag{6.7}$$

satisfied by the function (6.6). Notice that z_0 is a regular singular point of the differential equation (6.7) [98]. Expanding both sides of equation (6.7) in a Taylor series we easily obtain

$$z_0(j + 1) f_{j+1} + (a - j) f_j = 0 . \tag{6.8}$$

It follows from this recurrence relation that $\lim_{j\to\infty} |f_j/f_{j+1}| = |z_0|$ is the radius of convergence of the Taylor series.

We can rearrange (6.8) as a linear inhomogeneous equation with two unknowns: z_0 and a. Substituting $j - 1$ for j we obtain another linear inhomogeneous equation with the same two unknowns. Solving the resulting system of two equations for z_0 and a we obtain

$$z_0 = \frac{f_j f_{j-1}}{(j+1)f_{j+1}f_{j-1} - jf_j^2}, \quad a = \frac{(j^2-1)f_{j+1}f_{j-1} - (jf_j)^2}{(j+1)f_{j+1}f_{j-1} - jf_j^2}. \tag{6.9}$$

These expressions, which are exact only if f_j are the coefficients of the Taylor expansion of the function (6.6) about $z = 0$, are useful to estimate the position and exponent of an algebraic singular point of an unknown function $E(\lambda)$ if sufficient coefficients of its Taylor expansion (6.1) are available. To this end we simply substitute E_k for f_k ($k = j - 1, j, j + 1$) into equations (6.9) and estimate the limits of the right-hand sides as j increases. Such an improved ratio method and its variants prove useful for the estimation of the positions and exponents of the singular points of many functions of physical interest [99]–[103].

In most practical applications the function $E(z)$ is real for real values of z; therefore, for complex z we have $E(z)^* = E(z^*)$, and the singular points are either real or appear in complex conjugate pairs z_0, z_0^*. The expansion coefficients E_j are real and thereby equations (6.9) are not suitable for obtaining a complex singular point $z_0 = z_R + iz_I$. However, even in such a case we can apply the same method provided that we choose a slightly different ansatz $f(z)$. Taking into account that $(z - z_0)(z - z_0^*) = z^2 - 2z_R z + |z_0|^2$ we consider

$$f(z) = \left(z^2 - 2z_R z + |z_0|^2\right)^a \tag{6.10}$$

that satisfies the differential equation

$$\left(z^2 - 2z_R z + |z_0|^2\right) f'(z) = 2a\,(z - z_R)\,f(z). \tag{6.11}$$

Notice that z_0 and z_0^* are regular singular points of the differential equation (6.11) [98]. Expanding both sides of equation (6.11) in Taylor series we obtain a recurrence relation for the expansion coefficients f_j:

$$|z_0|^2(j+1)f_{j+1} + 2z_R(a-j)f_j - 2af_{j-1} = (1-j)f_{j-1}. \tag{6.12}$$

Substituting $j - 1$ and $j - 2$ for j we derive two additional equations, which together with equation (6.12) form a system of three inhomogeneous equations with three unknowns: z_R, $|z_0|^2$ and a. On solving it we obtain simple expressions for the unknowns in terms of the coefficients f_k. We then proceed as in the preceding case in order to estimate the exponent and position of a singular point of an unknown function $E(\lambda)$ from its Taylor expansion (6.1). This method is well known [99]–[103] and we apply it to some particular problems later in this chapter.

The two methods just outlined provide only the singular points closest to origin that cause the asymptotic behavior of the coefficients of the power series.

Because in all the practical examples discussed below we already know the exponent of the singular point, in what follows we concentrate on this simpler case in which it is sufficient to solve a system of two equations for the remaining unknowns z_R and $|z_0|^2$, thus obtaining

$$
\begin{aligned}
z_R &= \frac{(j+1)(j-2a-2)f_{j-2}f_{j+1} + j(2a-j+1)f_{j-1}f_j}{2[(j+1)(j-a-1)f_{j-1}f_{j+1} + j(a-j)f_j^2]}, \\
|z_0|^2 &= \frac{(j-a)(j-2a-2)f_{j-2}f_j + (j-a-1)(2a-j+1)f_{j-1}^2}{(j+1)(j-a-1)f_{j-1}f_{j+1} + j(a-j)f_j^2}.
\end{aligned}
\tag{6.13}
$$

The imaginary part of the singular points is given by $|z_I| = \sqrt{|z_0|^2 - z_R^2}$. These equations are exact only for the Taylor coefficients f_k of the ansatz (6.10). Substituting the coefficients E_k of an unknown function $E(\lambda)$ for f_k we can estimate the position of the pair of complex conjugate singular points closest to the origin of the λ plane provided the sequences (6.13) of z_R and $|z_0|$ values converge as j increases.

Equations (6.9) and (6.13) are suitable for obtaining the parameters that characterize algebraic singular points because the coefficients E_k of a power series carry information about the singular point closest to the origin. This information becomes more noticeable as k increases and determines the asymptotic behavior of E_k. If the coefficients E_k reveal the asymptotic behavior at moderate values of k, then the method gives accurate results with little computational effort; otherwise it converges slowly, requiring calculations of large order that may be time consuming.

6.2.2 Implicit Equations

In quantum mechanics we commonly derive exact or approximate quantization conditions of the form

$$Q(E, \lambda) = 0 \tag{6.14}$$

that give the allowed values of the energy E in terms of a model parameter λ (or of a set of such parameters). In what follows we concentrate on the case that λ is a perturbation parameter and obtain the perturbation expansion (6.1) either from the standard formulas given in preceding chapters or from the quantization condition (6.14).

The quantization condition (6.14) gives us either $E(\lambda)$ or $\lambda(E)$. In the latter case we can expand λ about a given point $E = E_b$:

$$\lambda = \lambda_b + \sum_{j=1}^{\infty} c_j (E - E_b)^j , \quad c_j = \frac{1}{j!} \frac{d^j \lambda}{dE^j} (E_b) , \tag{6.15}$$

where $\lambda_b = \lambda(E_b)$. If

$$\frac{d^j \lambda}{dE^j} (E_b) = 0, \; j = 1, 2, \ldots, n-1, \; \frac{d^n \lambda}{dE^n} (E_b) \neq 0 , \tag{6.16}$$

then $\lambda = \lambda_b + c_n(E - E_b)^n + \cdots$ and

$$E \approx E_b + [(\lambda - \lambda_b) / c_n]^{1/n} \tag{6.17}$$

in a neighborhood of λ_b. We see that equations (6.16) are sufficient conditions for a branch point of order n at $\lambda = \lambda_b$.

Here we are interested in the most usual case $n = 2$. Differentiating the quantization condition (6.14) with respect to E, and taking into account that λ depends on E, we have

$$\frac{\partial Q}{\partial E} + \frac{\partial Q}{\partial \lambda} \frac{d\lambda}{dE} = 0, \; \frac{\partial^2 Q}{\partial E^2} + 2\frac{\partial^2 Q}{\partial \lambda \partial E} \frac{d\lambda}{dE} + \frac{\partial^2 Q}{\partial \lambda^2} \left(\frac{d\lambda}{dE}\right)^2 + \frac{\partial Q}{\partial \lambda} \frac{d^2\lambda}{dE^2} = 0 . \tag{6.18}$$

Therefore, if

$$Q(E_b, \lambda_b) = 0, \; \frac{\partial Q}{\partial E}(E_b, \lambda_b) = 0, \; \frac{\partial Q}{\partial \lambda}(E_b, \lambda_b) \neq 0, \; \frac{\partial^2 Q}{\partial E^2}(E_b, \lambda_b) \neq 0 , \tag{6.19}$$

then $E(\lambda)$ exhibits a square-root branch point (for a more detailed discussion see reference [104]).

6.3   Radius of Convergence of the Perturbation Expansions

As discussed earlier in this book, perturbation theory provides approximate eigenvalues $E(\lambda)$ of a Hamiltonian operator $\hat{H} = \hat{H}_0 + \lambda \hat{H}'$ in the form of a power series (6.1). In order to understand the origin of the singular points that determine its radius of convergence we consider some simple trivial (exactly solvable) and nontrivial models below.

6.3.1   Exactly Solvable Models

In order to facilitate the study of the analytical properties of the eigenvalues $E(\lambda)$ in the complex λ plane, we first consider some exactly solvable time-independent Schrödinger equations.

Our first example is a Hamiltonian operator acting on a two-dimensional state space:

$$\hat{H} = \epsilon_1 |1><1| + \epsilon_2 |2><2| + \lambda (V|1><2| + V^*|2><1|) ,\qquad (6.20)$$

where the real numbers ϵ_1 and ϵ_2 are the only two eigenvalues of $\hat{H}_0 = \epsilon_1 |1><1| + \epsilon_2 |2><2|$ and V is a complex number. Notice that $\hat{H}_0$ and $\hat{H}' = V|1><2| + V^*|2><1|$ are, respectively, diagonal and off diagonal in the orthonormal basis set $\{|1>, |2>\}$. A particular form of this simple operator has already proved suitable to introduce concepts of finite-dimensional perturbation theory [105]. The secular determinant $|\mathbf{H} - E\mathbf{1}| = 0$, where

$$\mathbf{H} - E\mathbf{1} = \begin{pmatrix} \epsilon_1 - E & \lambda V \\ \lambda V^* & \epsilon_2 - E \end{pmatrix} \qquad (6.21)$$

is the matrix representation of $\hat{H} - E\hat{1}$, gives us the characteristic equation

$$E^2 - (\epsilon_1 + \epsilon_2)\, E + \epsilon_1 \epsilon_2 - \lambda^2 |V|^2 = 0 . \qquad (6.22)$$

The two roots of this equation are the two eigenvalues of $\hat{H}$:

$$E_1(\lambda) = \frac{\epsilon_1 + \epsilon_2}{2} - R(\lambda),\ \ E_2(\lambda) = \frac{\epsilon_1 + \epsilon_2}{2} + R(\lambda)$$

$$R(\lambda) = \frac{1}{2}\sqrt{\Delta\epsilon^2 + 4\lambda^2 |V|^2},\ \Delta\epsilon = |\epsilon_2 - \epsilon_1| . \qquad (6.23)$$

Each eigenvalue (6.23) exhibits a pair of complex conjugate square-root branch points λ_b, λ_b^* given by the zeros of $\Delta\epsilon^2 + 4\lambda^2 |V|^2$:

$$\lambda_b = i\,\frac{\Delta\epsilon}{2|V|} . \qquad (6.24)$$

Because there is no other singular point closer to the origin (in fact, there are no other singularities) the perturbation series

$$E_k(\lambda) = \sum_{j=0}^{\infty} E_{k,j}\lambda^{2j} \qquad (6.25)$$

for $k = 1, 2$ converge for all $|\lambda| < |\lambda_b|$.

The curves $E_1(\lambda)$ and $E_2(\lambda)$ cross at the branch points taking the common real value

$$E(\lambda_b) = E\left(\lambda_b^*\right) = E_b = \frac{\epsilon_1 + \epsilon_2}{2} . \qquad (6.26)$$

Alternatively, we can view $E_1(\lambda)$ and $E_2(\lambda)$ as the two branches of a two-valued function.

Because E is a function of λ^2, it is more convenient to choose $\xi = \lambda^2$ as a variable instead of λ itself. The characteristic equation (6.22) is a particular case of the quantization condition (6.14) that gives either $E(\xi)$ or $\xi(E)$. In the latter case we have

$$\xi = \frac{1}{4|V|^2}\left[4\left(E - E_b\right)^2 - \Delta\epsilon^2\right] ; \tag{6.27}$$

therefore

$$\frac{\partial \xi}{\partial E}(E_b) = 0 \Rightarrow \frac{\partial \lambda}{\partial E}(E_b) = 0 , \tag{6.28}$$

and

$$\frac{\partial^2 \xi}{\partial E^2}(E_b) \neq 0 \Rightarrow \frac{\partial^2 \lambda}{\partial E^2}(E_b) \neq 0 , \tag{6.29}$$

which are particular cases of the conditions (6.16) for $n = 2$.

It is also instructive to consider the eigenvectors of $\hat{H}$. A vector of the state space $\Psi = b_1|1 > +b_2|2 >$ is an eigenvector of $\hat{H}$ with eigenvalue E if the coefficients b_j satisfy $b_2 = (E-\epsilon_1)b_1/(\lambda V)$. The norm of the eigenvector reads

$$\|\Psi\| = \sqrt{<\Psi|\Psi>} = |b_1|\sqrt{1 + \frac{(E - \epsilon_1)^2}{\lambda^2|V|^2}} , \tag{6.30}$$

where we have explicitly assumed λ and E to be real. Notice that $\|\Psi\|$ vanishes when $b_1 = 0$ or when

$$(E - \epsilon_1)^2 + \lambda^2|V|^2 = 0 . \tag{6.31}$$

Real values of λ do not satisfy this equation, but $E_b, \pm\lambda_b$ are solutions. We conclude that the norm of an eigenvector vanishes at the branch points.

Another simple illustrative example is the harmonic oscillator with a harmonic perturbation:

$$\hat{H} = \hat{H}_0 + \lambda\hat{H}' = \frac{1}{2}\left(-\frac{d^2}{dx^2} + x^2 + \lambda x^2\right) , \tag{6.32}$$

because one easily obtains the eigenvalues as functions of the perturbation parameter λ:

$$E_k(\lambda) = \sqrt{1 + \lambda}\left(k + \frac{1}{2}\right), \quad k = 0, 1, \dots . \tag{6.33}$$

All the eigenvalues collapse at the branch point $\lambda = -1$ which corresponds to a zero force constant. In fact, when $\lambda = -1$ we have a free particle and no bound states; therefore, we do not expect a perturbation theory for the point spectrum to apply in such a case. When $\lambda < -1$ the potential-energy function is no longer a well but a barrier that does not support bound states. We will discuss barriers and other such problems later in this book.

The two trivial examples studied so far exhibit perturbation series with convergence radii determined by square-root branch points on the complex λ-plane where two or more eigenvalues cross. This situation is quite common in quantum mechanics as we will shortly see in other problems.

Many authors have resorted to exactly solvable models in their studies of perturbation series. In particular the delta function has proved useful as a one-dimensional model of Coulomb interaction in diatomic molecules with one electron, facilitating the understanding of the polarization expansion at large internuclear distances [106]–[109]. We have briefly discussed the application of perturbation theory to such problems in Chapter 4.

6.3.2 Simple Nontrivial Models

As stated earlier, we will not attempt a rigorous study of the convergence properties of the perturbation series in quantum mechanics. However, we outline some well-known mathematical results that are necessary for the discussions and applications in this chapter.

Many textbooks state that the perturbation series for the eigenvalues and eigenvectors of the operator $\hat{H}(\lambda) = \hat{H}_0 + \lambda \hat{H}'$ converge for a given value of λ if the perturbation $\lambda \hat{H}'$ is sufficiently smaller than the unperturbed part $\hat{H}_0$. However, it is not obvious what it means that one operator is smaller than another. In order to discuss this point briefly, in what follows $||\varphi|| = \sqrt{<\varphi|\varphi>}$ denotes the norm of a vector φ of the state space already introduced earlier. Suppose that E_0 is an isolated simple eigenvalue of $\hat{H}_0$, and that there are two real numbers a and b such that

$$\left\| \hat{H}' \Phi \right\| \leq a \left\| \hat{H}_0 \Phi \right\| + b \|\Phi\| \tag{6.34}$$

for all Φ in the state space. Under such conditions there is a unique eigenvalue $E(\lambda)$ of $\hat{H}$ near E_0, and $E(\lambda)$ is analytic in a neighborhood of $\lambda = 0$ in the complex λ plane [105, 110].

Before treating any nontrivial problem, we first show that the inequality (6.34) applies to the exactly solvable models discussed in the preceding subsection. For example, choosing an arbitrary vector $\Phi = c_1 |1> + c_2 |2>$ of the state space of the two-level model, we easily prove that $||\hat{H}'\Phi|| = |V| |||\Phi||$; therefore, any $a \geq 0$ and $b \geq |V|$ satisfies equation (6.34).

The proof that the inequality (6.34) also applies to the perturbed harmonic oscillator (6.32) is somewhat more laborious. For convenience we write $\hat{p} = -id/dx$, so that $\hat{H}_0 = (\hat{p}^2 + \hat{x}^2)/2$ and $\hat{H}' = \hat{x}^2/2$. Following the straightforward steps

$$
\begin{aligned}
\left(\hat{p}^2 + \hat{x}^2 \right)^2 &= \hat{p}^4 + \hat{x}^4 + \hat{p}^2 \hat{x}^2 + \hat{x}^2 \hat{p}^2 \\
&= \hat{p}^4 + \hat{x}^4 + \hat{p} \left(\left[\hat{p}, \hat{x}^2 \right] + \hat{x}^2 \hat{p} \right) + \hat{x} \left(\left[\hat{x}, \hat{p}^2 \right] + \hat{p}^2 \hat{x} \right) \\
&= \hat{p}^4 + \hat{x}^4 + \hat{p} \hat{x}^2 \hat{p} + \hat{x} \hat{p}^2 \hat{x} + 2i \left[\hat{x}, \hat{p} \right] \\
&= \hat{p}^4 + \hat{x}^4 + \hat{p} \hat{x}^2 \hat{p} + \hat{x} \hat{p}^2 \hat{x} - 2
\end{aligned}
\tag{6.35}
$$

we obtain

$$\left(\hat{p}^2 + \hat{x}^2 \right)^2 + 2 = \hat{p}^4 + \hat{x}^4 + \hat{p} \hat{x}^2 \hat{p} + \hat{x} \hat{p}^2 \hat{x} . \tag{6.36}$$

The commutator technique in equation (6.35) is well known and was used earlier by other authors [111]. Taking expectation values on both sides of equation (6.36), and realizing that

$$\left\langle \Psi \left| \hat{A} \hat{B}^2 \hat{A} \right| \Psi \right\rangle = \left\| \hat{B} \hat{A} \Psi \right\|^2 \geq 0 \tag{6.37}$$

for any two hermitian operators $\hat{A}$ and $\hat{B}$, we conclude that

$$\left\| \left(\hat{p}^2 + \hat{x}^2 \right) \Psi \right\|^2 + 2\|\Psi\|^2 \geq \left\| \hat{x}^2 \Psi \right\|^2 . \tag{6.38}$$

Making use of the well-known inequality

$$|\alpha| + |\beta| \geq \sqrt{\alpha^2 + \beta^2} \tag{6.39}$$

that holds for any two real numbers α and β, we rewrite equation (6.38) in the form of equation (6.34):

$$\left\| \hat{H}_0 \Psi \right\| + \frac{1}{\sqrt{2}} \|\Psi\| \geq \left\| \hat{H}' \Psi \right\| . \tag{6.40}$$

Therefore, the theorem enunciated above tells us that the Rayleigh–Schrödinger perturbation series for the perturbed harmonic oscillator (6.32) has a finite radius of convergence in agreement with the conclusion drawn earlier from the exact eigenvalues (6.33).

The perturbed rigid rotor discussed in Section 5.4 is a suitable example of a nontrivial quantum mechanical model that gives rise to perturbation series with finite radius of convergence. In this case we have

$$\hat{H}_0 = -\frac{1}{\sin(\theta)} \frac{d}{d\theta} \sin(\theta) \frac{d}{d\theta} + \frac{m^2}{\sin(\theta)^2}, \ \hat{H}' = v(\theta) \ . \tag{6.41}$$

Assuming that

$$|v(\theta)| \le v_M, \ 0 \le \theta \le \pi \ , \tag{6.42}$$

one easily proves that

$$\left\| \hat{H}' \Psi \right\| = \sqrt{< \Psi | v(\theta)^2 | \Psi >} \le v_M \| \Psi \| \ , \tag{6.43}$$

and the inequality (6.34) is satisfied for any $a \ge 0$ and $b \ge v_M$. According to the theorem above the perturbation series for every state of the perturbed rigid rotor exhibits a finite radius of convergence.

In Section 5.4 we developed two methods for the calculation of the perturbation corrections to the dimensionless energy ϵ in terms of $M = |m|$ and $\epsilon_0 = J(J+1)$. It was shown there that when $v(\theta) = -\cos(\theta)$ the energy coefficients of odd order vanish; therefore, we write

$$\epsilon = \sum_{j=0}^{\infty} \epsilon_{2j} \lambda^{2j} = \sum_{j=0}^{\infty} f_j z^j \ , \tag{6.44}$$

where $f_j = \epsilon_{2j}$ and $z = \lambda^2$. As argued above, we can obtain the position and exponent of the algebraic singular point closest to the origin by means of equation (6.9). Table 6.1 shows sequences of values of z_0 and a given by equation (6.9) for the ground-state energy ($M = 0$, $J = 0$). We see that the sequence of values of a appears to converge towards $a = 1/2$ suggesting that there is a square-root branch point at a negative value of $z = z_0 = \lambda_b^2$. The fourth column of Table 6.1 shows that a new sequence of values of z_0 obtained with $a = 1/2$ clearly approaches the same limit point estimated to be $z_0 \approx -3.6080$. We conclude that there are two complex conjugate branch points at $\pm \lambda_b$, where $\lambda_b = 1.8995i$ in agreement with previous calculations [99]–[101].

Table 6.2 shows similar results for the perturbation series with $M = 0$ and $J = 1$ ($\epsilon_0 = 2$). There is no doubt that this state shares a common branch point with the ground state where they match. As said above we can view them as two branches of the same two-valued function. What is surprising is that the sequences of singularity parameters z_0 and a for those states approach each other faster than they approach their common limits. Notice that after a relatively small value of j they agree up to the tenth digit.

Numerical calculations based on the perturbation series for the polar rigid rotor in a uniform electric field should be carried out carefully in order to remove vanishing denominators in the energy coefficients. One should first substitute the appropriate value of M and simplify the resulting expressions before substituting the value of ϵ_0. In this way zeros of the numerator and denominator cancel each other. The Maple command *simplify* produces perturbation corrections free from apparent poles.

The method for the estimation of singular points from the perturbation series just described applies only to the states with $J = M$ and $J = M + 1$, and diverges in other cases. We do not know the reason of such failure. Numerical calculation shows that each pair of states given by $J = M$ and $J = M + 1$, $M = 0, 1, \ldots$, exhibits a common branch point closest to the origin.

Table 6.1 Parameters of the Singular Point Closest to the Origin Calculated from the Perturbation Series for the Ground-State Energy of the Polar Linear Rigid Rotator in a Uniform Electric Field

j	z_0	a	$z_0(a = 1/2)$
1	−3.512417638	0.5707045109	−4.090909091
2	−3.556560626	0.5527415633	−3.686170213
3	−3.585821089	0.5326075288	−3.633209077
4	−3.596819366	0.5219724907	−3.619542326
5	−3.601477950	0.5161725703	−3.614468003
6	−3.603762887	0.5126933876	−3.612099210
7	−3.605027112	0.5104175963	−3.610814189
8	−3.605793807	0.5088247575	−3.610041506
9	−3.606292185	0.5076511427	−3.609541257
10	−3.606633893	0.5067517094	−3.609198974
11	−3.606878210	0.5060408867	−3.608954523
12	−3.607058873	0.5054651721	−3.608773879
13	−3.607196202	0.5049894780	−3.608636619
14	−3.607303017	0.5045898674	−3.608529882
15	−3.607387730	0.5042494624	−3.608445244
16	−3.607456041	0.5039560264	−3.608376998
17	−3.607511927	0.5037004735	−3.608321169
18	−3.607558228	0.5034759175	−3.608274917
19	−3.607597016	0.5032770442	−3.608236170
20	−3.607629833	0.5030996882	−3.608203387
21	−3.607657845	0.5029405378	−3.608175405
22	−3.607681946	0.5027969278	−3.608151329
23			−3.608130465

The Maple program for the calculation of branch points according to equation (6.9) is extremely simple and we do not show it here.

In order to verify the results just obtained, in what follows we compare them with those produced by the method of implicit equations outlined in Section 6.2.2. In order to apply this method we need an appropriate quantization condition; here we choose the well-known secular determinant. For brevity we denote the unperturbed eigenvectors by the kets $|J, M >$, where $M = |m|$ and $J = M, M + 1, \ldots$. Taking into account that M remains fixed during the calculation because m is a good quantum number, it is convenient to simplify the matrix notation by omitting M in the kets and writing $|i >= |M + i, M >, i = 0, 1, \ldots$.

Approximating the perturbed state Ψ as a finite linear combination of unperturbed states

$$\Psi = \sum_{i=0}^{N} c_i |i > \tag{6.45}$$

we obtain the secular equation

$$\sum_{j=0}^{N} \left(H_{i,j} - \epsilon \delta_{i,j} \right) c_j = 0, \quad H_{i,j} = \left\langle i \left| \hat{H} \right| j \right\rangle, \tag{6.46}$$

Table 6.2   Parameters of the Singular Point Closest to
the Origin Calculated from the Perturbation Series for
the State with $|m| = 0$ and $J = 1$ of the Polar Linear
Rigid Rotator in a Uniform Electric Field

j	z_0	a	$z_0(a = 1/2)$
1	−5.237205912	−0.09233151889	−2.397260274
2	−3.508036512	0.5984940281	−3.754571777
3	−3.542837739	0.5746700583	−3.651913167
4	−3.594244744	0.5249681057	−3.620069394
5	−3.602136537	0.5151423911	−3.614298564
6	−3.603865496	0.5125097594	−3.612081181
7	−3.605025052	0.5104223879	−3.610814793
8	−3.605792052	0.5088289160	−3.610041753
9	−3.606292071	0.5076514372	−3.609541268
10	−3.606633909	0.5067516623	−3.609198972
11	−3.606878213	0.5060408775	−3.608954523
12	−3.607058873	0.5054651721	−3.608773879
13	−3.607196202	0.5049894782	−3.608636619
14	−3.607303017	0.5045898674	−3.608529882
15	−3.607387730	0.5042494624	−3.608445244
16	−3.607456041	0.5039560264	−3.608376998
17	−3.607511927	0.5037004735	−3.608321169
18	−3.607558228	0.5034759175	−3.608274917
19	−3.607597016	0.5032770442	−3.608236170
20	−3.607629833	0.5030996882	−3.608203387
21	−3.607657845	0.5029405378	−3.608175405
22	−3.607681946	0.5027969278	−3.608151329
23			−3.608130465

where [78, 112]

$$
\begin{aligned}
H_{i,j} &= (M+i)(M+i+1)\delta_{i,j} - \lambda \left[\frac{i(i+2M)}{4(i+M)^2 - 1} \right]^{1/2} \delta_{i-1,j} \\
&\quad - \lambda \left[\frac{(i+1)(i+2M+1)}{4(i+M+1)^2 - 1} \right]^{1/2} \delta_{i+1,j} = H_{j,i} .
\end{aligned}
\tag{6.47}
$$

Because the matrix **H** of the Hamiltonian operator $\hat{H}$ is tridiagonal the secular equation (6.46)
becomes a three-term difference equation

$$
A_i c_{i-1} + B_i c_i + A_{i+1} c_{i+1} = 0, \ i = 0, 1, \ldots, N ,
\tag{6.48}
$$

where

$$
B_i = H_{i,i} - \epsilon, \ A_i = \begin{cases} 0 & \text{if } i \leq 0 \\ H_{i,i-1} & \text{if } 0 < i \leq N \\ 0 & \text{if } i > N \end{cases} .
\tag{6.49}
$$

The determinant of the homogeneous system of $N + 1$ equations (6.48) with $N + 1$ unknowns c_i is

$$
D_N = \begin{vmatrix}
B_0 & A_1 & 0 & 0 & \cdots & & & \\
A_1 & B_1 & A_2 & 0 & \cdots & & & \\
 & & \cdots & & & \ddots & & \\
 & & & \cdots & A_{N-1} & B_{N-1} & A_N & \\
 & & & \cdots & 0 & A_N & B_N &
\end{vmatrix}.
\tag{6.50}
$$

Expanding D_N by minors along the last row, and then the coefficient of A_N along the last column, we obtain the following three-term recurrence relation [113, 114]

$$
D_N = B_N D_{N-1} - A_N^2 D_{N-2}.
\tag{6.51}
$$

Taking into account the initial conditions $D_{-2} = 0$, and $D_{-1} = 1$, we easily calculate the secular determinants D_N, $N = 0, 1, \ldots$ to any desired dimension. It is well known that the approximate quantization condition $D_N(\epsilon, \lambda) = 0$ becomes increasingly accurate as N increases.

According to the discussion of Section 6.2.2, square-root branch points are simultaneous solutions of the equations

$$
D_N (\epsilon_b, \lambda_b) = 0, \quad \frac{\partial D_N}{\partial \epsilon} (\epsilon_b, \lambda_b) = 0,
\tag{6.52}
$$

provided that

$$
\frac{\partial D_N}{\partial \lambda} (\epsilon_b, \lambda_b) \neq 0, \quad \frac{\partial^2 D_N}{\partial \lambda^2} (\epsilon_b, \lambda_b) \neq 0.
\tag{6.53}
$$

We use Maple to obtain the determinants D_N as analytical functions of ϵ and λ, and to solve equations (6.52) numerically by means of the Newton–Raphson algorithm. Because Maple allows us to predetermine a sufficiently great floating-point precision through the variable *Digits*, the results of the final numerical step are supposed to be free from roundoff errors, and accurate up to the last digit reported. Table 6.3 shows the root closest to the origin of equations (6.52) for $m = 0$ and increasing values of N. Because of the remarkable rate of convergence we easily obtain accurate branch points from determinants of relatively small dimension. The converged value of λ_b^2 agrees with the value of z_0 obtained from the perturbation series by the method of Section 6.2.1 (cf. Tables 6.1 and 6.2). It is not surprising that the nonperturbative method based on the sequence of secular determinants D_N yields more accurate results. However, the method of the perturbation series is useful when there is no other approach to the studied physical property. The agreement of these two completely independent methods strongly supports the supposition that the radius of convergence of the eigenvalue $\epsilon(\lambda)$ is determined by a square-root branch point.

Table 6.4 shows singularity parameters $|\lambda_b|$ and $\epsilon_b = \epsilon(\lambda_b) = \epsilon(\lambda_b)^* = \epsilon(\lambda_b^*)$ for some states with $m = 0, 1, 2$ which are more accurate than earlier results [113, 114].

Figure 6.1 shows $\lambda(\epsilon)^2$ for selected intervals of ϵ, calculated by means of secular determinants and perturbation series. In the latter case we simply substitute the appropriate values of $M = |m|$ and J (in $\epsilon_0 = J(J + 1)$) for the two branches involved. We clearly see that each pair of energy eigenvalues with quantum numbers J, $J + 1$ share a common branch point (minimum of the curve). The curves that join states with $J = M$ and $J = M + 1$ look simpler than those joining states with $J > M + 1$. This noticeable difference may explain why the method of Section 6.2.1 diverges when $J > M + 1$. The curve connecting the states with $M = 0$, $J = 2$, and $J = 3$ was not shown in earlier discussions of the subject [113].

Maple greatly facilitates the calculation just discussed. The programs for the construction of the perturbation series were described in the preceding chapter. The procedure that builds the secular

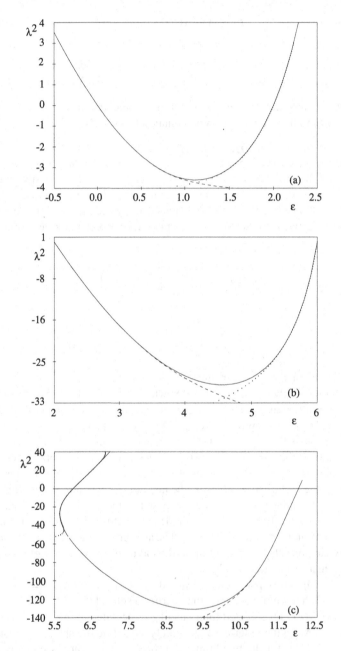

FIGURE 6.1

Curves λ^2 (ϵ) vs. ϵ connecting the states $(M = 0,\ J = 0) - (M = 0,\ J = 1)$ (a), $(M = 1, J = 1) - (M = 1, J = 2)$ (b), and $(M = 0, J = 2) - (M = 0, J = 3)$ (c). Continuous lines are results from secular determinants; broken lines and points come from perturbation theory.

Table 6.3 Approximate Branch Point Closest to the Origin for States with $m = 0$ Obtained from Determinants of Dimension N

N	ε_b	λ_b^2
2	1.121754911	−3.621793885
3	1.118493345	−3.607856626
4	1.118508634	−3.607916830
5	1.118508607	−3.607916730
6	1.118508607	−3.607916730

Table 6.4 Branch Points for the Linear Rigid Rotor in an Electric Field

$\|m\| = 0$		$\|m\| = 1$		$\|m\| = 2$	
ε_b	$\|\lambda_b\|$	ε_b	$\|\lambda_b\|$	ε_b	$\|\lambda_b\|$
1.118508607	1.899451692	4.558778867	5.413699680	10.32081667	10.42885501
9.182711108	11.44693732	16.13759075	19.03665394	25.42076233	28.15827404
24.27433747	29.15703642	34.74306246	40.82876537	47.54188835	54.04029322

determinants is straightforward and we do not show it here. We use the Maple command *fsolve* to obtain roots of one-variable functions, and a simple Newton–Raphson algorithm for two variables. We do not show our Newton–Raphson procedure here because it is extremely naive and there may be other more elaborate, automatic, and reliable Maple procedures available in the literature.

6.4 Divergent Perturbation Series

We say that a perturbation series is divergent when its radius of convergence is zero. Divergent perturbation series are so common in quantum mechanics that some authors have stated that they are more likely a rule than an exception. A rigorous discussion of divergent series is beyond the scope of this book. We merely summarize some well-known results, and briefly discuss methods for using divergent expansions in practical applications.

Some of the quantum-mechanical models discussed in earlier chapters lead to divergent series: for example, the anharmonic oscillators and the Zeeman and Stark effects in hydrogen. In this chapter we consider one-dimensional anharmonic oscillators as illustrative examples because they are the simplest models, and we can easily calculate sufficient perturbation coefficients for the application of summation methods.

We say that the series (6.1) is asymptotic to the function $E(\lambda)$ as $\lambda \to 0$ if [98]

$$\lim_{\lambda \to 0} \frac{E(\lambda) - \sum_{j=0}^{N} E_j \lambda^j}{\lambda^{N+1}} = E_{N+1} . \tag{6.54}$$

6.4.1 Anharmonic Oscillators

In earlier chapters we calculated perturbation corrections for some one-dimensional anharmonic oscillators. The most widely studied representative of this class of models is the quartic anharmonic oscillator

$$\hat{H} = -\frac{1}{2}\frac{d^2}{dx^2} + \frac{x^2}{2} + \lambda x^4 \,. \tag{6.55}$$

The analytic properties of the singular perturbation theory for this problem have been studied, first by means of approximate methods [115] and later in a more rigorous way [111]. It has been proved that any eigenvalue $E(\lambda)$ has a global third-order branch point at $\lambda = 0$. On the three-sheeted surface, $\lambda = 0$ is not an isolated singularity because there are infinitely many branch points of order two that accumulate towards origin. As a result the perturbation series is divergent, i.e., the radius of convergence is zero and the λ-power series does not converge for any value of λ (no matter how small it may be) [111, 115]. There are no real numbers a and b satisfying the inequality (6.34) for all Φ.

The asymptotic behavior of the energy perturbation coefficients E_j for the ground state is [115]

$$E_j \sim E_j^{asymp} = \frac{(-1)^{j+1}3^j\sqrt{6}}{\pi^{3/2}}\Gamma(j+1/2) \,. \tag{6.56}$$

It follows from this expression that the radius of convergence of the perturbation series is zero:

$$\lim_{j\to\infty}\frac{|E_j|}{|E_{j+1}|} = 0 \,. \tag{6.57}$$

Figure 6.2 shows the ratio $R_j = E_j/E_j^{asymp}$, where the exact perturbation coefficients E_j for the ground-state energy were calculated by means of the method of Swenson and Danforth described in Section 3.3.1. As j increases, this ratio slowly approaches unity.

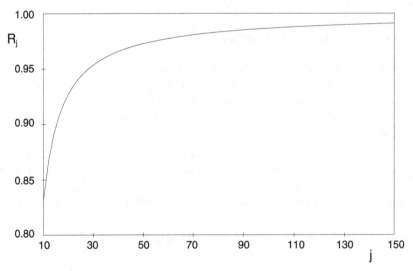

FIGURE 6.2

Ratio $R_j = E_j/E_j^{asymp}$ for the ground-state energy of the anharmonic oscillator.

If λ is sufficiently small the sequence of partial sums

$$S_N = \sum_{j=0}^{N} E_j \lambda^j, \ N = 0, 1, \ldots \tag{6.58}$$

for an asymptotic divergent series appears to converge as N increases, but after some value of N S_N clearly exhibits its divergent nature. However, it is commonly possible to obtain acceptable results by appropriate truncation. If we assume that the error $|E(\lambda) - S_N|$ is of the order of magnitude of the first neglected term $|E_{N+1}\lambda^{N+1}|$ then it is reasonable to choose the latter to be as small as possible [98].

Figure 6.3 shows that the minimum of $\log(|E_{j+1}\lambda^{j+1}|)$ vs. j increases and moves to smaller j values as λ increases. In other words, as λ increases we can sum less terms and obtain a less accurate estimation of the energy.

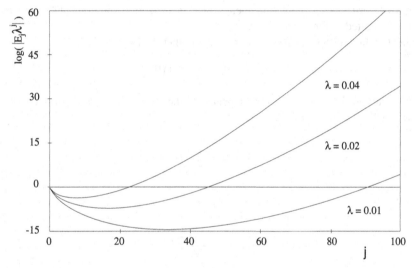

FIGURE 6.3
$\log(|E_j \lambda^j|)$ **for the ground state of the anharmonic oscillator.**

Table 6.5 shows the optimum value of N, the approximate energy calculated by perturbation theory S_N, the estimated error $|E_{N+1}\lambda^{N+1}|$, and the exact result obtained by means of a nonperturbative method [116]. Notice that $|E(\lambda) - S_N|$ is of the order of $|E_{N+1}\lambda^{N+1}|$ supporting the truncation rule suggested above. It is clear from the results of Table 6.5 that the perturbation series for the ground state of the anharmonic oscillator is useful only for sufficiently small λ values (say, $\lambda < 0.05$).

Table 6.5   Ground-State Energy of the Anharmonic Oscillator $\hat{H} = \dfrac{1}{2}\left(\hat{p}^2 + \hat{x}^2\right) + \lambda\hat{x}^4$

λ	N	Truncated Perturbation Series S_N	Estimated Absolute Error	Exact
0.01	32	0.5072562045246011	$0.3527780020 \ 10^{-14}$	0.5072562045246028
0.02	16	0.514086399	$0.5822940303 \ 10^{-7}$	0.5140864273
0.04	7	0.526837	$0.2060990198 \ 10^{-3}$	0.5267339644
0.06	4	0.5369	$0.2784570413 \ 10^{-2}$	0.5383192923

6.5   Improving the Convergence Properties of the Perturbation Series

We have just seen that divergent series are useful only for sufficiently small values of the perturbation parameter. However, in many cases it is possible to improve the convergence properties of
the perturbation series and obtain valuable results for greater values of the expansion variable. We
discuss some examples in what follows.

6.5.1   The Effect of $\hat{H}_0$

The convergence properties of the perturbation series depend dramatically on the reference model
$\hat{H}_0$. Therefore, a judicious choice of this operator is mandatory in difficult cases. In what follows
we illustrate this important point by means of simple examples.

The method that we present here is quite general and flexible. It has been suggested by a recent
application of perturbation theory by means of a factorization method, [117] and has not been
sufficiently exploited as far as we know.

Given a Hamiltonian operator $\hat{H}$ we construct another operator

$$\hat{\mathcal{H}}(\beta) = \hat{H} + (\beta - 1)\hat{W}(\beta) , \tag{6.59}$$

where $\hat{W}$ is an hermitian operator that depends on the new perturbation parameter β and can be
expanded in a Taylor series about $\beta = 0$:

$$\hat{W}(\beta) = \sum_{j=0}^{\infty} \hat{W}_j \beta^j . \tag{6.60}$$

If we write

$$\hat{\mathcal{H}}(\beta) = \sum_{j=0}^{\infty} \hat{\mathcal{H}}_j \beta^j , \tag{6.61}$$

where

$$\hat{\mathcal{H}}_0 = \hat{H} - \hat{W}_0, \ \hat{\mathcal{H}}_j = \hat{W}_{j-1} - \hat{W}_j, \ j = 1, 2 \dots , \tag{6.62}$$

and expand the eigenfunctions $\Psi(\beta)$ and eigenvalues $\mathcal{E}(\beta)$ of $\hat{\mathcal{H}}(\beta)$ in Taylor series about $\beta = 0$

$$\Psi(\beta) = \sum_{j=0}^{\infty} \Psi_j \beta^j, \ \mathcal{E}(\beta) = \sum_{j=0}^{\infty} \mathcal{E}_j \beta^j , \tag{6.63}$$

then we can apply perturbation theory in the way outlined in Chapter 1. The coefficients of those
series are determined by the perturbation equations

$$\left(\hat{\mathcal{H}}_0 - \mathcal{E}_0 \right) \Psi_j = \sum_{i=1}^{j} \left(\mathcal{E}_i - \hat{\mathcal{H}}_i \right) \Psi_{j-i} . \tag{6.64}$$

If the series (6.63) converge for $\beta = 1$, they give us the eigenfunctions and eigenvalues of $\hat{H}$
because $\hat{\mathcal{H}}(1) = \hat{H}$. Thus, the problem reduces to selecting an operator $\hat{W}$ that facilitates solving
the perturbation equations (6.64) and that leads to perturbation series that converge for $\beta = 1$.

The simplest case is given by $\hat{W}_j = 0$ for all $j > 0$, so that $\hat{W}(\beta) = \hat{W}_0$ is independent of β, and

$$\hat{\mathcal{H}}(\beta) = \hat{\mathcal{H}}_0 + \beta\hat{\mathcal{H}}_1, \ \hat{\mathcal{H}}_0 = \hat{H} - \hat{W}_0, \ \hat{\mathcal{H}}_1 = \hat{W}_0 . \tag{6.65}$$

As an illustrative example we choose the anharmonic oscillator (6.55) and

$$\hat{W}_0 = \frac{1-\omega^2}{2}x^2 + \lambda x^4 = \hat{\mathcal{H}}_1 \Rightarrow \hat{\mathcal{H}}_0 = -\frac{1}{2}\frac{d^2}{dx^2} + \frac{\omega^2 x^2}{2} . \tag{6.66}$$

The unperturbed model is a dimensionless harmonic oscillator with frequency ω that we may hopefully adjust in order to obtain perturbation series with better convergence properties.

This quite popular particular case of the method has proved to yield convergent renormalized series with partial sums [118]

$$S_N(\omega) = \sum_{j=0}^{N} \mathcal{E}_j(\omega) . \tag{6.67}$$

There are several ways to determine appropriate ω values. One of them is the principle of minimal sensitivity [119] which is based on the fact that the eigenvalues and eigenfunctions of $\hat{\mathcal{H}}(\beta = 1) = \hat{H}$ are independent of ω. Therefore, it is reasonable to look for ω values in the flattest part of the curve $S_N(\omega)$ vs. ω. From the Taylor series

$$S_N(\omega) = S_N(\omega_N) + \left.\frac{\partial S_N(\omega)}{\partial \omega}\right|_{\omega=\omega_N}(\omega-\omega_N) + \left.\frac{\partial^2 S_N(\omega)}{\partial \omega^2}\right|_{\omega=\omega_N}\frac{(\omega-\omega_N)^2}{2} + \cdots \tag{6.68}$$

we realize that ω_N may meet the above condition if

$$\left.\frac{\partial S_N(\omega)}{\partial \omega}\right|_{\omega=\omega_N} = 0 \tag{6.69}$$

and $|(\partial^2 S_N/\partial^2\omega)(\omega_N)|$ is small.

The calculation of the perturbation coefficients $\mathcal{E}_j(\omega)$ is straightforward by any of the methods described in earlier chapters. Here we choose the method of Swenson and Danforth discussed in Section 3.3.1 and easily calculate sufficient energy coefficients $\mathcal{E}_j$ by means of a program which is just a slight modification of the one for the dimensionless anharmonic oscillator already given in the program section. Table 6.6 shows results for the ground state of the anharmonic oscillator (6.55) with $\lambda = 0.06$ which is the maximum λ value considered in Table 6.5. For each value of N we see ω_N, the partial sum $S_N(\omega_N)$, and $\log(|(\partial^2 S_N/\partial\omega^2)(\omega_N)|)$. There is no doubt that the renormalized series converges towards the exact value indicated in Table 6.5. Moreover, the second derivative at the optimum value of ω decreases with N showing that the curve becomes flatter as N increases.

It is worth noticing that the renormalized series is divergent for each fixed ω value but the sequence $S_N(\omega_N)$ already converges towards the exact eigenvalue. An alternative and illustrative way of looking at this problem is through the truncation criterion discussed in Section 6.4.1. Given an ω value we choose $S_N(\omega)$ such that $\mathcal{E}_{N+1}(\omega)$ is the coefficient with the smallest absolute value, and we assume that the error $|\mathcal{E}(1) - S_N|$ must be of the order of $|\mathcal{E}_{N+1}|$. Obviously N is a function of ω which is equivalent to saying that ω depends on N.

Table 6.7 shows values of ω_N, N, and $S_N(\omega_N)$ for the anharmonic oscillator (6.55) with $\lambda = 0.06$. The convergence of the sequence of selected partial sums seems to be smoother than when we apply the principle of minimal sensitivity.

The λ value chosen in the two calculations above is rather small to convince someone that the appropriate choice of $\hat{H}_0$ may be the cure for a divergent perturbation series. For this reason, in what

Table 6.6 Renormalized Series and the Principle of Minimal Sensitivity for the Ground-State Energy of the Anharmonic Oscillator

$$\hat{H} = \frac{1}{2}\left(\hat{p}^2 + \hat{x}^2\right) + 0.06\hat{x}^4$$

N	ω_N	$S_N(\omega_N)$	$\log\left|\dfrac{\partial^2 S_N}{\partial\omega^2}(\omega_N)\right|$
1	1.146319568	0.5389144739	-0.311
2	No roots		
3	1.226408495	0.5383219630	-2.22
4	No roots		
5	1.300888395	0.5383192572	-3.87
6	1.292965589	0.5383192752	-4.49
7	1.400529907	0.5383192621	-4.45
8	1.324485680	0.5383192913	-5.48
9	1.495147674	0.5383192725	-4.76
10	1.357995526	0.5383192923	-6.52
11	1.583362711	0.5383192785	-5.00
12	1.392325919	0.5383192923	-7.56
13	1.666590777	0.5383192820	-5.21
14	1.431988690	0.5383192923	-8.50
15	1.426783737	0.5383192923	-8.98
16	1.481191656	0.5383192923	-8.92
17	1.446057835	0.5383192923	-9.50
18	1.531031398	0.5383192923	-9.14
19	1.467408165	0.5383192923	-10.14
20	1.579190627	0.5383192923	-9.33

Table 6.7 Renormalized Series and the Criterion of Minimal Error for the Ground States of Two Anharmonic Oscillators

ω_N	N	$S_N(\omega_N)$	$\log|\varepsilon_{N+1}|$
$\hat{H} = \dfrac{\hat{p}^2 + \hat{x}^2}{2} + 0.06\hat{x}^4$			
1.00	4	0.5369183938	-2.56
1.10	6	0.5382677955	-3.99
1.20	8	0.5383187776	-6.00
1.30	11	0.5383192937	-8.55
1.40	14	0.5383192923	-11.7
1.45	18	0.5383192923	-13.4
$\hat{H} = \dfrac{\hat{p}^2}{2} + \hat{x}^4$			
2.0	2	0.6679687500	-2.91
3.0	18	0.6679862617	-8.08
3.5	35	0.6679862591	-10.78
4.0	58	0.6679862592	-14.03

follows we consider the so-called strong-coupling limit of the anharmonic oscillator. As shown in Appendix C, when $\lambda \to \infty$ the quartic potential completely dominates and we are left with the pure quartic oscillator

$$\hat{H} = -\frac{1}{2}\frac{d^2}{dx^2} + x^4 . \tag{6.70}$$

If the renormalized series converges for this limit case, then we expect it to converge for all $0 < \lambda < \infty$. In order to construct the renormalized series for this model we choose

$$\hat{\mathcal{H}}_0 = -\frac{1}{2}\frac{d^2}{dx^2} + \frac{\omega^2 x^2}{2}, \ \hat{\mathcal{H}}_1 = \hat{W}_0 = x^4 - \frac{\omega^2 x^2}{2} \tag{6.71}$$

and obtain the perturbation series for the eigenvalues $\mathcal{E}(\beta)$ of $\hat{\mathcal{H}}(\beta) = \hat{\mathcal{H}}_0 + \beta \hat{\mathcal{H}}_1$ by means of the method of Swenson and Danforth. Table 6.7 shows that the sequence of selected partial sums converge towards the ground-state energy of the quartic anharmonic oscillator. As expected, the rate of convergence is smaller than in the case $\lambda = 0.06$ but the results in Table 6.7 suggest that the renormalized series is valid for all λ values.

The renormalized series is also suitable for excited states as shown in Table 6.8 for the quantum numbers $v = 10, 100, 1000,$ and $10,000$. When v is sufficiently large, a precise determination of ω is unnecessary because the renormalized series looks as if it were convergent for any value of ω in a neighborhood of ω_N. We have observed such behavior in the states with $v > 10$ shown in Table 6.8. The eigenvalues in Table 6.8 agree with those obtained by a nonperturbative method $E^{[BBCK]}$ [120] if we take into account that $E^{present} = 2^{-2/3} E^{[BBCK]}$ according to the scaling arguments in Appendix C.

Table 6.8 Renormalized Series for Some Excited States of the Quartic Oscillator $\hat{H} = \dfrac{\hat{p}^2}{2} + \hat{x}^4$

ω_N	N	$S_N(\omega_N)$	ω_N	N	$S_N(\omega_N)$
	$v = 10$			$v = 100$	
4.0	7	31.65942243	8.0	30	643.1833913
4.1	13	31.65945673		40	643.1833914
4.2	20	31.65945647		50	643.1833914
4.3	23	31.65945648		60	643.1833914
4.4	28	31.65945648			
4.5	34	31.65945648			
	$v = 1000$			$v = 10000$	
17.0	30	13774.25175	37.0	20	296579.3007
	40	13774.25198		30	296579.3010
	50	13774.25200		50	296579.3010
	60	13774.25200		60	296579.3010

It is a great advantage of perturbation theory that the treatment of highly excited states offers no more difficulty than the calculation of the ground state (at least for one-dimensional models). In fact, Table 6.8 suggests that the convergence properties of the renormalized series are better for the excited states than for the ground state. The reason for this behavior is well understood [121]. On the other hand, other approximate methods become increasingly demanding as the quantum number increases. For example, the Rayleigh–Ritz variational method requires enlarging the basis set (and thereby the matrix dimension), and numerical integration requires enlarging the variable interval (and probably augmenting the number of mesh points to account for increasing oscillation in the classical region).

It has been rigorously proved that the renormalized series for the eigenvalues of the anharmonic oscillator (6.55) converges uniformly for all values of λ [122]. However, such proof does not apply to other anharmonic oscillators of the form

$$\hat{H} = -\frac{1}{2}\frac{d^2}{dx^2} + \frac{x^2}{2} + \lambda x^{2K} \tag{6.72}$$

with $K \geq 3$. Numerical investigation shows that the convergence properties of the renormalized series deteriorate considerably as K increases. Table 6.9 shows results for some anharmonic oscillators

$$\hat{H} = -\frac{1}{2}\frac{d^2}{dx^2} + x^{2K} \tag{6.73}$$

that are the strong-coupling limit of the corresponding operators (6.72) as outlined in Appendix C.

Table 6.9 Renormalized Series for the Ground States of the Anharmonic Oscillators $\hat{H} = \dfrac{\hat{p}^2}{2} + \hat{x}^{2k}$

ω_N	N	$S_N(\omega_N)$	$\log\lvert E_{N+1}\rvert$
		$k = 3$	
5.0	13	0.6804483108	-4.51
6.0	19	0.6804140173	-4.87
7.0	26	0.6804021439	-4.96
8.0	35	0.6803871123	-5.37
9.0	44	0.6803859985	-6.46
10.0	55	0.6803834400	-5.58
Exact		0.6807036117	
		$k = 4$	
ω_N	N	$S_N(\omega_N)$	$\log\lvert E_{N+1}\rvert$
9.0	20	0.7165926122	-2.53
10.0	24	0.7187959806	-2.51
11.0	30	0.7161607879	-2.92
12.0	34	0.7201431753	-2.62
13.0	40	0.7199124049	-2.70
14.0	46	0.7206521523	-2.75
15.0	52	0.7220714811	-2.77
16.0	58	0.7240192116	-2.78
Exact		0.7040487741	

The truncated renormalized series for the ground states of the anharmonic oscillators (6.73) with $K = 3$ and $K = 4$ do not appear to converge as indicated by the fact that the estimated error $\lvert \mathcal{E}_{N+1} \rvert$ does not decrease sufficiently fast as N increases. The "exact" results added to that table for comparison were obtained by means of an accurate, reliable nonperturbative method [116]. In order to calculate accurate eigenvalues of the anharmonic oscillators (6.73) with $K \geq 3$ one has to resort to appropriate summation methods [123]–[127] that certainly perform better on the renormalized series than on the original perturbation series.

The choice $\hat{W}(\beta) = \hat{W}_0$ provides the simplest possible realization of the method outlined above. We have also tried a more general β-power series for $\hat{W}(\beta)$ which enables one to use a quasi-solvable anharmonic oscillator as unperturbed Hamiltonian operator. A quantum-mechanical problem is said to be quasi-solvable if one can solve the eigenvalue equation for just a few states. Because the unperturbed model is not completely solvable we cannot solve the perturbation equations exactly unless we choose the operator coefficients $\hat{W}_i$ conveniently. By means of the logarithmic perturbation theory described in Chapter 2, and aided by Maple, we calculated a perturbation series proposed recently [117] through larger order. As it did not appear to converge we decided not to show results here because, in our opinion, they would not add anything relevant to the present discussion.

Another way of building convergent perturbation series consists of splitting the Hamiltonian operator $\hat{H}$ into its diagonal $\hat{H}_D$ and off-diagonal $\hat{H}_N$ parts by means of a basis set of vectors $\{|j>, j = 0, 1, \dots\}$,

$$\hat{H} = \hat{H}_D + \hat{H}_N, \ \hat{H}_D = \sum_j |j><j|\hat{H}|j><j|, \ \hat{H}_N = \sum_i \sum_{j \neq i} |i><i|\hat{H}|j><j|. \quad (6.74)$$

Since $\hat{H}_D$ is exactly solvable we choose it to be the unperturbed model and $\hat{H}_N$ to be the perturbation: $\hat{H}(\beta) = \hat{H}_D + \beta \hat{H}_N$. At the end of the calculation we set the perturbation parameter β equal to unity. Moreover, we can introduce adjustable parameters into the basis set to improve the convergence properties of the resulting perturbation series, modifying the unperturbed part and the perturbation more favorably. A particular example of this strategy is the so-called operator method in which the splitting is carried out in terms of the generators of a Lie algebra [128]–[134].

To illustrate the application of this method we choose the anharmonic oscillators (6.73) which for convenience we write in operator form as $\hat{H} = \hat{p}^2/2 + \hat{x}^{2K}$, where $\hat{p} = -id/dx$. An appropriate basis set with an adjustable parameter α is

$$\{|n>_\alpha, n = 0, 1, \dots\}, \ |n>_\alpha = \hat{U}|n> , \quad (6.75)$$

where $|n>$ is an eigenvector of $\hat{p}^2 + \hat{x}^2$, and $\hat{U}$ is a unitary operator that generates a scaling transformation (see Appendix C):

$$\hat{U}^\dagger \hat{x} \hat{U} = \alpha^{1/2}\hat{x}, \ \hat{U}^\dagger \hat{p} \hat{U} = \alpha^{-1/2}\hat{p} . \quad (6.76)$$

We write the matrix elements of $\hat{H}$ in terms of matrix elements of powers of $\hat{p}$ and $\hat{x}$ in the basis set $\{|n>\}$ as follows from

$$H_{mn} = \ _\alpha \left\langle m \left| \hat{H} \right| n \right\rangle_\alpha = \left\langle m \left| \hat{U}^\dagger \hat{H} \hat{U} \right| n \right\rangle . \quad (6.77)$$

Notice that if we write

$$\hat{U}^\dagger \hat{H} \hat{U} = \frac{1}{2\alpha} \left(\hat{p}^2 + \hat{x}^2 \right) + \alpha^K \hat{x}^{2K} - \frac{\hat{x}^2}{2\alpha} \quad (6.78)$$

the problem reduces to the calculation of the matrix elements of powers of $\hat{x}$ which offers no difficulty if we resort to the recurrence relation (1.55). Therefore, by means of a slight modification of one of the Maple programs in the program section for Chapter 1, we easily carry out the calculation described in what follows.

We expand the eigenvectors Ψ of $\hat{H}$ in terms of the chosen basis set

$$\Psi = \sum_{j=0}^{\infty} c_j |j>_\alpha \quad (6.79)$$

and derive the perturbation equations for the energy E and the coefficients c_j as indicated in Chapter 1. For simplicity in this case we choose the intermediate normalization condition $c_n = 1$ when $E_0 = H_{nn}$ is the unperturbed energy of the state n, which we do not indicate explicitly by a subscript as we did in Chapter 1. The reader may easily verify that the perturbation equations are

$$E_p = \sum_{j \neq n} H_{nj} c_{j,p-1}, \ c_{n,p} = \delta_{p0}$$

$$c_{i,p} = \frac{1}{E_0 - H_{ii}} \left(\sum_{j \neq i} H_{ij} c_{j,p-1} - \sum_{s=1}^{p} E_s c_{i,p-s} \right), \ i \neq n, \tag{6.80}$$

where p indicates the perturbation order.

As a pedagogical illustration of how the operator method yields perturbation series with improved convergence properties we consider an exactly solvable problem already discussed above: the harmonic oscillator with a harmonic perturbation given by equation (6.32). This simple model was chosen some time ago for a test of the operator method, [130] and here we provide a more straightforward argument. We already know that the Rayleigh–Schrödinger perturbation series for the operator (6.32) converges only for $|\lambda| < 1$ in spite of the fact that there are bound states for all $\lambda > -1$. In order to separate the diagonal and off-diagonal parts of $\hat{H}$ for the application of the operator method we write $\hat{p}$ and $\hat{x}$ in terms of the creation and annihilation operators $\hat{a}^\dagger$ and $\hat{a}$, respectively. We easily obtain

$$\hat{H}_D = (1 + \lambda/2) \left(\hat{a}^\dagger \hat{a} + 1/2 \right), \ \hat{H}_N = \lambda \left[\hat{a}^2 + \left(\hat{a}^\dagger \right)^2 \right]/4, \tag{6.81}$$

so that

$$\hat{H}_D + \beta \hat{H}_N = (1 + \lambda/2 - \lambda\beta/2)\hat{p}^2/2 + (1 + \lambda/2 + \lambda\beta/2)\hat{x}^2/2. \tag{6.82}$$

Therefore, the eigenvalues are given by

$$E(\beta) = (v + 1/2)|1 + \lambda/2|\sqrt{1 - [\lambda\beta/(2 + \lambda)]^2}, \ v = 0, 1, \dots, \tag{6.83}$$

which shows that the operator method converges for all $|\lambda| < |2 + \lambda|$ ($\Rightarrow \lambda > -1$) when $\beta = 1$; that is to say, for all values of λ supporting bound states. At least for this trivial problem the operator method certainly improves the convergence properties of the perturbation series even when the scaling parameter α is arbitrarily chosen equal to unity.

As a more demanding test of the operator method, we consider the ground state of the anharmonic oscillators (6.73) with $K = 3$ and $K = 4$ for which the renormalized series diverges as shown above (for completeness we add the case $K = 2$). The results in Table 6.10 for several values of α do not clearly suggest convergence, but show that the operator method is preferable to the renormalized series in all those cases.

6.5.2   Intelligent Algebraic Approximants

There is a vast literature describing more or less successful methods for the summation of divergent series. Some of the most popular are Borel and Padé approximants, [135, 136] continued fractions, [135] and nonlinear transformations [137] among many others. Here we briefly discuss algebraic approximants that have produced results of unprecedented accuracy for anharmonic oscillators and other models [138].

Table 6.10   Operator Method for the Anharmonic Oscillators

$$\hat{H} = \frac{\hat{p}^2}{2} + \hat{x}^{2K}$$

α	N	$S_N(\alpha)$	$\log|E_{N+1}|$
		$K = 2$	
0.10	18	0.6874191200	-2.87
0.20	16	0.6679809935	-7.22
0.30	18	0.6679862592	-11.14
0.35	19	0.6679862592	-11.53
0.40	19	0.6679862596	-9.14
		$K = 3$	
0.10	16	0.6866545682	-2.62
0.15	16	0.6806778711	-4.36
0.20	17	0.6807037650	-6.61
0.23	17	0.6807036225	-7.91
0.25	14	0.6807038117	-6.97
		$K = 4$	
0.10	12	0.7112588258	-2.37
0.15	10	0.7034332514	-4.49
0.18	12	0.7040680092	-5.10
0.20	12	0.7040771959	-5.12
0.25	6	0.7050018759	-4.10

Suppose that we want to obtain meaningful values of the function $E(\lambda)$ from its asymptotic expansion (6.1). The simplest algebraic approximant is a rational function of the form

$$[M/N](\lambda) = \frac{A(\lambda)}{B(\lambda)} = \frac{\sum_{j=0}^{M} a_j \lambda^j}{\sum_{j=0}^{N} b_j \lambda^j}, \tag{6.84}$$

where we choose the coefficients a_j and b_j in order to obtain as many terms of the series (6.1) as possible. Notice that there are only $M + N + 1$ approximant coefficients at our disposal because we can always remove one of them by simply dividing numerator and denominator by it. Therefore, we can obtain $M + N + 1$ coefficients E_j:

$$[M/N](\lambda) = \sum_{j=0}^{M+N} E_j \lambda^j + \mathcal{O}\left(\lambda^{M+N+1}\right). \tag{6.85}$$

In some cases Padé approximants converge when we increase M and N conveniently, yielding reasonable approximate values of $E(\lambda)$ [135]. Notice that we can rewrite $E = [M/N]$ as a linear equation $A(\lambda) - B(\lambda)E = 0$ which we solve for the approximate value of E.

If the function $E(\lambda)$ is known to have branch points of order 2, then it is convenient to use a quadratic approximant of the form $A(\lambda)E^2 + B(\lambda)E + C(\lambda) = 0$, where $A(\lambda)$, $B(\lambda)$, and $C(\lambda)$ are polynomial functions of λ [138]. The two roots of the quadratic equation give us the two branches of the function $E(\lambda)$.

The linear and quadratic approximants just mentioned are particular cases of algebraic approximants of the form

$$\sum_{n=0}^{N} A_n(\lambda) E^n = 0 \,, \tag{6.86}$$

where $A_0(\lambda), A_1(\lambda), \ldots, A_n(\lambda)$ are polynomial functions of λ. They prove suitable for the accurate summation of divergent series to obtain the branches of multiple-valued functions [138].

In principle, one can construct many different algebraic approximants from the same set of perturbation coefficients E_j, so that there is great flexibility of choice for a given problem. This freedom may become a drawback because it may require some extensive numerical calculation to determine the most convenient sequence of approximants. For example, in the case of simple quantum-mechanical anharmonic oscillators, different sequences of algebraic approximants give different answers, and those with the correct large-coupling behavior (see Appendix C) prove to be the most accurate [138]. It appears to be most important to have clear directions of how to construct suitable algebraic approximants. Here we show how to address a wide class of problems.

Suppose that we can rewrite the unknown function $E(\lambda)$ as

$$E(\lambda) = \lambda^a W \left(\lambda^{-b} \right) \,, \tag{6.87}$$

where a and $b > 0$ are known rational numbers, and $W(\xi)$ is another unknown function. For concreteness and simplicity in what follows we restrict to the most usual case $a > 0$.

It is our purpose to obtain approximate values of $E(\lambda)$ by means of an implicit equation of the form $Q(E, \lambda) = 0$. We require that one of the roots of $Q(E, \lambda) = 0$ satisfies the expansion (6.1) through a given order. Moreover, in order to take into account equation (6.87) we build $Q(E, \lambda)$ to factorize as $Q(\lambda^a W, \lambda) = F(\lambda) G(W, \lambda^{-b})$, and assume that $G(W, \lambda^{-b}) = 0$ may give us approximate values of $W(\lambda^{-b})$. In this way we expect to obtain accurate values of $E(\lambda)$ for all λ having only the expansion (6.1) which is valid for sufficiently small values of λ.

Algebraic approximants (6.86) are particularly simple implicit equations which for convenience we rewrite as

$$A[M, N] = \sum_{m=0}^{M} \sum_{n=0}^{N} B_{mn} \lambda^m E^n \,. \tag{6.88}$$

In order to construct approximants that factorize in the way indicated above, we substitute $\lambda^a W$ for E in equation (6.88) and require that $m + an = aN - bj$, where $j = 0, 1, \ldots J_m \leq (aN - m)/b$. More precisely, from all the possible algebraic approximants we choose those of the form

$$A[M, N] = \sum_{m=0}^{M} \sum_{j=0}^{J_m} A_{mj} \lambda^m E^{N-(m+bj)/a}, \quad J_m = \left[\frac{aN - m}{b} \right] \,, \tag{6.89}$$

where $[u]$ stands for the greatest integer smaller than or equal to the real number u. The new coefficients A_{mj} are related to the original ones B_{mn} by $A_{mj} = B_{m\,N-(m-bj)/a}$. The integers M and N are independent except for the restriction $N \geq M/a$ necessary to have $J_m \geq 0$ for all m. In order to remove one degree of freedom we arbitrarily choose N to be the smallest integer greater than or equal to M/a.

Substituting $E = \lambda^a W$ into $A[M, N] = 0$ and dividing by λ^{aN}, the resulting implicit equation

$$\sum_{m=0}^{M} \sum_{j=0}^{J_m} A_{mj} \lambda^{-bj} W^{N-(m+bj)/a} = 0 \tag{6.90}$$

gives us $W(\lambda^{-b})$. If

$$W(0) = \lim_{\lambda \to \infty} \lambda^{-a} E(\lambda) \tag{6.91}$$

exists, then we obtain it approximately as a root of the implicit equation

$$\sum_{m=0}^{M} A_{m0} W(0)^{N-m/a} = 0 \tag{6.92}$$

easily derived from (6.90). Moreover, if $W(\xi)$ satisfies a formal Taylor expansion $W_0 + W_1\xi + \cdots + W_n\xi^n + \cdots$ about $\xi = 0$, then we expect appropriate partial sums for

$$E(\lambda) = \lambda^a \sum_{j=0}^{\infty} W_j \lambda^{-bj} \tag{6.93}$$

to be valid for sufficiently large λ values. We say that the approximants (6.89) are intelligent because their roots also satisfy (6.87) as follows from (6.90).

As said earlier we require that one of the roots of $A[M, N] = 0$ satisfies the Taylor expansion (6.1) through a given order. Because one of the approximant coefficients is not independent we arbitrarily set $A_{00} = 1$. Choosing the total number

$$\sum_{m=0}^{M} \sum_{j=0}^{J_m} 1 - 1 = \sum_{m=0}^{M} (J_m + 1) - 1 = M + \sum_{m=0}^{M} J_m \tag{6.94}$$

of independent adjustable parameters appropriately we can force a root of the approximant to give the λ-power series (6.1) exactly through order

$$P(M) = M - 1 + \sum_{m=0}^{M} J_m . \tag{6.95}$$

The derivation of the expressions that give the approximant coefficients A_{mj} in terms of the series coefficients E_k is rather tedious, but we show it here for completeness. We first rewrite the approximant as

$$A = \sum_{m=0}^{M} \lambda^m P_m(E), \quad P_m(E) = \sum_{n=0}^{N_m} B_{mn} E^n . \tag{6.96}$$

It is convenient to introduce a cutoff function $\theta(x)$ which is zero if $x < 0$ and unity otherwise, and rewrite the approximant as

$$A = \sum_{m=0}^{\infty} \theta(M - m)\lambda^m P_m(E) . \tag{6.97}$$

Writing

$$E^n = \sum_{j=0}^{\infty} C_{nj} \lambda^j , \tag{6.98}$$

where $C_{0j} = \delta_{0j}$ and $C_{1j} = E_j$, we have

$$P_m(E) = \sum_{j=0}^{\infty} p_{mj}\lambda^j, \quad p_{mj} = \sum_{n=0}^{N_m} B_{mn}C_{nj} . \tag{6.99}$$

Therefore

$$A = \sum_{k=0}^{\infty} \lambda^k \sum_{m=0}^{k} \theta(M-m) p_{m\,k-m} . \tag{6.100}$$

$A = 0$ for all λ provided that

$$\sum_{m=0}^{k} \theta(M-m) \sum_{n=0}^{N_m} B_{mn}C_{n\,k-m} = 0 . \tag{6.101}$$

Finally, taking into account the relation between n, m, and j, and between the approximant coefficients B_{mn} and A_{mj} we obtain

$$\sum_{m=0}^{\min(k,M)} \sum_{j=0}^{J_m} A_{mj}C_{N-(m+bj)/a\,k-m} = 0 . \tag{6.102}$$

Substituting the expansion (6.1) into $E^n = EE^{n-1}$ we obtain a recursion relation for the coefficients C_{nj}:

$$C_{nj} = \sum_{i=0}^{j} E_{j-i}C_{n-1\,i} . \tag{6.103}$$

If $P(M)$ is the perturbation order, then we are left with a system of $P(M) + 1$ linear inhomogeneous equations (6.102) $k = 0, 1, \ldots, P(M)$ with $P(M) + 1$ unknowns A_{mj} (remember that $A_{00} = 1$).

As simple illustrative examples we consider one-dimensional anharmonic oscillators

$$\hat{H} = -\frac{1}{2}\frac{d^2}{dx^2} + \frac{x^2}{2} + \lambda x^K . \tag{6.104}$$

As shown in Appendix C the eigenvalues $E(\lambda)$ of $\hat{H}$ satisfy equation (6.87) with $a = 2/(K+2) = b/2$. If K is even we arbitrarily choose $N = (K+2)M/2$.

We first consider the quartic anharmonic oscillator ($K = 4$) for which $a = 1/3$ and $b = 2/3$ that lead to $J_m = [(N - 3m)/2]$ and $N = 3M$. From the perturbation series of order 5 for the ground state we construct the approximant

$$\begin{aligned}
A[2, 6] &= E^6 - \frac{4921}{14852}E^4 - \frac{2317}{59408}E^2 - \frac{19497}{7426}\lambda E^3 + \frac{10131}{14852}\lambda E \\
&\quad + \frac{164997}{237632}\lambda^2 + \frac{75}{5056} .
\end{aligned} \tag{6.105}$$

For each λ value the equation $A[2, 6] = 0$ has 6 roots, two of them real. The smallest real root gives an approximation to the energy of the quartic anharmonic oscillator. We discard the other root because it does not give the exact result $E(0) = 1/2$. Figure 6.4 shows a satisfactory agreement between the smallest real root of the simple implicit equation $A[2, 6] = 0$, and the eigenvalue

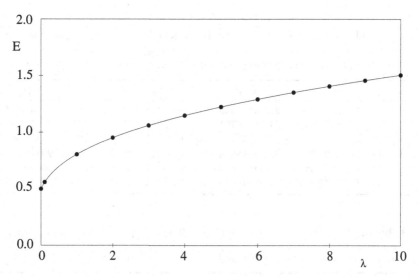

FIGURE 6.4

Ground-state energy of the anharmonic oscillator (6.55) calculated by means of the intelligent approximant (6.105) (continuous line) and by a nonperturbative method (points).

calculated accurately by a nonperturbative method [116]. In what follows we show that the simple intelligent approximant (6.105) gives reasonable results even for λ values much larger than those in Figure 6.4.

Substituting $\lambda^{1/3}W$ for E in equation (6.105) and dividing the result by λ^2 we obtain

$$W^6 - \frac{4921}{14852}W^4\xi - \frac{2317}{59408}W^2\xi^2 - \frac{19497}{7426}W^3 + \frac{10131}{14852}\xi W$$
$$+ \frac{75}{5056}\xi^3 + \frac{164997}{237632} = 0 , \qquad (6.106)$$

where $\xi = \lambda^{-2/3}$. As argued earlier in this chapter, when $\xi = 0$ ($\lambda \to \infty$) the implicit expression resulting from (6.106) gives us an approximant for the ground state energy of the pure quartic oscillator (equation (6.73) with $K = 2$). One of the roots of this approximant $W_0 = 0.668215$ provides the ground-state energy with an error of 0.034%, clearly showing that equation (6.105) is a good approach to the corresponding eigenvalue of the quartic anharmonic oscillator for all λ values (even for $\lambda \to \infty$). Moreover, the simpler intelligent approximant A[1,6] proves to be more accurate than other expressions built from perturbation series of order four [139, 140].

In order to test the convergence properties of the intelligent approximants we calculate the leading coefficient W_0 of the strong-coupling expansion for the anharmonic oscillators (6.104) with $K = 4$, 6, 8, given by a root of equation (6.92) with $a = 2/(K + 2)$.

Table 6.11 shows that the rate of convergence of the intelligent approximants is remarkable for the ground as well as highly excited states of the pure quartic oscillator. The converged eigenvalues agree with those in Table 6.8 obtained earlier by means of the renormalized series. By simple inspection of both tables one concludes that the rate of convergence is considerably greater for the intelligent approximants than for the renormalized series.

Table 6.12 shows results for the pure sextic and octic anharmonic oscillators. The intelligent approximants appear to converge in the former case but not in the latter one. We have been unable to identify a convergent sequence from the roots of equation (6.92) (see 5th, 6th, and 7th columns in Table 6.12) for the octic anharmonicity.

Table 6.11 Convergence Rate of the Intelligent Approximants $A[M, 3M]$ for

Several States of the Anharmonic Oscillator $\hat{H} = \dfrac{\hat{p}^2}{2} + \hat{x}^4$

M	P	$v = 0$	$v = 10$	$v = 100$	$v = 1000$	$v = 10000$
1	1	0.7211247852	33.18593389	673.9051114	14432.11270	310743.9659
2	5	0.6682150331	31.65916791	643.1772376	13774.12177	296576.4973
3	10	0.6679536006	31.65945459	643.1833908	13774.25199	296579.3007
4	17	0.6679862615	31.65945648	643.1833914	13774.25200	296579.3010
5	25	0.6679862592	31.65945648	643.1833914	13774.25200	296579.3010
6	35	0.6679862592	31.65945648	643.1833914	13774.25200	296579.3010

Table 6.12 Convergence Rate of the Intelligent Approximants for the Ground

States of the Anharmonic Oscillators $\hat{H} = \dfrac{\hat{p}^2}{2} + \hat{x}^K$ with $K = 6$ and $K = 8$

	$K = 6$	$A[M, 4M]$	$K = 8$	$A[M, 5M]$	
M	P	E	P	E	
1	2	0.5993203292	2	0.5897586004	
2	7		8	0.4554075304	
3	14	0.6593272643	16	0.4277561778	0.6561496492
4	23	0.6807661937	27		
5	34	0.6807031149	40	0.4663678730	0.6864498896
6	47	0.6807048473	56	0.4592557841	0.5707878070	0.8107106475
7	62	0.6807036615			

The intelligent approximants are suitable for obtaining the strong-coupling expansion (6.93). For example, we can rewrite the Hamiltonian operator (6.55) as

$$\hat{U}^\dagger \hat{H} \hat{U} = \lambda^{1/3} \left[-\frac{1}{2} \frac{d^2}{dx^2} + \frac{1}{2} \lambda^{-2/3} x^2 + x^4 \right] \tag{6.107}$$

by means of the equivalent transformation discussed in Appendix C. Choosing the quadratic term as a perturbation, and $\xi = \lambda^{-2/3}$ as a perturbation parameter we obtain the expansion

$$E(\lambda) = \lambda^{1/3} \sum_{j=0}^{\infty} W_j \lambda^{-2j/3} . \tag{6.108}$$

The calculation of the coefficients W_j proceeds as follows: first substitute $W\lambda^{1/3}$ for E in the approximant $A[M, N]$, then divide it by λ^M, and, finally, substitute ξ for $\lambda^{-2/3}$, in order to obtain a polynomial function of W and ξ:

$$\sum_{m=0}^{\min\{M,[N/3]\}} \sum_{j=0}^{[(N-3m)/2]} A_{mj} \xi^j W^{N-3m-2j} = 0 . \tag{6.109}$$

Equation (6.106) is a particular case of this expression which is a particular case of (6.90). If we substitute the truncated expansion $W_0 + W_1 \xi + W_2 \xi^2 + \cdots + W_n \xi^n$ for W we obtain a polynomial function of ξ. Setting each of its coefficient equal to zero we extract the expansion coefficients W_j term by term.

Table 6.13 shows the coefficients W_j, $1 \leq j \leq 5$, for the ground and excited states; the leading coefficient W_0 is given in Table 6.11. The coefficients W_0 and W_1 increase with the quantum number because they are respectively an eigenvalue of the pure quartic oscillator and the expectation value $\langle x^2 \rangle /2$. The next coefficient W_2 appears to approach a finite nonzero value as the quantum number increases. This curious and interesting behavior of W_2 is not well known because most of the calculations of the strong coupling series for the anharmonic oscillators have been restricted to the ground state [116, 125, 127], [141]–[144]. The absolute values of the remaining coefficients decrease with the quantum number suggesting an increasing radius of convergence. Present coefficients W_j are related to earlier ones W_j' [141, 144, 145] by means of the transformation $W_j = 2^{-2(j+1)/3} W_j'$ which follows from the scaling arguments discussed in Appendix C.

Table 6.13 Strong-Coupling Series for the Anharmonic Oscillator $\hat{H} = \frac{\hat{p}^2}{2} + \xi \frac{\hat{x}^2}{2} + \hat{x}^4$

P	W_1	W_2	W_3	W_4	W_5
			$v = 0$		
	0.1247530653	0	$-0.1450431970\ 10^{-2}$	$0.2708825696\ 10^{-3}$	0
5	0.1434473426	$-0.8477670177\ 10^{-2}$	$0.7551679354\ 10^{-3}$	$-0.6175925863\ 10^{-4}$	$0.2377454709\ 10^{-5}$
10	0.1437221947	$-0.8785810015\ 10^{-2}$	$0.1190188239\ 10^{-2}$	$-0.9174251401\ 10^{-3}$	$0.1870527091\ 10^{-2}$
17	0.1436687775	$-0.8627560060\ 10^{-2}$	$0.8182047111\ 10^{-3}$	$-0.8242677907\ 10^{-4}$	$0.8068340550\ 10^{-5}$
25	0.1436687832	$-0.8627565683\ 10^{-2}$	$0.8182088855\ 10^{-3}$	$-0.8242919997\ 10^{-4}$	$0.8069484300\ 10^{-5}$
35	0.1436687830	$-0.8627565841\ 10^{-2}$	$0.8182088741\ 10^{-3}$	$-0.8242921887\ 10^{-4}$	$0.8069494581\ 10^{-5}$
			$v = 10$		
1	1.160790617	0	$-0.5201558690\ 10^{-3}$	$0.1907145983\ 10^{-4}$	0
5	1.285488559	$-0.1170313526\ 10^{-1}$	$0.9055091205\ 10^{-4}$	$0.1827544018\ 10^{-5}$	$-0.1106049773\ 10^{-6}$
10	1.285402347	$-0.1168532870\ 10^{-1}$	$0.8833622063\ 10^{-4}$	$0.2034299068\ 10^{-5}$	$-0.1261223438\ 10^{-6}$
17	1.285401386	$-0.1168501615\ 10^{-1}$	$0.8827429329\ 10^{-4}$	$0.2043545272\ 10^{-5}$	$-0.1272375932\ 10^{-6}$
25	1.285401385	$-0.1168501552\ 10^{-1}$	$0.8827411300\ 10^{-4}$	$0.2043586236\ 10^{-5}$	$-0.1272451861\ 10^{-6}$
35	1.285401385	$-0.1168501553\ 10^{-1}$	$0.8827411537\ 10^{-4}$	$0.2043585982\ 10^{-5}$	$-0.1272451412\ 10^{-6}$
46	1.285401384	$-0.1168501550\ 10^{-1}$	$0.8827411803\ 10^{-4}$	$0.2043585800\ 10^{-5}$	$-0.1272451427\ 10^{-6}$
			$v = 100$		
1	5.234510168	0	$-0.1155679066\ 10^{-3}$	$0.9405426049\ 10^{-6}$	0
5	5.794717030	$-0.1169307698\ 10^{-1}$	$0.2008458603\ 10^{-4}$	$0.8882862735\ 10^{-7}$	$-0.1197338502\ 10^{-8}$
10	5.794317915	$-0.1167509893\ 10^{-1}$	$0.1959442465\ 10^{-4}$	$0.9895203040\ 10^{-7}$	$-0.1367717608\ 10^{-8}$
17	5.794317840	$-0.1167509278\ 10^{-1}$	$0.1959411962\ 10^{-4}$	$0.9896348006\ 10^{-7}$	$-0.1368068196\ 10^{-8}$
25	5.794317849	$-0.1167509313\ 10^{-1}$	$0.1959412671\ 10^{-4}$	$0.9896336791\ 10^{-7}$	$-0.1368066423\ 10^{-8}$
35	5.794317844	$-0.1167509284\ 10^{-1}$	$0.1959411923\ 10^{-4}$	$0.9896349738\ 10^{-7}$	$-0.1368068317\ 10^{-8}$
46	5.794317842	$-0.1167509276\ 10^{-1}$	$0.1959411788\ 10^{-4}$	$0.9896350056\ 10^{-7}$	$-0.1368068043\ 10^{-8}$
			$v = 1000$		
1	24.22394697	0	$-0.2497338533\ 10^{-4}$	$0.4391918791\ 10^{-7}$	0
5	26.81631188	$-0.1169279518\ 10^{-1}$	$0.4339177809\ 10^{-5}$	$0.4150472515\ 10^{-8}$	$-0.1208946558\ 10^{-10}$
10	26.81448429	$-0.1167499026\ 10^{-1}$	$0.4234161483\ 10^{-5}$	$0.4619742513\ 10^{-8}$	$-0.1379841062\ 10^{-10}$
17	26.81448396	$-0.1167498447\ 10^{-1}$	$0.4234099352\ 10^{-5}$	$0.4620247175\ 10^{-8}$	$-0.1380175162\ 10^{-10}$
25	26.81448390	$-0.1167498565\ 10^{-1}$	$0.4234065495\ 10^{-5}$	$0.4619314054\ 10^{-8}$	$-0.1382752428\ 10^{-10}$
35	26.81448396	$-0.1167498444\ 10^{-1}$	$0.4234099260\ 10^{-5}$	$0.4620247302\ 10^{-8}$	$-0.1380175139\ 10^{-10}$
			$v = 10000$		
1	112.4038887	0	$-0.5381967605\ 10^{-5}$	$0.2039771776\ 10^{-8}$	0
5	124.4329670	$-0.1169279230\ 10^{-1}$	$0.9351259981\ 10^{-6}$	$0.1927647361\ 10^{-9}$	$-0.1210043192\ 10^{-12}$
10	124.4244875	$-0.1167498913\ 10^{-1}$	$0.9124960890\ 10^{-6}$	$0.2145578100\ 10^{-9}$	$-0.1381081866\ 10^{-12}$
17	124.4244859	$-0.1167498327\ 10^{-1}$	$0.9124826986\ 10^{-6}$	$0.2145811763\ 10^{-9}$	$-0.1381415151\ 10^{-12}$
25	124.4244857	$-0.1167498262\ 10^{-1}$	$0.9124807799\ 10^{-6}$	$0.2145864061\ 10^{-9}$	$-0.1381556080\ 10^{-12}$
35	124.4244858	$-0.1167498314\ 10^{-1}$	$0.9124825886\ 10^{-6}$	$0.2145812126\ 10^{-9}$	$-0.1381414873\ 10^{-12}$

It has been proved that the perturbation series for any eigenvalue $W(\xi)$ of the Hamiltonian operator

$$\hat{H} = -\frac{1}{2}\frac{d^2}{dx^2} + \frac{1}{2}\xi x^2 + x^4 \tag{6.110}$$

has a finite radius of convergence determined by a pair of the so-called Bender and Wu branch points [111, 115]. Two eigenvalues become degenerate at every one of those square-root branch points, some of which have been accurately calculated by means of nonperturbative methods [146, 147]. It is our purpose to show that intelligent approximants are suitable for obtaining Bender and Wu branch points ξ_b. Such a calculation is an even more demanding test of a perturbation method than the accurate determination of the coefficients of the strong-coupling expansion. We start from the transformation of an intelligent approximant $A[M, N]$ into a polynomial function of W and ξ equation (6.109). Consequently, $A[M, 3M] = 0$ becomes a quantization condition of the form $Q(W, \xi) = 0$ from which we obtain square-root branch points as indicated in Section 6.2.2.

Table 6.14 shows Bender and Wu branch points calculated from intelligent approximants of increasing perturbation order P. They correspond to crossings between pairs of eigenvalues ($v = 0, v = 2$), ($v = 1, v = 3$), ($v = 2, v = 4$), and ($v = 3, v = 5$). Although convergence is rather slow, we clearly see that our sequences approach the branch points calculated by accurate nonperturbative methods [146, 147]. The latter values of W_b and ξ_b were multiplied by $2^{-2/3}$ and $2^{2/3}$, respectively, in order to compare them with the present ones. We could not obtain all the branch points reported by Shanley [146, 147]; it is not clear to us whether the failure was due to the algebraic approximants or to our rather primitive numerical root finding algorithm. Suffice to say that we do not try to give accurate Bender and Wu branch points but to test intelligent approximants built from divergent series at singular points of $E(\lambda)$.

Another interesting application of the intelligent approximants is the calculation of the energies of metastable states. For the sake of concreteness consider the anharmonic oscillator (6.104) with $K = 4$ and $\lambda < 0$. The potential-energy function is unbounded from below as shown in Figure 6.5,

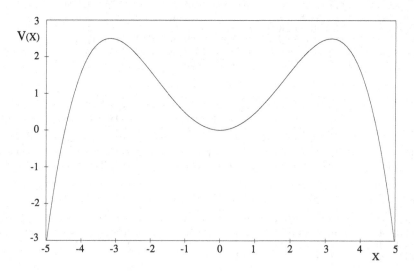

FIGURE 6.5
Potential-energy function $V(x) = x^2/2 - x^4/40$.

and therefore supports no bound state. The eigenfunctions satisfy the boundary conditions

$$\Psi(x) \to \exp\left(i\frac{\sqrt{2|\lambda|}}{3}|x|^3\right) \ as \ |x| \to \infty, \tag{6.111}$$

which correspond to outgoing waves in both channels ($x < 0$ and $x > 0$), only for discrete complex values of the energy E. The real and imaginary parts of E are commonly interpreted as the position and width, respectively, of a resonance, the latter being related to the lifetime of the metastable state.

Table 6.14   Bender and Wu Branch Points of the Anharmonic Oscillator $\hat{H} = \dfrac{\hat{p}^2}{2} + \xi \dfrac{\hat{x}^2}{2} + \hat{x}^4$ for Pairs of States with Quantum Numbers v and $v + 2$

	$v = 0$	
P	W_b	ξ_b
35	$0.2630428496 + 1.624594658i$	$-6.624501579 + 3.576330437i$
46	$0.2889000800 + 1.876309615i$	$-6.669345055 + 3.445029671i$
59	$0.2617508958 + 1.852538472i$	$-6.658672632 + 3.444746306i$
73	$0.2547451096 + 1.849390225i$	$-6.657061544 + 3.444280112i$
89	$0.2545953206 + 1.849466123i$	$-6.657055939 + 3.444246366i$
106	$0.2546068447 + 1.849473621i$	$-6.657058696 + 3.444247072i$
Shanley	$0.2546072144 + 1.849472732i$	$-6.657058631 + 3.444247225i$

	$v = 1$	
P	W_b	ξ_b
35	$1.200465709 + 4.979841203i$	$-7.909214963 + 6.369208765i$
46	$1.263430140 + 4.974064781i$	$-7.916591583 + 6.386014949i$
59	$1.265137461 + 4.973918388i$	$-7.916826785 + 6.386354390i$
73	$1.265141090 + 4.973873844i$	$-7.916820032 + 6.386360110i$
89	$1.265136142 + 4.973871821i$	$-7.916819326 + 6.386359686i$
106	$1.265135846 + 4.973873892i$	$-7.916819519 + 6.386359449i$
Shanley	$1.265135800 + 4.973873648i$	$-7.916819487 + 6.386359463i$

	$v = 2$	
P	W_b	ξ_b
35	$2.530479106 + 8.281701739i$	$-8.980053030 + 8.859695343i$
46	$2.658342737 + 8.439401379i$	$-9.065073467 + 8.847819213i$
59	$2.640928270 + 8.450766476i$	$-9.064697852 + 8.842465234i$
73	$2.649060188 + 8.454160591i$	$-9.066174055 + 8.843304710i$
106	$2.648977207 + 8.454181256i$	$-9.066163981 + 8.843293687i$
Shanley	$2.649031393 + 8.454111313i$	$-9.066162551 + 8.843307413i$

	$v = 3$	
P	W_b	ξ_b
59	$5.177125194 + 11.67311958i$	$-10.09069392 + 12.75585928i$
73	$4.304386970 + 12.23652612i$	$-10.13632596 + 11.02995502i$
89	$4.299380545 + 12.23424907i$	$-10.13550858 + 11.02965413i$
106	$4.299887277 + 12.23487423i$	$-10.13563744 + 11.02961991i$
Shanley	$4.299828498 + 12.23480418i$	$-10.13562445 + 11.02962218i$

In order to obtain resonance energies of the anharmonic oscillator mentioned above, we simply select convergent sequences of roots of $A[M, 3M] = 0$ for negative values of λ. Table 6.15 shows the remarkable rate of convergence of the one obtained from the perturbation series for the ground state ($v = 0$) for $\lambda = -0.025$.

Table 6.15 Convergence of Intelligent Approximants to a Resonance of the Anharmonic Oscillator
$$\hat{H} = \frac{1}{2}(\hat{p}^2 + \hat{x}^2) - \frac{1}{40}\hat{x}^4$$

P	$E\,(v = 0)$
1	0.4800747778
5	0.4791200094
10	0.4791144844
17	$0.4791167889 - 0.7289308712 \times 10^{-5}\,i$
25	$0.4791168180 - 0.7282280620 \times 10^{-5}\,i$
35	$0.4791168182 - 0.7282385091 \times 10^{-5}\,i$
46	$0.4791168182 - 0.7282386678 \times 10^{-5}\,i$
59	$0.4791168182 - 0.7282386684 \times 10^{-5}\,i$
73	$0.4791168182 - 0.7282386684 \times 10^{-5}\,i$

In the program section we show a collection of simple procedures for the construction of intelligent approximants for the anharmonic oscillator (6.104) with $K = 4$. One must keep in mind that they are just the starting point of the calculations described in this section, which must be explicitly carried out according to the equations given above.

There are many methods for the summation of divergent or slowly convergent perturbation series that have not been mentioned in this section. The reader may look up some of them in the literature cited. It is not our purpose to be exhaustive on this subject which we leave at this point.

Chapter 7

Polynomial Approximations

7.1 Introduction

We call polynomial approximation a particular form of perturbation theory based on the expansion of a nonpolynomial potential-energy function in a Taylor series about a conveniently chosen coordinate point, in a way similar to the approach known as small-amplitude oscillation in classical mechanics. This approximate method commonly gives more accurate results for deep wells and energies close to the minimum of the potential-energy function. Throughout this chapter we discuss several polynomial approximations, including the celebrated large-N expansion and its variants [148].

7.2 One-Dimensional Models

For simplicity we begin our discussion with a simple one-dimensional model in the coordinate representation

$$\hat{H} = -\frac{\hbar^2}{2m}\frac{d^2}{dx^2} + V(x), \ -\infty < x < \infty \,, \tag{7.1}$$

where the potential-energy function $V(x)$ exhibits a single minimum V_e at $x = x_e$ and supports bound states. Except for these conditions, the potential-energy function is arbitrary.

It is convenient to work with a dimensionless Hamiltonian operator as in preceding chapters. To this end we define a dimensionless coordinate $q = x/\gamma$, where γ is an arbitrary length unit, a dimensionless energy $\epsilon = m\gamma^2 E/\hbar^2$ and a dimensionless potential-energy function $v(q) = m\gamma^2 V(\gamma q)/\hbar^2$. The dimensionless Hamiltonian operator is

$$\hat{\mathcal{H}} = \frac{m\gamma^2}{\hbar^2}\hat{H} = -\frac{1}{2}\frac{d^2}{dq^2} + v(q) \,. \tag{7.2}$$

7.2.1 Deep-Well Approximation

In order to apply the polynomial approximation we define a new dimensionless coordinate $z = (q - q_0)/\beta$, where β is an arbitrary parameter. On expanding $v(q)$ in a Taylor series around q_0

$$v(q) = \sum_{j=0}^{\infty} v_j (q - q_0)^j, \quad v_j = \frac{1}{j!} \frac{d^j v}{dq^j}\bigg|_{q=q_0} \tag{7.3}$$

the Hamiltonian operator becomes

$$\hat{\mathcal{H}} = \frac{1}{\beta^2} \left(-\frac{1}{2} \frac{d^2}{dz^2} + \beta^2 v_0 + \beta^3 v_1 z + \beta^4 v_2 z^2 + \sum_{j=3}^{\infty} v_j \beta^{j+2} z^j \right). \tag{7.4}$$

Notice that

$$v_j = \frac{m\gamma^{j+2}}{\hbar^2} \frac{1}{j!} \frac{d^j V}{dx^j}\bigg|_{x=x_0}, \quad x_0 = \gamma q_0. \tag{7.5}$$

If $v_2 > 0$ we can view equation (7.4) as a harmonic oscillator with a power-series perturbation. Although the exact eigenvalues are independent of q_0 and β, these parameters affect the rate of convergence of the perturbation series. Another degree of freedom that is relevant to the construction of the perturbation series is the way we collect and group the terms of the perturbation into polynomial contributions. We will discuss these points later; for the time being we keep the approach as straightforward and simple as possible, setting the adjustable parameters beforehand.

At first sight it seems most reasonable to choose q_0 to be the value of the coordinate at the minimum of the well so that the linear term of the Hamiltonian operator (7.4) vanishes because $v_1(q_0) = v'(q_0) = 0$. We also set

$$\beta = \left(\frac{1}{2v_2} \right)^{1/4} = \left(\frac{\hbar^2}{2m\gamma^4 V_2} \right)^{1/4}, \tag{7.6}$$

and define a new dimensionless Hamiltonian operator

$$\hat{h} = \beta^2 \left(\hat{\mathcal{H}} - v_0 \right) = -\frac{1}{2} \frac{d^2}{dz^2} + \frac{1}{2} z^2 + \sum_{j=1}^{\infty} b_j \beta^j z^{j+2}, \tag{7.7}$$

where β plays the role of a perturbation parameter and

$$b_j = \frac{v_{j+2}}{2v_2} = \frac{V_{j+2}\gamma^j}{2V_2}. \tag{7.8}$$

The eigenvalue equation $\hat{h}\Phi = e\Phi$ gives us the dimensionless energies

$$e = \frac{m\gamma^2 \beta^2 (E - V_0)}{\hbar^2}, \tag{7.9}$$

from which we easily recover the eigenvalues E of $\hat{\mathcal{H}}$. Notice that $\hbar^2/(m\gamma^2\beta^2) = \hbar\omega$, where

$$\omega = \sqrt{\frac{2V_2}{m}} \tag{7.10}$$

is the classical frequency of a harmonic oscillator with force constant $2V_2$.

The perturbation coefficients e_j of the dimensionless energy series

$$e = \sum_{j=0}^{\infty} e_j \beta^j \tag{7.11}$$

depend on the potential coefficients b_j and on the unperturbed energy

$$e_0 = v + \frac{1}{2}, \ v = 0, 1, \dots . \tag{7.12}$$

The net effect of the scaling transformation described in Appendix C that substitutes $-z$ for z is the substitution of $-\beta$ for β in the dimensionless Hamiltonian operator $\hat{h}$ as follows from $\beta^j(-z)^{j+2} = (-\beta)^j z^{j+2}$. Since the states of one-dimensional models are nondegenerate we conclude that $e(-\beta) = e(\beta)$, from which it follows that $e_{2j+1} = 0$. Therefore, the energy series for the eigenvalues of $\hat{\mathcal{H}}$ are of the form

$$\epsilon = v_0 + \frac{1}{\beta^2} \sum_{j=0}^{\infty} e_{2j} \beta^{2j} \tag{7.13}$$

and those for $\hat{H}$ read

$$E = V_0 + \hbar\omega \left(e_0 + e_2\beta^2 + e_4\beta^4 + \cdots \right) . \tag{7.14}$$

The first term in this equation is the classical energy of a particle at rest at the bottom of the well, the second term is the quantum-mechanical energy of a harmonic oscillation about this minimum, and the remaining contributions are anharmonic corrections to this oscillatory motion. In other words, the perturbation series (7.14) looks very much like successive quantum-mechanical corrections to a classical approach. This point of view is reinforced by the fact that the perturbation parameter β is proportional to $\hbar^{1/2}$ and decreases with the mass of the particle. However, this seemingly semiclassical approach differs markedly from the WKB method [149] in that the accuracy of the perturbation series (7.14) decreases with the vibrational quantum number.

The dimensionless potential coefficients b_j, and, consequently, the dimensionless energy corrections e_j, are invariant under the substitution of $CV(x)$ for $V(x)$. Only V_0, ω, and β depend on C in equation (7.14). As C increases, β decreases and the convergence properties of the perturbation series (7.14) improve. If $V(x)$ describes a finite potential well, then the depth of $CV(x)$ increases with C. For this reason, and that given earlier, we decided to call the present approach "deep-well approximation" instead of, say, "semiclassical expansion" in spite of the fact that in some cases the well may become increasingly shallower as the perturbation parameter decreases.

One easily calculates the dimensionless energy coefficients e_{2j} by means of any of the methods outlined in earlier chapters. For example, the application of the method of Swenson and Danforth discussed in Section 3.3 is straightforward and we choose it for the calculations described below because we are not interested in the eigenfunctions. The reader may easily derive the necessary equations following the lines indicated in Section 3.3 and write a simple Maple program by a straightforward modification of that one shown in the program section. Table 7.1 shows the first energy coefficients in terms of e_0 and the potential coefficients b_j.

As shown in Section 3.3, the method of Swenson and Danforth gives us perturbation series for the expectation values

$$Z_k = \left\langle z^k \right\rangle = \sum_{j=0}^{\infty} Z_{k,j} \beta^j . \tag{7.15}$$

Table 7.1 Energy Coefficients of the Deep-Well Expansion for an Arbitrary Potential-Energy Function

$$e_2 = -\frac{15}{4} e_0^2 b_1^2 - \frac{7}{16} b_1^2 + \frac{3}{8} b_2 + \frac{3}{2} b_2 e_0^2$$

$$e_4 = -\frac{705}{16} e_0^3 b_1^4 - \frac{1155}{64} e_0 b_1^4 + \frac{459}{16} e_0 b_1^2 b_2 + \frac{225}{4} b_1^2 b_2 e_0^3 - \frac{95}{8} b_1 e_0 b_3 - \frac{35}{2} b_1 b_3 e_0^3$$
$$\quad - \frac{67}{16} e_0 b_2^2 - \frac{17}{4} b_2^2 e_0^3 + \frac{25}{8} b_4 e_0 + \frac{5}{2} b_4 e_0^3$$

$$e_6 = \frac{7335}{16} b_1 b_2 e_0^2 b_3 + \frac{116325}{64} b_1^4 b_2 e_0^4 - \frac{23865}{32} b_1^3 e_0^2 b_3 - \frac{9765}{16} b_1^3 b_3 e_0^4$$
$$\quad - \frac{62013}{64} b_1^2 e_0^2 b_2^2 - \frac{24945}{32} b_1^2 b_2^2 e_0^4 + \frac{8535}{32} b_1^2 b_4 e_0^2 + \frac{2715}{16} b_1^2 b_4 e_0^4$$
$$\quad - \frac{1365}{16} b_1 b_5 e_0^2 - \frac{315}{8} b_1 b_5 e_0^4 + \frac{5667}{128} b_1 b_2 b_3 - \frac{885}{16} b_2 b_4 e_0^2 - \frac{165}{8} b_2 b_4 e_0^4$$
$$\quad - \frac{209055}{256} e_0^2 b_1^6 - \frac{115755}{128} e_0^4 b_1^6 + \frac{131817}{1024} b_1^4 b_2 - \frac{14777}{256} b_1^3 b_3 - \frac{40261}{512} b_1^2 b_2^2$$
$$\quad + \frac{6055}{256} b_1^2 b_4 - \frac{1155}{128} b_1 b_5 + \frac{1707}{32} e_0^2 b_2^3 + \frac{375}{16} b_2^3 e_0^4 - \frac{945}{128} b_2 b_4 - \frac{315}{16} b_3^2 e_0^4$$
$$\quad - \frac{1085}{32} b_3^2 e_0^2 - \frac{1107}{256} b_3^2 + \frac{245}{16} b_6 e_0^2 + \frac{35}{8} b_6 e_0^4 + \frac{315}{128} b_6 - \frac{101479}{2048} b_1^6$$
$$\quad + \frac{1539}{256} b_2^3 + \frac{2415}{8} b_1 b_2 b_3 e_0^4 + \frac{239985}{128} e_0^2 b_1^4 b_2$$

According to the scaling transformation discussed in Appendix C we have

$$\left\langle (-z)^k \right\rangle (\beta) = (-1)^k Z_k(\beta) = Z_k(-\beta) \tag{7.16}$$

so that $Z_{k,j} = 0$ if $j + k$ is odd. This property is a useful test for the program.

In the case of a parity-invariant potential-energy function $V(-x) = V(x)$ it is convenient to proceed in a slightly different way exploiting the fact that only even powers of x appear in the Taylor expansion of $V(x)$. We define $q = \sqrt{\beta} z$ and expand $v(q)$ around $q = 0$

$$v(q) = \sum_{j=0}^{\infty} v_j \beta^j z^{2j} , \tag{7.17}$$

where

$$v_j = \frac{1}{(2j)!} \frac{d^{2j} v}{dq^{2j}} \bigg|_{q=0} = \frac{m\gamma^{2j+2}}{\hbar^2} V_j, \quad V_j = \frac{1}{(2j)!} \frac{d^{2j} V}{dx^{2j}} \bigg|_{x=0} . \tag{7.18}$$

We apply perturbation theory to

$$\hat{h} = \beta(\hat{\mathcal{H}} - v_0) = -\frac{1}{2} \frac{d^2}{dz^2} + \frac{1}{2} z^2 + \sum_{j=1}^{\infty} b_j \beta^j z^{2j+2} , \tag{7.19}$$

where $b_j = v_{j+1}/(2v_1)$. From the eigenvalues of $\hat{\mathcal{H}}$

$$\epsilon = v_0 + \frac{1}{\beta} \sum_{j=0}^{\infty} e_j \beta^j , \tag{7.20}$$

we obtain the energies

$$E = V_0 + \hbar\omega \sum_{j=0}^{\infty} e_j \beta^j \tag{7.21}$$

exactly as in the preceding case. Table 7.2 shows the first perturbation corrections e_j in terms of e_0 and the dimensionless potential coefficients b_j.

Table 7.2 Energy Coefficients of the Deep-Well Expansion for a Parity-Invariant Potential-Energy Function

$e_1 = \frac{3}{8} b_1 (1 + 4 e_0^2)$

$e_2 = -\frac{67}{16} e_0 b_1^2 - \frac{17}{4} e_0^3 b_1^2 + \frac{25}{8} e_0 b_2 + \frac{5}{2} b_2 e_0^3$

$e_3 = \frac{1707}{32} e_0^2 b_1^3 + \frac{375}{16} e_0^4 b_1^3 - \frac{885}{16} b_1 b_2 e_0^2 - \frac{165}{8} b_1 b_2 e_0^4 + \frac{1539}{256} b_1^3 - \frac{945}{128} b_1 b_2$

$\quad + \frac{315}{128} b_3 + \frac{245}{16} b_3 e_0^2 + \frac{35}{8} b_3 e_0^4$

$e_4 = -\frac{89165}{128} e_0^3 b_1^4 - \frac{10521}{64} b_1 e_0 b_3 - \frac{2205}{8} b_1 b_3 e_0^3 + \frac{29555}{32} b_1^2 b_2 e_0^3 + \frac{117281}{256} e_0 b_1^2 b_2$

$\quad - \frac{305141}{1024} e_0 b_1^4 - \frac{10689}{64} e_0^5 b_1^4 - \frac{189}{4} b_1 b_3 e_0^5 + \frac{3129}{16} b_1^2 b_2 e_0^5 - \frac{19277}{256} e_0 b_2^2$

$\quad - \frac{4145}{32} b_2^2 e_0^3 - \frac{393}{16} b_2^2 e_0^5 + \frac{5607}{128} b_4 e_0 + \frac{945}{16} b_4 e_0^3 + \frac{63}{8} b_4 e_0^5$

As a first illustrative example we consider the Morse potential-energy function [150]

$$V(x) = D \left[1 - \exp(-\alpha x) \right]^2 , \tag{7.22}$$

where $D > 0$ is the depth of the potential well and α determines the range of the interaction (greater α shorter range and vice versa). This anharmonic oscillator proves to be a simple two-parameter model for the study of vibrational properties of diatomic molecules. Figure 7.1 shows the shape of $V(x)$ for two values of D and α. The Schrödinger equation for this model is exactly solvable when $-\infty < x < \infty$, and the energies are given by [150]

$$E = \hbar\omega \left(e_0 - \frac{\hbar\omega e_0^2}{4D} \right) , \tag{7.23}$$

where

$$\omega = \sqrt{\frac{2D}{m}} \alpha . \tag{7.24}$$

Choosing $\gamma = 1/\alpha$ the dimensionless potential-energy function turns out to be $v(q) = B(1 - e^{-q})^2$, where $B = mD/(\hbar^2\alpha^2)$, and the energy is $E = D\epsilon(B)/B$. According to equation (7.6) the perturbation parameter is $\beta = (2B)^{-1/4}$. Moreover, since $V_2 = D\alpha^2$ the frequency (7.10) appearing in the perturbation expansion (7.14) agrees with equation (7.24). A straightforward calculation shows that all the perturbation corrections to the dimensionless energy vanish except e_0 and

$$e_2 = -\frac{1}{2} e_0^2 . \tag{7.25}$$

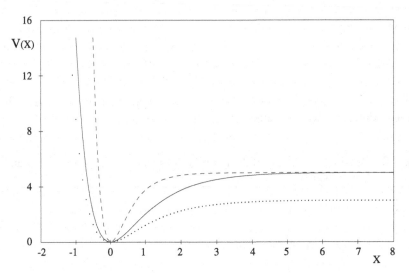

FIGURE 7.1
Morse potential-energy function for $D = 5, \alpha = 1$ **(continuous line),** $D = 5, \alpha = 2$ **(broken line), and** $D = 3, \alpha = 1$ **(points).**

Therefore, the resulting perturbation series $E = \hbar\omega(e_0 + e_2\beta^2)$ gives the exact result (7.23) because $\beta^2 = \hbar\omega/(2D)$. Notice that the energies of the Morse oscillator approach those of the harmonic oscillator as the well depth increases because the perturbation parameter β is proportional to $D^{-1/4}$.

One may think that perturbation theory also gives the exact eigenfunctions, but it is not the case because the perturbation series for them do not terminate. In order to illustrate this point Table 7.3 shows the first coefficients of the perturbation series for the expectation values $\langle z \rangle$ and $\langle z^2 \rangle$. None of these series terminate although some coefficients vanish as discussed above.

Another interesting exactly solvable model is given by the potential-energy function

$$V(x) = -\frac{A}{\cosh(x/\alpha)^2}, \ A, \alpha > 0 \tag{7.26}$$

shown in Figure 7.2 for two values of A and α. The exact eigenvalues are given by [151]

$$E = -4A\left(\frac{1}{2}\sqrt{1 + u^2} - e_0 u\right)^2, \ u = \frac{\hbar}{2\alpha}\sqrt{\frac{1}{2mA}} . \tag{7.27}$$

In this case we choose $\gamma = \alpha$ so that $v(q) = -B/\cosh(q)^2$, where $B = m\alpha^2 A/\hbar^2$, and $E = A\epsilon(B)/B$. According to the general equations given above for a parity-invariant potential-energy function, the perturbation parameter is $\beta = 1/\sqrt{2B} = 2u$ and perturbation theory yields

$$E = -A + A\sqrt{\frac{2}{B}} \sum_{j=0}^{\infty} e_j(2B)^{-j/2} . \tag{7.28}$$

It is not difficult to verify that the perturbation series (7.28) agrees with the Taylor expansion of the exact energy about $u = 0$. Table 7.4 shows perturbation coefficients obtained by means of the method of Swenson and Danforth.

The function $\sqrt{1 + z^2}$ exhibits a pair of complex-conjugate square-root branch points at $z = \pm i$. Consequently, the Taylor expansion of the exact energy (7.27) about $u = 0$ converges for $u < 1$, and

Table 7.3 Perturbation Corrections to the Expectation Values $\langle z \rangle$ and $\langle z^2 \rangle$ of the Morse Oscillator

$$Z_{1,0} = 0$$

$$Z_{1,1} = \tfrac{3}{2}\, e_0$$

$$Z_{1,2} = 0$$

$$Z_{1,3} = \tfrac{5}{96} + \tfrac{7}{8}\, e_0{}^2$$

$$Z_{1,4} = 0$$

$$Z_{1,5} = \tfrac{3}{32}\, e_0 + \tfrac{5}{8}\, e_0{}^3$$

$$Z_{2,0} = e_0$$

$$Z_{2,1} = 0$$

$$Z_{2,2} = \tfrac{7}{32} + \tfrac{23}{8}\, e_0{}^2$$

$$Z_{2,3} = 0$$

$$Z_{2,4} = \tfrac{43}{72}\, e_0 + \tfrac{109}{36}\, e_0{}^3$$

$$Z_{2,5} = 0$$

Table 7.4 Perturbation Coefficients for the Energies Supported by the Potential-Energy Function $V(x) = -\dfrac{A}{\cosh(\frac{x}{\alpha})^2}$

$$e_1 = -\tfrac{1}{8} - \tfrac{e_0{}^2}{2}$$

$$e_2 = \tfrac{e_0}{8}$$

$$e_3 = 0$$

$$e_4 = -\tfrac{e_0}{128}$$

$$e_5 = 0$$

$$e_6 = \tfrac{e_0}{1024}$$

the perturbation series (7.20) for $\beta < 2$. Surprisingly, this radius of convergence is large enough for the calculation of critical constants; that is to say, particular values of β such that $E = 0$. It follows from the exact expression for the energy (7.27) that the critical constants are given by

$$u_v = \frac{\beta_v}{2} = \sqrt{\frac{1}{(2v + 1)^2 - 1}}, \quad v = 1, 2, \ldots . \tag{7.29}$$

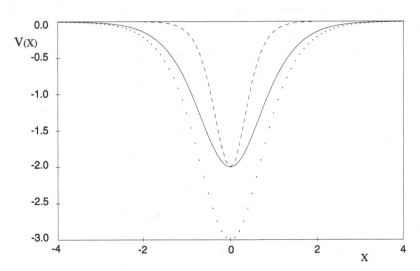

FIGURE 7.2
Potential-energy function $V(x) = -A/\cosh(x/\alpha)^2$ for $A = 2, \alpha = 1$ (continuous line), $A = 2$,
$\alpha = 1/2$ **(broken line), and $A = 3, \alpha = 1$ (points).**

Since $\beta_\nu < 2$ for all $\nu \geq 1$ we can calculate them by means of the perturbation series. However, taking into account that each root of $E(\beta) = 0$ is double, it is advisable to look for a root of $dE/d\beta = 0$ instead.

Table 7.5 shows that a positive root of

$$\frac{d}{d\beta}\left(2\beta \sum_{j=0}^{N} e_j \beta^j - 1\right) = 0, \; N = 2, 3, \ldots \tag{7.30}$$

converges towards the exact value $\beta_1 = 1/\sqrt{2}$ when $e_0 = 3/2$. This is the most unfavorable case because β_ν decreases with ν.

The radius of convergence of the deep-well series for the model (7.26) is unusually large. We have applied the method to a Gaussian well of the form

$$V(x) = -A \exp\left(-\alpha x^2\right), \; A, \alpha > 0, \tag{7.31}$$

obtaining poorer results for the energies and failing to estimate the critical constants. However, in principle the deep-well approximation can be improved by a more judicious choice of the arbitrary parameters. We will discuss this point later in this chapter.

7.2.2 Weak Attractive Interactions

The reader may wonder why we obtained critical constants of the model (7.26) with $\nu = 1, 2, \ldots$, but not with $\nu = 0$. The reason is that such a critical constant does not exist because there is a ground state with negative energy for all values of $B > 0$. In fact, the Taylor series for $\epsilon(B)$ about $B = 0$ clearly shows that the dimensionless ground-state energy approaches zero from below as B tends to zero:

$$\epsilon(B) = -2B^2 + 8B^3 - 40B^4 + 224B^5 + \cdots. \tag{7.32}$$

Table 7.5 Critical
Constant β_1 Obtained
from the Perturbation
Series for the
Hyperbolic Well

N	β_1
2	0.7150377815
3	0.7150377815
4	0.7064548587
5	0.7064548587
6	0.7071722752
7	0.7071722752
8	0.7070998079
9	0.7070998079
10	0.7071075521
11	0.7071075521
12	0.7071066940
13	0.7071066940
14	0.7071067912
15	0.7071067912
16	0.7071067800
17	0.7071067800
18	0.7071067813
19	0.7071067813
20	0.7071067812
Exact	0.70710678120

It is not possible to obtain this expansion by means of a polynomial approximation, and we have to resort to a different form of perturbation theory that we show in what follows for the sake of completeness.

It is our purpose to outline a method that produces perturbation expansions for weak attractive interactions. Consider the dimensionless Schrödinger equation

$$\Phi''(q) + 2\epsilon\Phi(q) = 2\lambda v(q)\Phi(q) . \tag{7.33}$$

It has been proved that if

$$\int_{-\infty}^{\infty} \left(1 + q^2\right) |v(q)| \, dq < \infty , \tag{7.34}$$

then there is a bound state for all small positive λ if and only if

$$\int_{-\infty}^{\infty} v(q) \, dq \le 0 . \tag{7.35}$$

Moreover, if

$$\int_{-\infty}^{\infty} \exp(\alpha|q|)|v(q)| \, dq < \infty \tag{7.36}$$

for some $\alpha > 0$ then the energy $\epsilon(\lambda)$ is analytic at $\lambda = 0$ [152].

Taking into account the asymptotic behavior of the square-integrable eigenfunctions

$$\Phi(q) \to \begin{cases} C_L \exp(kq), & q \to -\infty \\ C_R \exp(-kq), & q \to \infty \end{cases}, \quad k = \sqrt{-2\epsilon} \tag{7.37}$$

we conclude that

$$\lim_{|q| \to \infty} \Phi'(q) = 0 \tag{7.38}$$

is the appropriate boundary condition when $\epsilon = 0$. Since $\epsilon \to 0^-$ when $\lambda \to 0^+$ the general solution of the Schrödinger equation in this limit is $\Phi_0(q) = c_{00} + c_{01}q$ which satisfies equation (7.38) provided that $c_{01} = 0$. Without loss of generality we choose $c_{00} = 1$.

In order to obtain an appropriate expression for the solution of the Schrödinger equation we consider the general integration formulas developed in Appendix B. Viewing equation (7.33) as an ordinary differential equation with an inhomogeneous term $f(q) = 2\lambda v(q)\Phi(q)$ we have

$$\begin{aligned}
\Phi(q) = \ & c_1 \exp(-kq) + c_2 \exp(kq) \\
& + \frac{\lambda}{k} \int_{q_i}^{q} \left\{ \exp\left[k\left(q - q'\right)\right] - \exp\left[-k\left(q - q'\right)\right] \right\} v\left(q'\right) \Phi\left(q'\right) dq',
\end{aligned} \tag{7.39}$$

where q_i is an arbitrary coordinate point. When $q \to -\infty$ we require that the coefficient of $\exp(-kq)$ vanishes, and obtain

$$c_1 = \frac{\lambda}{k} \int_{q_i}^{-\infty} \exp\left(kq'\right) v\left(q'\right) \Phi\left(q'\right) dq'; \tag{7.40}$$

analogously, when $q \to \infty$ we have

$$c_2 = -\frac{\lambda}{k} \int_{q_i}^{\infty} \exp\left(-kq'\right) v\left(q'\right) \Phi\left(q'\right) dq'. \tag{7.41}$$

Substituting equations (7.40) and (7.41) into (7.39) and rearranging the result we finally obtain

$$\Phi(q) = -\frac{\lambda}{k} \int_{-\infty}^{\infty} \exp\left(-k\left|q - q'\right|\right) v\left(q'\right) \Phi\left(q'\right) dq'. \tag{7.42}$$

In particular, when $q = 0$ we have an appropriate expression for the energy:

$$\frac{k}{\lambda}\Phi(0) = -\int_{-\infty}^{\infty} \exp\left(-k\left|q'\right|\right) v\left(q'\right) \Phi\left(q'\right) dq'. \tag{7.43}$$

We easily obtain the perturbation series for the eigenfunction and energy from equations (7.42) and (7.43), respectively. We solve equation (7.42) iteratively:

$$\Phi_j(q) = -\frac{\lambda}{k} \int_{-\infty}^{\infty} \exp\left(-k\left|q - q'\right|\right) v\left(q'\right) \Phi_{j-1}\left(q'\right) dq', \tag{7.44}$$

where $j = 1, 2, \ldots$ and $\Phi_0(q) \equiv 1$ which is an appropriate starting point for sufficiently small λ. At each iteration step we expand

$$\Phi_j(q) = \sum_{i=0}^{j} \Phi_{j,i}(q)\lambda^i + \cdots, \tag{7.45}$$

$$k = \sum_{i=1}^{j} k_i \lambda^i + \cdots, \tag{7.46}$$

where we explicitly indicate those perturbation coefficients calculated previously. Notice that equation (7.46) takes into account that $k = 0$ when $\lambda = 0$. We then substitute these series into equation (7.43), expand both sides to order j, and solve for the next coefficient k_{j+1}.

For example, substituting $\Phi(q) = \Phi_0(q)$ into equation (7.43) and expanding both sides to λ^0 we obtain

$$k_1 = -\int_{-\infty}^{\infty} v\left(q'\right) dq' . \tag{7.47}$$

The first iteration of equation (7.42) yields

$$
\begin{aligned}
\Phi_1(q) &= -\frac{\lambda}{k} \int_{-\infty}^{\infty} \exp\left(-k\left|q - q'\right|\right) v\left(q'\right) dq' \\
&= 1 + \lambda \left[\int_{-\infty}^{\infty} \left|q - q'\right| v\left(q'\right) dq' - \frac{k_2}{k_1} \right] + \cdots .
\end{aligned}
\tag{7.48}
$$

From the expansions of the left- and right-hand sides of equation (7.43) to order λ, respectively,

$$
\begin{aligned}
\frac{k}{\lambda}\Phi_1(0) &= k_1 + k_1 \lambda \int_{-\infty}^{\infty} \left|q'\right| v\left(q'\right) dq' \\
&\quad - \int_{-\infty}^{\infty} \exp\left(-k\left|q'\right|\right) v\left(q'\right) \Phi_1\left(q'\right) dq' + \cdots \\
&= k_1 + k_1 \lambda \int_{-\infty}^{\infty} \left|q'\right| v\left(q'\right) dq' \\
&\quad - k_2 \lambda - \lambda \int_{-\infty}^{\infty} \int_{-\infty}^{\infty} \left|q - q'\right| v(q) v\left(q'\right) dq\, dq'
\end{aligned}
\tag{7.49}
$$
$$
\tag{7.50}
$$

we conclude that

$$k_2 = -\int_{-\infty}^{\infty} \int_{-\infty}^{\infty} \left|q - q'\right| v(q) v\left(q'\right) dq\, dq' . \tag{7.51}$$

We obtain contributions of higher order exactly in the same way. However, we leave the discussion at this point because the procedure soon becomes tedious. The perturbation expansion for weakly attractive potentials has received some attention [152]–[155] and a few more energy terms are already available [154].

We easily derive weak- and strong-coupling expansions for the Gaussian well $v(q) = -\exp(-q^2)$:

$$\epsilon(\lambda) = -\pi\lambda^2 + 8.885765874\lambda^3 + \cdots , \tag{7.52}$$

$$\epsilon(\lambda) = -\lambda + \frac{\sqrt{2\lambda}}{2} - \frac{3}{16} - \frac{\sqrt{2}}{256\sqrt{\lambda}} + \frac{7}{2048\lambda} + \cdots , \tag{7.53}$$

respectively, by means of the method just described and the deep-well approach given earlier. We have obtained analytical and numerical expressions of k_2 by means of Maple, but we only show the latter here.

7.3 Central-Field Models

In what follows we consider a particle of mass m under the effect of a spherically symmetric potential $V(r)$. Arguing as in Section 3.3.2 we first separate the Schrödinger equation in spherical coordinates, then define dimensionless coordinate $q = r/\gamma$, potential-energy function $v(q) = m\gamma^2 V(\gamma q)/\hbar^2$, and energy $\epsilon = m\gamma^2 E/\hbar^2$ in the usual way, and finally make the radial part look like a one-dimensional model:

$$\hat{\mathcal{H}}\Phi(q) = \epsilon\Phi(q), \ \hat{\mathcal{H}} = -\frac{1}{2}\frac{d^2}{dq^2} + u(q), \ u(q) = \frac{l(l+1)}{2q^2} + v(q) \tag{7.54}$$

in order to apply the method of the preceding section. As usual, $l = 0, 1, \ldots$ is the angular momentum quantum number. In order to apply the polynomial approximation we expand the effective potential-energy function $u(q)$ about a coordinate point q_0:

$$u(q) = \sum_{j=0}^{\infty} u_j (q - q_0)^j \tag{7.55}$$

The variable domain $0 \leq q < \infty$ and the boundary condition at origin $\Phi(0) = 0$ do not pose a problem for the application of the method developed for one-dimensional models because the change of variable $z = (q - q_0)/\beta$ maps $q = 0$ onto $z_0 = -q_0/\beta$ which tends to $-\infty$ as $\beta \to 0$. In this case it is convenient to select the perturbation parameter $\lambda = \beta/q_0$, where q_0 is the minimum of $u(q)$ and, thereby, a root of

$$q_0^3 v'(q_0) = l(l+1) . \tag{7.56}$$

Notice that a perturbation expansion about $\lambda = 0$ is consistent with the preceding discussion on the left boundary condition, and that we can choose $\beta = 1/(2u_2)^{1/4}$ as in the nonsymmetric one-dimensional model.

The operator $\hat{h}$ is given by equation (7.7) except that

$$b_j = \frac{u_{j+2}q_0^j}{2u_2} , \tag{7.57}$$

and the series for the eigenvalues of $\hat{\mathcal{H}}$ read

$$\epsilon = u_0 + \frac{1}{\beta^2}\sum_{j=0}^{\infty} e_{2j}\lambda^{2j} . \tag{7.58}$$

When q_0 is the minimum of $u(q)$ as in equation (7.56), the resulting polynomial approximation proves suitable for the treatment of Lennard–Jones potentials by means of perturbation theory [156].

The calculation of analytical perturbation coefficients e_j through the methods discussed in Chapters 2 and 3 is straightforward, specially if one resorts to Maple. The first three nonzero corrections are exactly those in Table 7.1 provided that the potential coefficients b_j are given by equation (7.57).

We have already seen that the application of approximate methods to exactly solvable models is most instructive. Here we choose the Kratzer oscillator [157, 158]

$$V(r) = -\frac{C_1}{r} + \frac{C_2}{r^2}, \ C_1, C_2 > 0 , \tag{7.59}$$

which resembles the potential-energy function of a diatomic molecule. Choosing the length unit $\gamma = \hbar^2/(mC_1)$, and defining $B = mC_2/h^2$, we obtain

$$u(q) = -\frac{1}{q} + \frac{l(l+1)+2B}{2q^2} . \tag{7.60}$$

One can solve the Schrödinger equation for the Kratzer oscillator in many different ways [157, 158]. To facilitate the discussion below we define an effective real quantum number L satisfying $L(L+1) = l(l+1) + 2B$, so that we can resort to the solutions of the hydrogen atom. It is well known that every eigenfunction behaves as $\Phi(q) \approx q^{L+1}$ sufficiently close to origin. Because $L + 1$ has to be positive for all $l \geq 0$, the only acceptable root is

$$L = -\frac{1}{2} + \sqrt{(l + 1/2)^2 + 2B} \ . \tag{7.61}$$

The eigenvalues of the Kratzer oscillator are given by [157, 158]

$$\epsilon = -\frac{1}{2(\nu + L + 1)^2} \ , \tag{7.62}$$

where $\nu = 0, 1, \dots$ is the radial quantum number.

The parameters of the deep-well approximation for this simple model are $q_0 = L(L+1)$, $\beta = q_0^{3/4}$, and $\lambda = q_0^{-1/4}$; therefore, the accuracy of the series is expected to increase with L. Taking into account that $u_0 = -1/(2q_0)$ we obtain

$$\epsilon = -\frac{\lambda^4}{2} + \lambda^6 \sum_{j=0}^{\infty} e_{2j} \lambda^{2j} \ , \tag{7.63}$$

where the coefficients e_{2j} are given in Table 7.1 with

$$b_j = \frac{(-1)^j (j + 1)}{2} \ . \tag{7.64}$$

Figure 7.3 shows the effective potential-energy function $u(q)$ for three values of $L(L + 1)$. Notice that the well becomes shallower as L increases so that the name "deep-well approximation" is not the most appropriate in this case.

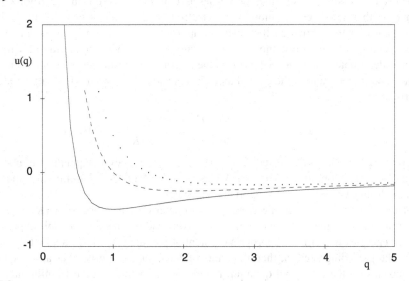

FIGURE 7.3
Effective potential-energy function $u(q) = -1/q + L(L+1)/(2q^2)$ for $L(L+1) = 1$ (continuous line), 2 (broken line), and 3 (points).

In order to obtain the perturbation expansion (7.63) from the exact result (7.62) we define the parameter $\xi = 1/(L + 1)$ and solve $L(L + 1) = \lambda^{-4} = (\xi^{-1} - 1)\xi^{-1}$ for ξ:

$$\xi = \frac{2\lambda^2}{\lambda^2 + \sqrt{4 + \lambda^4}} . \tag{7.65}$$

Substituting this result into

$$\epsilon = -\frac{\xi^2}{2(1 + \nu\xi)^2} \tag{7.66}$$

we obtain ϵ as a function of λ. Its Taylor expansion about $\lambda = 0$ gives us exactly the perturbation series (7.63) in terms of $e_0 = \nu + 1/2$.

This example allows us to study the convergence properties of the deep-well approximation. The exact dimensionless energy ϵ exhibits a pole at $L_s = -\nu - 1$ [see equation (7.66)], and square-root branch points at $\lambda_s^2 = \pm 2i$ [see equation (7.65)]; therefore, we expect the perturbation series converge for all $\lambda^2 < 1/\sqrt{\nu(\nu + 1)}$ if $\nu > 0$, and for all $\lambda^2 < 2$ when $\nu = 0$. It is clear that the radius of convergence decreases with the radial (vibrational) quantum number ν. Fortunately, there is considerable room for improving the convergence properties of the perturbation series. For example, if we choose ξ to be the perturbation parameter, then the radius of convergence of the perturbation series is $1/\nu > 1/\sqrt{\nu(\nu + 1)}$ if $\nu > 0$, and we obtain the exact result when $\nu = 0$. Moreover, if $C_2 = 0$ in the Kratzer oscillator (7.59), then the expressions given above are unsuitable to treat s states because $L = l = 0$ and the perturbation parameter λ is undefined. In such a case the variable ξ is certainly more convenient. Even better choices are possible as discussed later in this chapter.

7.4   Vibration-Rotational Spectra of Diatomic Molecules

The theoretical study of molecular properties is commonly based on the Born–Oppenheimer approximation that separates the motions of electrons and nuclei because of their considerably different masses [159]. Provided that such an approach is valid, one can model the vibration-rotational spectrum of a diatomic molecule by means of a Schrödinger equation for the motion of the nuclei under a potential $V(R)$, where R is the internuclear distance. This equation is separable in spherical coordinates, and in the case of an electronic state $^1\Sigma^+$ we are left with a radial equation of the form [160]

$$\hat{H}\Phi = E\Phi, \quad \hat{H} = -\frac{\hbar^2}{2m}\frac{d^2}{dR^2} + \frac{\hbar^2 J(J + 1)}{2mR^2} + V(R) , \tag{7.67}$$

where m is the reduced mass of the nuclei, $J = 0, 1, \ldots$ is the rotational quantum number, and $\Phi(R)$ satisfies the boundary condition $\Phi(0) = 0$. We assume that $V(R)$ has a minimum at the equilibrium internuclear distance $R = R_e$, and supports bound states.

One of the aims of molecular spectroscopy is to determine the form of $V(R)$ as accurately as possible from the vibration-rotational spectrum. To this end one needs a suitable expression for the bound-state energies in terms of appropriate potential parameters. If we applied the perturbation method of the preceding section, then the value of $V(R)$ and its derivatives at a coordinate point R_J that depends on the rotational quantum number would appear in the Hamiltonian operator $\hat{h}$. Because spectroscopists are more interested in what they call molecular parameters at equilibrium (that is to say, values of $V(R)$ and its derivatives at R_e), we then resort to a different approach in what follows.

The molecular parameters at equilibrium appear naturally in the perturbation method if we expand the potential-energy function in a Taylor series about R_e:

$$V(R) = \sum_{j=0}^{\infty} V_j (R - R_e)^j \ . \tag{7.68}$$

In order to facilitate the calculation we define dimensionless coordinate $q = (R - R_e)/(\lambda R_e)$, potential-energy function $v(q) = m R_e^2 \lambda^2 V(R)/\hbar^2$, and energy $\epsilon = m R_e^2 \lambda^2 E/\hbar^2$, where

$$\lambda = \left(\frac{\hbar^2}{2m R_e^4 V_2} \right)^{1/4} \tag{7.69}$$

is a dimensionless parameter. We can rewrite λ as

$$\lambda = \sqrt{\frac{2B}{\hbar\omega}} \ , \tag{7.70}$$

where the rotational constant B and the oscillator frequency ω are, respectively, given by

$$B = \frac{\hbar^2}{2m R_e^2}, \quad \omega = \sqrt{\frac{2V_2}{m}} \ . \tag{7.71}$$

For most molecules λ is sufficiently small to be chosen as perturbation parameter.

If we substitute the definitions above into the Taylor expansion of the dimensionless potential-energy function we finally obtain

$$v(q) = v_0 + \frac{1}{2}q^2 + \sum_{j=1}^{\infty} a_j \lambda^j q^{j+2}, \quad v_0 = \frac{V_0}{\hbar\omega}, \quad a_j = \frac{V_{j+2} R_e^j}{2V_2} \ . \tag{7.72}$$

We apply perturbation theory to the eigenvalue equation $\hat{h}\Phi = e\Phi$ for the dimensionless Hamiltonian operator

$$\hat{h} = \frac{1}{\hbar\omega}\left(\hat{H} - V_0 \right) - J_B = -\frac{1}{2}\frac{d^2}{dq^2} + \frac{1}{2}q^2 + \sum_{j=1}^{\infty} \lambda^j p_j(q) \ , \tag{7.73}$$

where

$$J_B = \frac{J(J+1)\lambda^2}{2} = \frac{J(J+1)B}{\hbar\omega} \ , \tag{7.74}$$

and

$$p_j(q) = (-1)^j (j+1)J_B q^j + a_j q^{j+2} \ . \tag{7.75}$$

The first term in the right-hand side of this equation comes from the Taylor expansion of the centrifugal part of the radial Schrödinger equation (7.67). This example shows a different way of grouping the anharmonic terms into perturbation contributions. We mentioned earlier that this practice is an additional degree of freedom in the construction of perturbation series by means of polynomial approximations, and we will come back to it later in this chapter.

Straightforward application of perturbation theory to $\hat{h}\Phi = e\Phi$ gives us the perturbation series $e = e_0 + e_2\lambda^2 + \cdots$, where $e_0 = v + 1/2$, $v = 0, 1, \ldots$, and $e_{2j+1} = 0$, as argued before. The perturbation series for the dimensionless energy $\epsilon = E/(\hbar\omega)$ reads

$$\epsilon = v_0 + J_B + \sum_{j=0}^{\infty} e_{2j}\lambda^{2j} . \tag{7.76}$$

Table 7.6 shows e_2 and e_4 obtained by means of the method of Swenson and Danforth discussed in Section 3.3. We do not provide more corrections because their length increases considerably with the perturbation order and because one easily obtains as many of them as desired by means of a set of simple Maple procedures similar to those shown in the program section for anharmonic oscillators. Many energy coefficients were obtained before in order to study molecular spectra [160].

Table 7.6 Perturbation Corrections to the Vibration-Rotational Energies of Diatomic Molecules

$$e_2 = -2 J_B{}^2 + (6 a_1 + 3) e_0 J_B + \left(-\tfrac{15}{4} a_1{}^2 + \tfrac{3}{2} a_2\right) e_0{}^2 - \tfrac{7}{16} a_1{}^2 + \tfrac{3}{8} a_2$$

$$\begin{aligned}
e_4 = \;& (12 + 8 a_1) J_B{}^3 + \left(-54 a_1 - \tfrac{57}{2} - 54 a_1{}^2 + 24 a_2\right) e_0 J_B{}^2 \\
& + \left[\left(90 a_1{}^3 - 9 a_2 + 15 a_3 + 30 a_1 - 78 a_1 a_2 + 45 a_1{}^2 + \tfrac{15}{2}\right) e_0{}^2 + \tfrac{15}{8} - \tfrac{9}{4} a_2 + \tfrac{21}{4} a_1{}^2 \right. \\
& \left. + \tfrac{15}{4} a_3 + \tfrac{21}{2} a_1{}^3 + \tfrac{7}{2} a_1 - \tfrac{23}{2} a_1 a_2 \right] J_B \\
& + \left(-\tfrac{35}{2} a_1 a_3 + \tfrac{225}{4} a_1{}^2 a_2 - \tfrac{705}{16} a_1{}^4 - \tfrac{17}{4} a_2{}^2 + \tfrac{5}{2} a_4\right) e_0{}^3 \\
& + \left(\tfrac{25}{8} a_4 - \tfrac{95}{8} a_1 a_3 + \tfrac{459}{16} a_1{}^2 a_2 - \tfrac{67}{16} a_2{}^2 - \tfrac{1155}{64} a_1{}^4\right) e_0
\end{aligned}$$

The first three terms $v_0 + J_B + e_0$ in equation (7.76) give the electronic, rotational, and vibrational energies (in units of $\hbar\omega$) according to the simplest model of a rigid rotor and a harmonic oscillator. The remaining polynomial function of e_0 and J_B accounts for anharmonic effects, vibration-rotation coupling, and centrifugal stretching [160].

In order to compare present perturbation expansion with those in the preceding section we consider the same exactly solvable model studied earlier: the Kratzer oscillator, which for convenience we write as

$$V(R) = D\left(-2\frac{R_e}{R} + \frac{R_e^2}{R^2}\right) . \tag{7.77}$$

The exact dimensionless energies are

$$\epsilon = -\frac{\lambda^2 C^2}{2(v + L + 1)^2} , \tag{7.78}$$

where $C = 2m R_e^2 D/\hbar^2$ and

$$L = \sqrt{(J + 1/2)^2 + C} - \frac{1}{2} . \tag{7.79}$$

The reader may easily verify that the perturbation parameter is $\lambda = C^{-1/4}$.

There are two types of singular points in the complex C plane: a pole of order two at $C_{s,1} = v(v+1) - J(J+1)$ and a branch point at $C_{s,2} = -(J+1/2)^2$. Comparing $|C_{s,1}|$ and $|C_{s,2}|$ we conclude that the radius of convergence of the λ^2-power series is $1/(J+1/2)$ if $v < \sqrt{2}(J+1/2)-1/2$ and $1/\sqrt{v(v+1) - J(J+1)}$ otherwise.

Taking into account that $v_0 = -1/(2\lambda^2)$ we obtain the perturbation series

$$\epsilon = -\frac{1}{2\lambda^2} + J_B + \sum_{j=0}^{\infty} e_{2j}\lambda^{2j} . \tag{7.80}$$

One easily derives the coefficients e_j of the Kratzer oscillator by substitution of the appropriate potential parameters a_j into the general expressions in Table 7.6 (Maple facilitates it). In order to test the results of perturbation theory we expand the exact expression

$$\lambda^2 \epsilon = -\frac{1}{2\left(e_0\lambda^2 + \sqrt{1+2J_B\lambda^2 + \lambda^4/4}\right)^2} \tag{7.81}$$

in a Taylor series about $\lambda = 0$ and compare the coefficients.

7.5 Large-N Expansion

The most popular polynomial approximation is the large-N expansion which consists of expanding the eigenvalues of the Schrödinger equation for N dimensions in powers of $1/N$ (or a related variable) and then substituting the required value of N [148]. In order to illustrate this approach we consider the Schrödinger equation for a central-field model which is separable in hyperspherical coordinates. After removal of the $N - 1$ angular variables and appropriate transformation of the resulting radial equation we are left with a dimensionless eigenvalue problem of the form $\hat{H}\Phi = E\Phi$, where [148]

$$\hat{H} = -\frac{1}{2}\frac{d^2}{dr^2} + u(r), \ u(r) = \frac{(k-1)(k-3)}{8r^2} + V(r) , \tag{7.82}$$

$k = N + 2l$, and $l = 0, 1, \ldots$ is the angular momentum quantum number. The boundary condition at origin is $\Phi(0) = 0$.

Following the polynomial approximation discussed above we define a new variable $q = (r - r_0)/(\beta r_0)$, where β and r_0 are to be determined, and consider

$$\beta^2 r_0^2 \hat{H} = -\frac{1}{2}\frac{d^2}{dq^2} + \frac{(k-1)(k-3)\beta^2}{8(1+\beta q)^2} + \beta^2 r_0^2 V(r) . \tag{7.83}$$

We write the Taylor expansion of the potential-energy function as follows

$$\begin{aligned}
\beta^2 r_0^2 V(r) &= \beta^2 r_0^2 \sum_{j=0}^{\infty} V_j (r-r_0)^j \\
&= \beta^2 r_0^2 V_0 + \beta^3 r_0^3 V_1 q + \beta^4 r_0^4 V_2 q^2 + \sum_{j=1}^{\infty} V_{j+2} (r_0\beta)^{j+4} q^{j+2} . \tag{7.84}
\end{aligned}$$

If k is sufficiently large the terms proportional to k^2 dominate in the centrifugal term. Therefore we write its Taylor expansion as follows:

$$
\begin{aligned}
\beta^2 r_0^2 \frac{(k-1)(k-3)}{8r^2} &= \frac{(k-1)(k-3)\beta^2}{8} \sum_{j=0}^{\infty} (-1)^j (j+1)(\beta q)^j \\
&= \frac{(k-1)(k-3)\beta^2}{8} - \frac{k^2\beta^3 q}{4} + \frac{3k^2\beta^4 q^2}{8} + \frac{k^2}{8} \sum_{j=1}^{\infty} (-1)^j (j+3)\beta^{j+4} q^{j+2} \\
&\quad + \frac{(3-4k)\beta^2}{8} \sum_{j=1}^{\infty} (-1)(j+1)\beta^j q^j .
\end{aligned}
\tag{7.85}
$$

We choose r_0 in such a way that the coefficient of q in the potential plus centrifugal term is zero to the leading term in k:

$$
\frac{4r_0^3 V_1}{k^2} = 1 .
\tag{7.86}
$$

It gives the location of the minimum of the effective potential-energy function

$$
V_{eff}(r) = V(r) + \frac{k^2}{8r^2} .
\tag{7.87}
$$

From the coefficient of the quadratic term to the leading order in k we define an oscillator frequency ω as

$$
\beta^4 \left(\frac{3k^2}{8} + r_0^4 V_2 \right) = \left(\frac{3}{8} + r_0^4 \frac{V_2}{k^2} \right) = \frac{\omega^2}{2} ,
\tag{7.88}
$$

provided that $\beta = 1/\sqrt{k}$.

In this way we obtain the following operator

$$
\begin{aligned}
\hat{\mathcal{H}} &= r_0^2 \beta^2 \hat{H} - \frac{(k-1)(k-3)\beta^2}{8} - r_0^2 \beta^2 V_0 \\
&= -\frac{1}{2} \frac{d^2}{dq^2} + \frac{\omega^2}{2} q^2 + \sum_{j=1}^{\infty} \left(a_j q^{j+2} + b_j q^j \right) \beta^j + \sum_{j=1}^{\infty} c_j q^j \beta^{j+2} ,
\end{aligned}
\tag{7.89}
$$

where

$$
\begin{aligned}
a_j &= \frac{(-1)^j}{8}(j+3) + \frac{V_{j+2}}{k^2} r_0^{j+4} , \\
b_j &= -\frac{(-1)^j}{2}(j+1), \quad c_j = \frac{3}{8}(-1)^j (j+1) .
\end{aligned}
\tag{7.90}
$$

Notice that we have substituted $1/\sqrt{k}$ for β only in some places, leaving β unchanged in others where it will play the role of a perturbation parameter. (We finally set $\beta = 1/\sqrt{k}$ in the resulting perturbation series.) In this way the Hamiltonian operator (7.89) exactly agrees with the one commonly used by other authors [148].

If we apply perturbation theory to the eigenvalue equation $\hat{\mathcal{H}}\Phi = \mathcal{E}\Phi$ with perturbation parameter β, we obtain a perturbation series of the form

$$
\mathcal{E} = \sum_{j=0}^{\infty} \mathcal{E}_{2j} \beta^{2j}, \quad \mathcal{E}_0 = (\nu + 1/2)\omega
\tag{7.91}
$$

because the coefficients of odd order vanish as argued earlier in this book. Finally, the energy reads

$$E = V_0 + \frac{(k-1)(k-3)}{8r_0^2} + \frac{1}{r_0^2\beta^2}\sum_{j=0}^{\infty}\mathcal{E}_{2j}\beta^{2j} . \tag{7.92}$$

A simple scaling argument shows that the actual perturbation parameter is not β but $\lambda = \beta/\sqrt{\omega}$. In fact, the change of variable $q = z/\sqrt{\omega}$ enables us to rewrite the Hamiltonian operator as $\hat{\mathcal{H}} = \omega\hat{h}$, where

$$\hat{h} = -\frac{1}{2}\frac{d^2}{dz^2} + \frac{1}{2}z^2 + \sum_{j=1}^{\infty}\lambda^j\left(\frac{a_j}{\omega^2}z^{j+2} + \frac{b_j}{\omega}z^j\right) + \sum_{j=1}^{\infty}c_j\lambda^{j+2}z^j . \tag{7.93}$$

If we apply perturbation theory to the eigenvalue equation $\hat{h}\Phi = e\Phi$ we obtain the series

$$e = \sum_{j=0}^{\infty}e_{2j}\lambda^{2j} \tag{7.94}$$

in terms of which the energy reads

$$E = V_0 + \frac{(k-1)(k-3)}{8r_0^2} + \frac{1}{r_0^2\lambda^2}\sum_{j=0}^{\infty}e_{2j}\lambda^{2j} . \tag{7.95}$$

Table 7.7 shows the coefficients $\mathcal{E}_2$ and $\mathcal{E}_4$ in terms of the potential parameters a_j, b_j, and c_j. Coefficients of higher order are increasingly more complicated to be shown there, but one easily obtains as many of them as desired by means of a simple Maple program similar to those described earlier. Here, we have chosen the method of Swenson and Danforth discussed in Section 3.3. The perturbation corrections in Table 7.7 account for most of the results obtained earlier by other authors [148].

Table 7.7 General Energy Coefficients of the Large-N Expansion

$$\mathcal{E}_2 = \frac{-\frac{1}{2}b_1^2 + \frac{3}{8}a_2 + b_2 e_0}{\omega^2} + \frac{-3a_1 b_1 e_0 - \frac{7}{16}a_1^2 + \frac{3}{2}a_2 e_0^2}{\omega^4} - \frac{15}{4}\frac{a_1^2 e_0^2}{\omega^6}$$

$$\mathcal{E}_4 = \frac{-b_1 c_1 + c_2 e_0 + \frac{3}{8}b_4}{\omega^2} + (-\frac{7}{8}a_1 b_3 - 3b_1 b_3 e_0 - \frac{15}{8}b_1 a_3 + b_1^2 b_2 + \frac{25}{8}a_4 e_0 - \frac{1}{2}b_2^2 e_0$$

$$+\frac{3}{2}b_4 e_0^2 - \frac{3}{4}a_2 b_2 - 3a_1 c_1 e_0)/\omega^4 + (-\frac{15}{2}a_1 b_3 e_0^2 + 9a_1 e_0 b_2 b_1 + 6b_1^2 a_2 e_0$$

$$-\frac{15}{2}b_1 a_3 e_0^2 - 3a_2 b_2 e_0^2 - \frac{95}{8}a_1 e_0 a_3 + \frac{5}{2}a_4 e_0^3 + \frac{23}{4}a_1 a_2 b_1 - a_1 b_1^3 + \frac{7}{4}a_1^2 b_2$$

$$-\frac{67}{16}a_2^2 e_0)/\omega^6 + (-\frac{27}{2}a_1^2 b_1^2 e_0 - \frac{35}{2}a_1 a_3 e_0^3 + \frac{459}{16}a_2 a_1^2 e_0 + 39a_1 a_2 e_0^2 b_1$$

$$+15b_2 e_0^2 a_1^2 - \frac{17}{4}a_2^2 e_0^3 - \frac{21}{4}a_1^3 b_1)/\omega^8$$

$$+\frac{-45a_1^3 b_1 e_0^2 + \frac{225}{4}a_1^2 a_2 e_0^3 - \frac{1155}{64}e_0 a_1^4}{\omega^{10}} - \frac{705}{16}\frac{a_1^4 e_0^3}{\omega^{12}}$$

In order to understand the relevant features of the large-N expansion we apply it to an exactly solvable model as we did before with other approximations. In this case we choose the hydrogen atom in N dimensions. When $V(r) = -1/r$ we have $V_j = (-1/r_0)^{j+1}$, $r_0 = k^2/4$, and $\omega = 1/2$. Therefore, equation (7.92) becomes

$$E = -2\beta^4 - 8\beta^6 + 6\beta^8 + 16\beta^6 \sum_{j=0}^{\infty} \mathcal{E}_{2j}\beta^{2j} , \qquad (7.96)$$

where $\mathcal{E}_0$ is given in equation (7.91), and $\mathcal{E}_2$ and $\mathcal{E}_4$ follows from straightforward substitution of the particular values of the potential coefficients a_j, b_j, and c_j into the expressions of Table 7.7. It is not difficult to verify that the perturbation series (7.96) agrees with the Taylor expansion of the exact result

$$E = -\frac{2}{[2(e_0 - 1) + k]^2} = -\frac{2\beta^4}{[1 + 2(e_0 - 1)\beta^2]^2} \qquad (7.97)$$

where $e_0 = 2\mathcal{E}_0 = \nu + 1/2$.

It is not difficult to prove that the large-N expansion for the hydrogen atom converges for all

$$\beta^2 < \frac{1}{|2\nu - 1|} \Rightarrow k > |2\nu - 1| ; \qquad (7.98)$$

that is to say, the radius of convergence decreases with the vibrational quantum number $\nu = 0, 1, \ldots$ as in previous examples. The radius of convergence of the large-N expansion for a Kratzer oscillator (which we can view as a generalization of the Coulomb interaction) was already discussed some time ago [161].

In order to apply the approach developed in Section 7.3 to the radial equation in N dimensions, we write $r = r_0(1 + \xi q)$ and choose r_0 to be the minimum of $U(r) = (k - 1)(k - 3)/(8r^2) + V(r)$, which is given by

$$\frac{4r_0^3 V_1}{(k - 1)(k - 3)} = 1 . \qquad (7.99)$$

If

$$\xi = \left[\frac{3(k - 1)(k - 3)}{4} + 2V_2 r_0^4\right]^{-1/4} = \left(3r_0^3 V_1 + 2V_2 r_0^4\right)^{-1/4} , \qquad (7.100)$$

then the coefficient of q^2 equals unity. Notice that the large-N expansion and the deep-well approximation lead to different expressions of both r_0 and the perturbation parameter, which agree when $k \to \infty$.

In order to appreciate the difference between both expansions more clearly we apply the deep-well approximation to the hydrogen atom in N dimensions. A straightforward calculation shows that $r_0 = (k - 1)(k - 3)/4$ and $\xi = r_0^{-1/4}$, and it follows from equation (7.98) that the perturbation series converges for all

$$\xi^2 < \frac{2}{\sqrt{|2\nu - 1|(|2\nu - 1| + 1)}} < 2 \qquad (7.101)$$

provided that there is no other singular point. There is, however, another singularity coming from the transformation between the perturbation parameters:

$$k = -\frac{1}{2} + \frac{2}{\xi^2}\sqrt{1 + \xi^4/4} . \qquad (7.102)$$

Since $\xi = 0$ is not a singular point of E as one easily verifies by substituting equation (7.102) into equation (7.97), then we are left with the branch point $\xi_b^4 = -4$ which tells us that the Taylor expansion of the square root converges for all $\xi^2 < 2$. Since this radius of convergence is greater than the one in equation (7.101), we conclude that both perturbation series converge for the same values of k (although the rate of convergence may be different).

In order to compare the deep-well approximation and the large-N expansion we apply both to a quantum-mechanical model with potential-energy function $V(r) = r$. When $N = 3$ we have to solve a radial equation of the form (7.82) with

$$u(r) = r + \frac{l(l+1)}{2r^2} . \tag{7.103}$$

Notice that the deep-well approximation does not apply to s states because $u(r)$ has no minimum when $l = 0$. On the other hand, the large-N series is based on an expansion about the minimum of the effective potential-energy function (7.87) which already exists for all values of l because $k = 3 + 2l$. In order to compare the results of both approaches we choose $l = 1$. Figure 7.4 shows $u(r)$ for three values of l. In the selected case $l = 1$, $u(r)$ is a single well that supports bound states for all $E > u_0 = 2^{1/3} + 2^{-2/3}$.

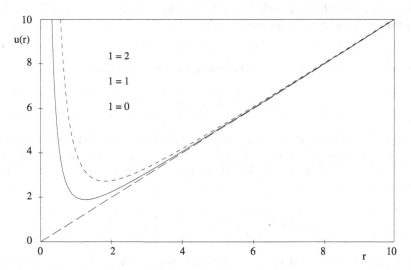

FIGURE 7.4
Effective potential-energy function $u(r) = r + l(l+1)/(2r^2)$.

We compare the convergence properties of different perturbation series by means of a logarithmic error defined as the logarithm of the absolute value of the first term neglected in the partial sum, which we calculate by means of the method of Swenson and Danforth discussed in Section 3.3. For concreteness consider the eigenvalue with $\nu = 0$. Figure 7.5 shows that neither the deep-well approximation nor the large-N expansion converge, and that the latter gives better results if we apply the truncation criterion discussed in Chapter 6. In fact, from a deep-well approximation of order 16 and a large-N expansion of order 34 we obtain the best estimates $E = 2.66782$ and $E = 2.66782947$, respectively, while the exact result provided by the Riccati–Padé method [162] is $E = 2.667829483$.

The conclusion just drawn appears to be at variance with earlier calculations on Lennard–Jones potentials which suggest that at low perturbation orders the deep-well approximation is more accurate even than the shifted large-N expansion (an improved version of the large-N expansion to be discussed later in this chapter) [156]. In order to investigate whether the relative accuracy of those approaches is model dependent, we consider the Lennard–Jones interactions in what follows.

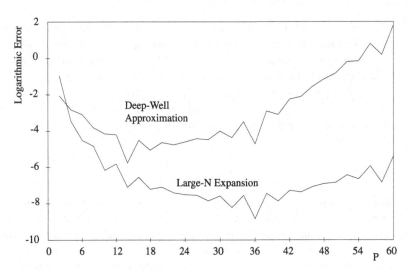

FIGURE 7.5

Logarithmic error for the deep-well approximation and large-N expansion for the bound-state energy with $\nu = 0$ and $l = 1$ supported by $V(r) = r$.

A general $j - n$ Lennard–Jones potential-energy function is of the form

$$V(r) = \frac{D}{j-n} \left[n \left(\frac{r_e}{r} \right)^j - j \left(\frac{r_e}{r} \right)^n \right] , \qquad (7.104)$$

where $j > n$, r_e is the equilibrium distance, and $D > 0$ is the well depth. This potential-energy function may support bound states for $-D < E < 0$. Notice that the Kratzer oscillator is a particular case of (7.104) with $j = 2$ and $n = 1$. The dimensionless energy and potential-energy function read $\epsilon = mr_e^2 E / \hbar^2$ and

$$v(q) = \frac{mr_e^2 V(r_e q)}{\hbar^2} = \frac{\eta^2}{j-n} \left(\frac{n}{q^j} - \frac{j}{q^n} \right) , \qquad (7.105)$$

respectively, where $\eta^2 = mr_e^2 D / \hbar^2$. For comparison purposes, here we choose one of the examples considered by the authors who proposed the deep-well approximation [156]: $j = 12$, $n = 6$, and $\eta = 50\sqrt{2}$; its shape can be seen in Figure 7.6.

Figure 7.7 shows the rate of convergence of the deep-well approximation and large-N expansion for several s states which we have purposely chosen because $l = 0$ is expected to be the most unfavorable case for both methods. The behavior of the perturbation series agrees with our earlier investigation on the much simpler Kratzer oscillator: both perturbation series appear to converge with an almost identical rate for all the states considered, and their convergence rate decreases with the radial quantum number ν. Table 7.8 shows dimensionless energies $E/D = \epsilon/\eta^2$ for the selected states. It is worth noticing that the smallest perturbation order P (also given in Table 7.8) at which the last digit becomes stable is exactly the same for both series, and that the perturbation eigenvalues agree with the most accurate numerical calculation available [163].

From the results above one may be tempted to conjecture that the deep-well approximation and the large-N expansion for a given state of a Lennard–Jones model exhibit the same nonzero radius of convergence which decreases with the radial quantum number ν; assumption that already applies to the particular case of the Kratzer oscillator as shown above. However, the methods for the estimation of the convergence radius of a power series discussed in Section 6.2.1 fail to predict the location

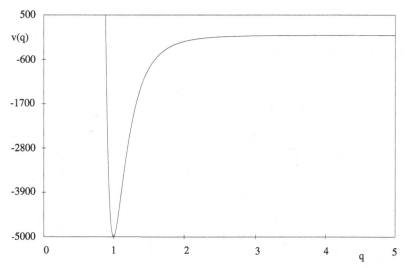

FIGURE 7.6
Dimensionless Lennard–Jones potential energy function $v(q) = (\eta^2/6)(6/q^{12} - 12/q^6)$ **for** $\eta = 50\sqrt{2}$.

Table 7.8   Eigenvalues $\dfrac{\epsilon}{\eta^2}$ of the Schrödinger Equation with the Dimensionless

Lennard–Jones Potential-Energy Function $v(q) = \dfrac{\eta^2}{6}(\dfrac{6}{q^{12}} - \dfrac{12}{q^6}),\ \eta = 50\sqrt{2}$

	$v = 0$			$v = 5$	
P	Polynomial Approximation	Large-N Expansion	P	Polynomial Approximation	Large-N Expansion
6	−0.9410460320	−0.9410460320	18	−0.4698229102	−0.4698229102
Exact		−0.941046	Exact		−0.469823
	$v = 10$			$v = 15$	
P	Polynomial Approximation	Large-N Expansion	P	Polynomial Approximation	Large-N Expansion
28	−0.1857237018	−0.1857237018	44	−0.04646991136	−0.04646991136
Exact		−0.185724	Exact		−0.046470

and exponent of the singular points. It may therefore happen that the deep-well and large-N series for the Lennard–Jones potential just discussed are slowly divergent. A more detailed and rigorous investigation is necessary in order to draw a convincing conclusion on this point.

7.6   Improved Perturbation Series

It is worth summarizing the most noticeable differences between the deep-well approximation and the large-N expansion. First, one expands the potential-energy function around different coordinate

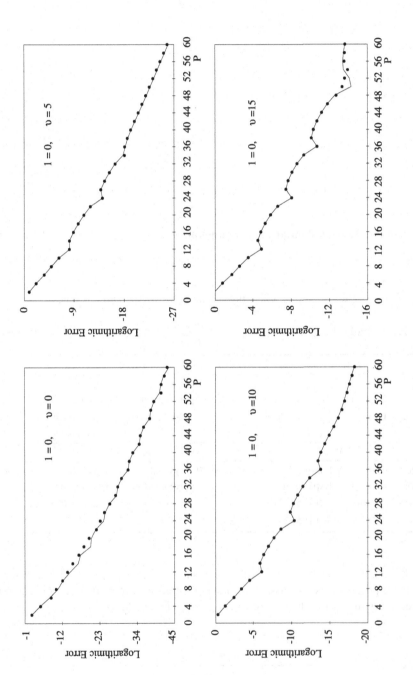

FIGURE 7.7
Logarithmic error for the deep-well approximation (line) and large-N expansion (points) for several s states supported by the dimensionless Lennard–Jones potential of Figure 7.6.

points: the minimum of $u(r)$ (equation (7.82)) in the deep-well approximation, and the minimum of $V_{eff}(r)$ (equation (7.87)) in the large-N expansion. For this reason it is possible to apply the latter and not the former when $u(r)$ does not exhibit a minimum. Second, although in both approaches the reference model or unperturbed Hamiltonian is a harmonic oscillator, the corresponding oscillator frequencies are different. Third, the grouping of the perturbation terms that is noticeably different comes from the use of different perturbation variables.

All in all, the difference between the deep-well approximation and the large-N expansion comes from the three degrees of freedom available for the construction of perturbation series by means of a polynomial approximation. In the most general case we may try to obtain the eigenvalues of a quantum-mechanical problem by application of perturbation theory to a Hamiltonian operator of the form

$$\hat{\mathcal{H}} = -\frac{1}{2}\frac{d^2}{dq^2} + \frac{\omega^2}{2}q^2 + \sum_{j=0}^{\infty}\lambda^j p_j(q) \,, \tag{7.106}$$

where $p_0(q)$, $p_1(q)$, ... are polynomial functions of q that one conveniently chooses in order to fit the potential-energy function and centrifugal term for a given value of λ.

Those appear to be the main degrees of freedom at our disposal if we restrict to the harmonic oscillator as the unperturbed model. However, the lack of clear guidelines for setting them conveniently makes the task a complicated and tedious process of trial and error. For this reason the methods commonly used to improve the convergence properties of the polynomial approximations try to reduce the degrees of freedom to one adjustable parameter.

In what follows we briefly consider straightforward ways of improving the convergence properties of the polynomial approximation. We simply give the main ideas and show some results, but it is not our purpose to provide rigorous proofs of convergence which in most cases are not available.

7.6.1 Shifted Large-N Expansion

In the preceding section we obtained the radius of convergence of the large-N expansion for the hydrogen atom in N spatial dimensions. It follows from equation (7.97) that if we expand the energy in powers of $1/[2(e_0 - 1) + k]$ instead of in powers of $1/k$ we obtain the exact result with just one term and all the corrections are zero. This is the basis for the celebrated shifted large-N expansion [148, 164] that follows from the change of expansion parameter from $\beta = 1/\sqrt{k}$ to $\overline{\beta} = 1/\sqrt{\overline{k}}, \overline{k} = k - a$, where a is a real number. If in the case of the hydrogen atom we choose $a = -2(e_0 - 1)$, then $E = -2\overline{\beta}^4$ and all the remaining coefficients of the $\overline{\beta}^2$-power series vanish. Since the actual variable of the large-N expansion for the energy is β^2, we conclude that the net effect of shifting k is the Euler transformation

$$\beta^2 = \frac{\overline{\beta}^2}{1 + a\overline{\beta}^2}, \, \overline{\beta}^2 = \frac{\beta^2}{1 - a\beta^2} \,. \tag{7.107}$$

In order to apply the shifted large-N expansion to a general central-field model we rewrite the radial Hamiltonian operator (7.82) as

$$\hat{H} = -\frac{1}{2}\frac{d^2}{dr^2} + u(r), u(r) = \frac{(\overline{k} + a - 1)(\overline{k} + a - 3)}{8r^2} + V(r) \,, \tag{7.108}$$

and expand $u(r)$ in powers of $\overline{\beta} = 1/\sqrt{\overline{k}}$ exactly as we did before for the large-N expansion. The resulting effective potential

$$V_{eff}(r) = V(r) + \frac{\overline{k}^2}{8r^2} \tag{7.109}$$

exhibits a minimum at r_0 given by

$$\frac{4r_0^3 V_1}{\overline{k}^2} = 1 , \qquad (7.110)$$

and the Hamiltonian operator reads

$$\hat{\mathcal{H}} = r_0^2 \overline{\beta}^2 \hat{H} - \frac{(k-1)(k-3)\overline{\beta}^2}{8} - r_0^2 \overline{\beta}^2 V_0 = -\frac{1}{2}\frac{d^2}{dq^2} + \frac{\omega^2}{2} q^2$$

$$+ \sum_{j=1}^{\infty}(a_j q^{j+2} + b_j q^j)\overline{\beta}^j + \sum_{j=1}^{\infty} c_j q^j \overline{\beta}^{j+2} , \qquad (7.111)$$

where

$$\omega^2 = \left(\frac{3}{4} + 2r_0^4 \frac{V_2}{\overline{k}^2}\right) , \qquad (7.112)$$

and

$$a_j = \frac{(-1)^j}{8}(j+3) + \frac{V_{j+2}}{\overline{k}^2} r_0^{j+4} ,$$

$$b_j = -\frac{(-1)^j(a-2)(j+1)}{4}, \quad c_j = \frac{(a-1)(a-3)(-1)^j(j+1)}{8} . \qquad (7.113)$$

Notice that when $a = 0$ these equations reduce to those above for the large-N expansion.

Straightforward application of perturbation theory to $\hat{\mathcal{H}}\Phi = \mathcal{E}\Phi$ gives us a series similar to equation (7.91) with $\overline{\beta}$ instead of β. The coefficients $\mathcal{E}_{2j}$ are functions of the potential parameters a_j, b_j, and c_j identical to those for the case $a = 0$, the first of which are given in Table 7.7. The optimum value of a is customarily chosen in such a way that the coefficient of $\overline{\beta}^{-2}$ that appears when we rewrite the expansion of the energy

$$E = V_0 + \frac{\overline{k}^2 + (2a-4)\overline{k} + (a-1)(a-3)}{8r_0^2} + \frac{1}{r_0^2 \overline{\beta}^2}\sum_{j=0}^{\infty} \mathcal{E}_{2j}\overline{\beta}^{2j} \qquad (7.114)$$

in terms of $\overline{\beta}$

$$E = V_0 + \frac{1}{8r_0^2 \overline{\beta}^4} + \frac{2a-4+8\mathcal{E}_0}{8r_0^2 \overline{\beta}^2} + \frac{(a-1)(a-3)+8\mathcal{E}_2}{8r_0^2} + \frac{\mathcal{E}_4}{r_0^2}\overline{\beta}^2 + \cdots \qquad (7.115)$$

vanishes. There is a good reason for this prescription: if a is given by

$$a = 2 - 4\mathcal{E}_0 = 2 - 2(2\nu + 1)\omega , \qquad (7.116)$$

then the shifted large-N expansion gives the exact energies of the hydrogen atom and harmonic oscillator in N dimensions [148, 164].

In order to realize how much the shifting just mentioned improves the convergence properties of the perturbation series we consider a case in which the large-N expansion fails badly. For this purpose it is sufficient to choose the state $l = 0$, $\nu = 5$ supported by the potential-energy function $V(r) = r$. Figure 7.8 shows the logarithmic error defined above for the large-N expansion and for its shifted version. It is evident that the former series diverges whereas the latter appears to converge. Even if the shifted large-N expansion proved to be divergent, we would still obtain reasonable results from it by means of the truncation criterion discussed in Chapter 6.

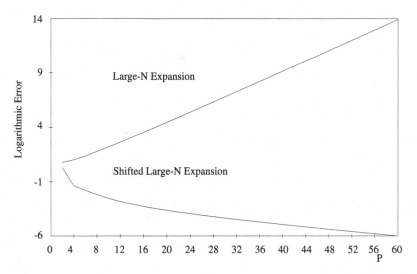

FIGURE 7.8
Logarithmic error for the large-N and shifted large-N perturbation series for the state $l = 0$, $v = 5$ supported by $V(r) = r$.

7.6.2 Improved Shifted Large-N Expansion

The shift given by equation (7.116) has become quite popular and has been routinely used to obtain satisfactory results for many quantum-mechanical models [148]. However it is suitable only for perturbation approximations of low order. When one takes into account perturbation terms of sufficiently large order, then the shift given by equation (7.116) leads to divergent series, and a different criterion for selecting the optimum shift is necessary. A successful prescription is to set a so that $\mathcal{E}_P = 0$, and calculate the eigenvalue by means of a partial sum of order P [165]–[167]. This choice of a, which is inspired in the truncation criterion discussed in Chapter 6, corrects most of the problems encountered in earlier applications of the shifted large-N expansion [148].

In order to compare results for different values of the free parameter a, we consider the ground-state energy of the Schrödinger equation with the potential-energy function $V(r) = 2^{7/2}r$, already chosen in earlier discussions of the shifted large-N expansion [165]. Figure 7.9 shows results for the shifted large-N expansion and for its improved version with a given by $\mathcal{E}_{10} = 0$, $\mathcal{E}_{40} = 0$, and $\mathcal{E}_{60} = 0$. The peaks at $P = 10, 40$, and 60 are finite (instead of $-\infty$) simply because the logarithm of the perturbation terms has been purposefully calculated with low numerical precision. Notice that the shifted large-N expansion is the most accurate at sufficiently low perturbation order. However, as we add more perturbation terms the value of a given by $\mathcal{E}_P = 0$ is preferable.

In Table 7.9 we give the best estimate of the eigenvalue obtained from a partial sum of order M determined according to the truncation criterion proposed in Chapter 6. In the case of the improved shifted large-N expansion we omit the term of order P forced to be zero by the choice of a. Notice that the greater the value of P, the more negative the value of a, and the greater the order M of the optimum partial sum. When there are more than one real root of $\mathcal{E}_P = 0$ we arbitrarily choose the smallest (most negative) one. By means of the smallest root of $\mathcal{E}_{60} = 0$ and the partial sum of order 58 we obtain the ground-state energy with ten-digit accuracy (with respect to the exact result obtained by means of the powerful Riccati–Padé method [162]).

It is worth noticing that the shifted large-N expansion and its variants are based on just one adjustable parameter that clearly modifies the expansion point r_0 and the frequency ω of the harmonic oscillator. Allowing more degrees of freedom will certainly result in perturbation series with better

FIGURE 7.9
Logarithmic error for the perturbation series for the ground state of $V(r) = 2^{7/2}r$.

Table 7.9   Ground-State Energy Supported by the
Potential-Energy Function $V(r) = 2^{\frac{7}{2}}r$

	P	M	a	E
Shifted Large-N Expansion		20	0.267949192	9.3526
Improved Shifted Large-N Expansion	10	28	−1.656887641	9.3524301
	20	26	−2.212808322	9.3524295
	30	32	−2.618371270	9.3524297
	40	42	−2.922963203	9.35242966
	50	44	−3.211017542	9.35242965
	60	58	−4.848680852	9.352429642
Exact				9.352429641839

convergence properties but will at the same time make the calculation more complicated, especially if we lack sound criteria for setting the optimum values of the adjustable parameters.

7.7   Born–Oppenheimer Perturbation Theory

The starting point of a typical quantum-mechanical treatment of a molecule is the Born–Oppenheimer approximation that consists of the separation of the electronic and nuclear degrees of freedom. In this way one simplifies the problem considerably and makes it computationally more tractable. The first such approach was based on perturbation theory [168] but later a more convenient strategy was proposed [169] which was rapidly adopted by most researchers, becoming an important part of the routine theoretical treatment of molecules [53, 159].

The perturbation approach to the separation of electronic and nuclear degrees of freedom resembles the polynomial approximation discussed in this chapter. For this reason we briefly discuss it in the present section under the name of Born–Oppenheimer perturbation theory. The original perturbation method is rather cumbersome making it considerably difficult for the derivation of moderately large perturbation orders [168]. This fact motivated the development of a more systematic procedure to facilitate the calculation [170].

The application of the Born–Oppenheimer perturbation method to a system composed of an arbitrary number of light and heavy particles is straightforward [170]. However, in order to concentrate on the main ideas and avoid the diversion which may possibly arise from the notation required for the description of many degrees of freedom, we consider here a system of only two coordinates:

$$\hat{H} = -\frac{\hbar^2}{2m}\frac{\partial^2}{\partial x^2} - \frac{\hbar^2}{2M}\frac{\partial^2}{\partial X^2} + V(x, X) \,. \tag{7.117}$$

This model describes the one-dimensional motion of a light particle of mass m and position x and a heavy particle of mass $M \gg m$ and position X.

In the first step we derive a dimensionless Hamiltonian operator $\hat{\mathcal{H}} = mL^2\hat{H}/\hbar^2$, where L is a length unit. Defining dimensionless coordinates $q = x/L$ and $Q = X/L$, and a dimensionless potential-energy function $v(q, Q) = mL^2 V(Lq, LQ)/\hbar^2$, we obtain

$$\hat{\mathcal{H}} = -\frac{1}{2}\frac{\partial^2}{\partial q^2} - \frac{m}{2M}\frac{\partial^2}{\partial Q^2} + v(q, Q) \,. \tag{7.118}$$

The dimensionless Schrödinger equation reads $\hat{\mathcal{H}}\Psi = \mathcal{E}\Psi$, where $\mathcal{E} = mL^2 E/\hbar^2$ and E is an eigenvalue of $\hat{H}$. In order to apply the polynomial approximation to the slow motion of the heavy particle define

$$Q = Q_0 + \beta y \,, \tag{7.119}$$

where y is a new coordinate and the parameters Q_0 and β of the transformation will be determined later on.

Expanding the potential-energy function about Q_0

$$v(q, Q) = \sum_{j=0}^{\infty} v_j(q)(Q - Q_0)^j = \sum_{j=0}^{\infty} v_j(q)(\beta y)^j \tag{7.120}$$

the dimensionless Hamiltonian operator becomes

$$\hat{\mathcal{H}} = -\frac{1}{2}\frac{\partial^2}{\partial q^2} - \frac{m}{2M\beta^2}\frac{\partial^2}{\partial y^2} + v_0 + v_1\beta y + v_2\beta^2 y^2 + \cdots \,. \tag{7.121}$$

In order to simplify the notation we define the kinetic energy operators

$$\hat{T}_q = -\frac{1}{2}\frac{\partial^2}{\partial q^2}, \ \hat{T}_y = -\frac{1}{2}\frac{\partial^2}{\partial y^2} \,. \tag{7.122}$$

The kinetic energy and the harmonic term for the slow motion will appear at the same perturbation order if $m/(M\beta^2) = \beta^2$ that leads to

$$\beta = (m/M)^{1/4} \,. \tag{7.123}$$

In order to apply perturbation theory we expand the dimensionless Hamiltonian operator as

$$\hat{\mathcal{H}} = \sum_{j=0}^{\infty} \hat{\mathcal{H}}_j \beta^j , \tag{7.124}$$

where

$$\begin{aligned}
\hat{\mathcal{H}}_0 &= \hat{T}_q + v_0, \; \hat{\mathcal{H}}_1 = v_1 y, \; \hat{\mathcal{H}}_2 = \hat{T}_y + v_2 y^2, \\
\hat{\mathcal{H}}_j &= v_j y^j, \; j > 2 .
\end{aligned} \tag{7.125}$$

Straightforward application of perturbation theory to the Schrödinger equation with this expansion of the Hamiltonian operator proves to be rather lengthy requiring great ingenuity to figure out how to combine the perturbation equations to obtain suitable results [168]. For this reason, in what follows we develop more convenient equations for the application of perturbation theory.

We first define the Hamiltonian operator

$$\hat{\mathcal{H}}_q = \hat{T}_q + v(q, Q) \tag{7.126}$$

that depends on the coordinate Q parametrically. Strictly speaking equation (7.126) represents a family of Hamiltonian operators, one for each value of Q. In the Born–Oppenheimer approximation the eigenvalues of $\hat{\mathcal{H}}_q$ are effective potential-energy functions for the slow motion

$$\hat{\mathcal{H}}_q \Phi(q, Q) = U(Q) \Phi(q, Q) . \tag{7.127}$$

For the sake of simplicity we assume that the chosen eigenvalue $U(Q)$ is nondegenerate and normalize $\Phi(q, Q)$ to unity for all values of Q:

$$\langle \Phi | \Phi \rangle_q = \int \Phi(q, Q)^2 \, dq = 1 , \tag{7.128}$$

where the subscript in the ket $| \rangle_q$ indicates integration over q. Notice that without loss of generality we choose $\Phi(q, Q)$ to be real.

We define a function of the coordinate for the slow motion

$$f(Q) = \langle \Phi | \Psi \rangle_q \tag{7.129}$$

and a correlation function

$$F(q, Q) = \Psi(q, Q) - f(Q) \Phi(q, Q) \tag{7.130}$$

which is orthogonal to Φ with respect to q:

$$\langle \Phi | F \rangle_q = 0 . \tag{7.131}$$

This correlation function vanishes identically if the state function $\Psi(q, Q)$ is exactly factorizable as $f(Q)\Phi(q, Q)$, which, in general, does not occur. The function $F(q, Q)$ is therefore a measure of such a factorization and satisfies the differential equation $(\hat{\mathcal{H}} - \mathcal{E})F = f(\mathcal{E} - U)\Phi - \beta^2 \hat{T}_y \Phi f$. We rewrite $\hat{T}_y \Phi f = \Phi \hat{T}_y f - [\Phi, \hat{T}_y]f$, where $[\Phi, \hat{T}_y]$ is a formal notation for the commutator

$$\left[\Phi, \hat{T}_y \right] = \Phi \hat{T}_y - \hat{T}_y \Phi = \frac{1}{2} \left(\frac{\partial^2 \Phi}{\partial y^2} + 2 \frac{\partial \Phi}{\partial y} \frac{\partial}{\partial y} \right) . \tag{7.132}$$

The differential equation for $F(q, Q)$ becomes

$$\left(\hat{\mathcal{H}} - \mathcal{E}\right) F = \Phi \left(\mathcal{E} - U - \beta^2 \hat{T}_y\right) f + \beta^2 \left[\Phi, \hat{T}_y\right] f \,. \tag{7.133}$$

Applying the bra $_q < \Phi|$ to equation (7.133) from the left, and taking into account equations (7.127), (7.128), and (7.131) we conclude that $(\mathcal{E} - U - \beta^2 \hat{T}_y + \beta^2 \langle\Phi|[\Phi, \hat{T}_y]\rangle_q)f = \beta^2 \langle\Phi|\hat{T}_y|F\rangle_q$. Differentiating equation (7.128) with respect to y once and twice we obtain

$$\left\langle\Phi|\frac{\partial\Phi}{\partial y}\right\rangle_q = 0, \ \left\langle\Phi|\frac{\partial^2\Phi}{\partial y^2}\right\rangle_q = -\left\langle\frac{\partial\Phi}{\partial y}|\frac{\partial\Phi}{\partial y}\right\rangle_q \,, \tag{7.134}$$

respectively. Therefore, the differential equation for $f(Q)$ reads

$$\left(\hat{\mathcal{H}}_y - \mathcal{E}\right) f(Q) = -R(Q) \,, \tag{7.135}$$

where

$$\hat{\mathcal{H}}_y = \beta^2 \hat{T}_y + U(Q) + W(Q) \tag{7.136}$$

and

$$W(Q) = \frac{\beta^2}{2}\left\langle\frac{\partial\Phi}{\partial y}|\frac{\partial\Phi}{\partial y}\right\rangle_q \geq 0, \ R(Q) = \beta^2 \left(\frac{1}{2}\left\langle\frac{\partial^2\Phi}{\partial y^2}|F\right\rangle_q + \left\langle\frac{\partial\Phi}{\partial y}|\frac{\partial F}{\partial y}\right\rangle_q\right) \,. \tag{7.137}$$

By means of equation (7.135) we can rewrite equation (7.133) as follows:

$$\left(\hat{\mathcal{H}} - \mathcal{E}\right) F = \Phi W f + R + \beta^2 \left[\Phi, \hat{T}_y\right] f \,. \tag{7.138}$$

In order to apply perturbation theory to the equations above we need the following expansions:

$$\hat{\mathcal{H}}_q = \sum_{j=0}^{\infty} \hat{\mathcal{H}}_{q,j}\beta^j, \ \hat{\mathcal{H}}_{q,j} = \hat{\mathcal{H}}_j - \hat{T}_y\delta_{j2} \,, \tag{7.139}$$

$$\Phi(q, Q) = \sum_{j=0}^{\infty} \Phi_j(q)\beta^j y^j \,, \tag{7.140}$$

$$U(Q) = \sum_{j=0}^{\infty} U_j\beta^j y^j \,, \tag{7.141}$$

$$W(Q) = \sum_{j=0}^{\infty} W_j\beta^{j+2} y^{j-2} \,, \tag{7.142}$$

$$R(Q) = \sum_{j=0}^{\infty} R_j(y)\beta^{j+2} \,, \tag{7.143}$$

$$F(q, Q) = \sum_{j=0}^{\infty} F_j(q, y)\beta^j ,$$ (7.144)

$$\hat{\mathcal{H}}_y = \sum_{j=0}^{\infty} \hat{\mathcal{H}}_{y,j}\beta^j, \quad \hat{\mathcal{H}}_{y,j} = \hat{T}_y\delta_{j2} + U_j y^j + W_{j-2}y^{j-4} .$$ (7.145)

Notice that $\partial\Phi/\partial y$ and $\partial^2\Phi/\partial y^2$ are of order β and β^2, respectively. Therefore, $W_0 = W_1 = R_0 = 0$, so that W and R are at least of order β^4 and β^3, respectively. Because the right-hand side of equation (7.138) is at least of order β^3 we can choose $F_0 = F_1 = F_2 = 0$. It therefore follows from equation (7.137) that $R_1 = R_2 = R_3 = 0$, and

$$R_4 = \left\langle \Phi_1 \middle| \frac{\partial F_3}{\partial y} \right\rangle_q$$ (7.146)

may be the first nonzero coefficient of the expansion (7.143).

Consider the perturbation coefficients of equation (7.135):

$$\sum_{j=0}^{k} \left(\hat{\mathcal{H}}_{y,j} - \mathcal{E}_j \right) f_{k-j} = -R_{k-2} ,$$ (7.147)

where $R_{k-2} = 0$ if $k < 6$. The solution of the perturbation equation of order zero $(U_0 - \mathcal{E}_0)f_0 = 0$ is

$$U_0 = \mathcal{E}_0 .$$ (7.148)

From the equation of first order $(U_1 y - \mathcal{E}_1)f_0 = 0$ we obtain

$$U_1 = \mathcal{E}_1 = 0 ,$$ (7.149)

which states that Q_0 is a stationary point of $U(Q)$:

$$\frac{\partial U}{\partial Q}(Q_0) = 0 .$$ (7.150)

The equation of second order $(\hat{\mathcal{H}}_{y,2} - \mathcal{E}_2)f_0 = 0$ is the eigenvalue equation for the harmonic oscillator $\hat{\mathcal{H}}_{y,2} = \hat{T}_y + U_2 y^2$; therefore

$$\mathcal{E}_2 = (v + 1/2)\sqrt{2U_2}, v = 0, 1, \ldots ,$$ (7.151)

and f_0 is the corresponding eigenfunction. It is clear that the present approach applies provided that $U_2 > 0$; that is to say, if Q_0 is a minimum of $U(Q)$.

It is not difficult to verify that the perturbation equations (7.147) of order $k = 3$, 4, and 5 are identical to those for a harmonic oscillator perturbed by anharmonic terms of the form $\beta U_3 y^3 + \beta^2(U_4 y^4 + W_2) + \beta^3(U_5 y^5 + W_3 y)$. The inhomogeneous term in equation (7.147) appears at sixth order. According to equation (7.146) we need F_3 in order to obtain R_4. It follows from equation (7.138) that F_3 is a solution of

$$\left(\hat{\mathcal{H}}_0 - \mathcal{E}_0 \right) F_3 = \left(\hat{\mathcal{H}}_{q,0} - U_0 \right) F_3 = \Phi_1 \frac{\partial f_0}{\partial y} .$$ (7.152)

Since $< \Phi_0|\Phi_1 >_q = 0$ we can formally write

$$F_3 = \left(\hat{\mathcal{H}}_{q,0} - U_0\right)^{-1} \Phi_1 \frac{\partial f_0}{\partial y} , \qquad (7.153)$$

so that

$$R_4 = \left\langle \Phi_1 \middle| \left(\hat{\mathcal{H}}_{q,0} - U_0\right)^{-1} \Phi_1 \right\rangle_q \frac{\partial^2 f_0}{\partial y^2} = -2\mathcal{K}\hat{T}_y f_0 , \qquad (7.154)$$

where

$$\mathcal{K} = \left\langle \Phi_1 \middle| \left(\hat{\mathcal{H}}_{q,0} - U_0\right)^{-1} \Phi_1 \right\rangle_q . \qquad (7.155)$$

The reader can easily verify that the perturbation equations (7.147) with $k \leq 6$ follow from the eigenvalue equation

$$\hat{\mathcal{H}}_{eff} f = \mathcal{E}_{eff} f , \qquad (7.156)$$

where the effective Hamiltonian operator $\hat{\mathcal{H}}_{eff}$ given by

$$\begin{aligned}
\hat{\mathcal{H}}_{eff} &= \left(1 - 2\mathcal{K}\beta^4\right)\hat{T}_y + U_2 y^2 + \beta U_3 y^3 + \beta^2 \left(U_4 y^4 + W_2\right) \\
&\quad + \beta^3 \left(U_5 y^5 + W_3 y\right) + \beta^4 \left(U_6 y^6 + W_4 y^2\right)
\end{aligned} \qquad (7.157)$$

is accurate through order β^4. Finally, the eigenvalue of $\hat{\mathcal{H}}$ through order six results to be

$$\mathcal{E} \approx U_0 + \beta^2 \mathcal{E}_{eff} . \qquad (7.158)$$

The coefficients of the perturbation series for the eigenfunction

$$\Psi = \sum_{j=0}^{\infty} \Psi_j \beta^j \qquad (7.159)$$

are given by

$$\Psi_j = \sum_{k=0}^{j} \Phi_k y^k f_{j-k} + F_j . \qquad (7.160)$$

Suppose that we want to derive an approximate expression for the transition integral $\langle \Psi|d|\Psi'\rangle$, where $d = c_1 x + c_2 X$ is the dipole moment of the system and Ψ and Ψ' are initial and final states, respectively. In order to carry out this calculation by means of perturbation theory we rewrite $d = c_1 L(q + \alpha\beta y) + c_2 L Q_0$, where $\alpha = c_2/c_1$. Therefore

$$\langle \Psi|d|\Psi'\rangle = \sum_{j=0}^{\infty} \langle \Psi|d|\Psi'\rangle_j \beta^j , \qquad (7.161)$$

where

$$\langle \Psi|d|\Psi'\rangle_j = c_1 L \left(\sum_{k=0}^{j} \langle \Psi_k|q|\Psi'_{j-k}\rangle + \alpha \sum_{k=0}^{j-1} \langle \Psi_k|y|\Psi'_{j-1-k}\rangle \right) . \qquad (7.162)$$

The leading term

$$\langle \Psi | d | \Psi' \rangle_0 = c_1 L \langle \Phi_0 | q | \Phi_0' \rangle_q \langle f_0 | f_0' \rangle_y \tag{7.163}$$

is proportional to the well-known Frank–Condon overlap integral $\langle f_0 | f_0' \rangle_y$.

It is worth noticing that we could approximately separate the eigenvalue equation with two degrees of freedom into two one-dimensional eigenvalue equations so that the energies result to be eigenvalues of an effective Hamiltonian operator for the slow motion. In the case of an actual molecular system we obtain equations for the electrons (light particles, rapid motion) and the nuclei (heavy particles, slow motion), and the molecular energies are the eigenvalues of a nuclear effective Hamiltonian operator [170]. The treatment of systems with many degrees of freedom closely parallels the much simpler example considered here, which we have solved in detail for pedagogical purposes. We point out that the approach just discussed is of practical value only in the case of many degrees of freedom [170].

One of the main reasons for discussing the early Born–Oppenheimer approximation here is that it does not appear to be widely known. Although the polynomial approximations in perturbation theory (specially the large-N expansion and its variants) are quite popular, nobody seems to think of the early Born–Oppenheimer method as an example in which the polynomial approximation is restricted to some properly selected degrees of freedom.

It is instructive to apply the Born–Oppenheimer method to an exactly solvable model. The eigenvalues of the dimensionless Hamiltonian operator

$$\hat{\mathcal{H}} = -\frac{1}{2} \frac{\partial^2}{\partial q^2} - \frac{\beta^4}{2} \frac{\partial^2}{\partial Q^2} + \frac{1}{2} \left(q^2 + Q^2 \right) + \lambda q Q \tag{7.164}$$

are

$$\mathcal{E} = (v + 1/2) \sqrt{\xi_+} + \left(v' + 1/2 \right) \sqrt{\xi_-} , \tag{7.165}$$

where v, $v' = 0, 1, \ldots$, and

$$\xi_{\pm} = \frac{1 + \beta^4}{2} \pm \frac{1}{2} \sqrt{(1 - \beta^4)^2 + 4\lambda^2 \beta^4} . \tag{7.166}$$

In order to verify if the Born–Oppenheimer perturbation theory applies, we solve the eigenvalue equation (7.127) with

$$v(q, Q) = \frac{1}{2} \left(q^2 + Q^2 \right) + \lambda q Q = \frac{1}{2} (q + \lambda Q)^2 + \frac{1 - \lambda^2}{2} Q^2 , \tag{7.167}$$

which gives us

$$U(Q) = v + \frac{1}{2} + \frac{1 - \lambda^2}{2} Q^2 . \tag{7.168}$$

We see that $U(Q)$ will have a minimum at $Q = Q_0 = 0$ if $\lambda^2 < 1$. In agreement with this result notice that the dimensionless force constant ξ_- is positive only if $\lambda^2 < 1$.

The convergence radius of the Born–Oppenheimer perturbation series is determined by a pair of complex conjugate branch points that are roots of $(1 - \beta^4)^2 + 4\lambda^2 \beta^4 = 0$. Solving this equation for β^4 we obtain

$$\beta_b^4 = 1 - 2\lambda^2 \pm 2\lambda i \sqrt{1 - \lambda^2} \tag{7.169}$$

and the radius of convergence is $|\beta_b|^4 = 1$; that is to say, the series converges for all $\beta^4 = m/M < 1$.

We can exactly express the solutions of most of the equations of the Born–Oppenheimer method for the present trivial problem in terms of the eigenvalues $e = v + 1/2$ and eigenfunctions χ of the harmonic oscillator

$$\hat{h} = -\frac{1}{2}\frac{d^2}{du^2} + \frac{u^2}{2}, \ \hat{h}\chi(u) = e\chi(u) . \tag{7.170}$$

For example,

$$\Phi(q, Q) = \chi(q + \lambda\beta y) , \tag{7.171}$$

from which it follows that $\partial\Phi/\partial y = \lambda\beta(d\chi/du)$, and

$$W(Q) = \frac{\lambda^2\beta^4}{2}\left\langle\frac{d\chi}{du}\Big|\frac{d\chi}{du}\right\rangle_u = -\frac{\lambda^2\beta^4}{2}\left\langle\chi\Big|\frac{d^2\chi}{du^2}\right\rangle_u = \frac{\lambda^2\beta^4}{2}\left(v + \frac{1}{2}\right) . \tag{7.172}$$

In the same way $\partial\Phi/\partial\beta = \lambda y(d\chi/du)$ so that

$$\Phi_1 = \lambda\frac{d\chi_v(q)}{dq} = \frac{\lambda}{\sqrt{2}}\left(\sqrt{v}\chi_{v-1} - \sqrt{v+1}\chi_{v+1}\right) , \tag{7.173}$$

where we have explicitly indicated the harmonic oscillator quantum number $v = 0, 1, \ldots$. One easily obtains the result in equation (7.173) writing d/dq in terms of creation and annihilation boson operators [49]. Since $\hat{\mathcal{H}}_{q,0} = \hat{h}$ with $u = q$ and $U_0 = v + 1/2$, we obtain $\mathcal{K} = \lambda^2/2$, and the effective Hamiltonian operator

$$\hat{\mathcal{H}}_{eff} = \left(1 - \lambda^2\beta^4\right)\hat{T}_y + \frac{1 - \lambda^2}{2}y^2 + \frac{\lambda^2\beta^2}{2}\left(v + \frac{1}{2}\right) \tag{7.174}$$

is a harmonic oscillator with mass $1/(1 - \lambda^2\beta^4)$ and force constant $1 - \lambda^2$; consequently

$$\mathcal{E}_{eff} = \left(v' + \frac{1}{2}\right)\sqrt{(1 - \lambda^2)(1 - \lambda^2\beta^4)} + \frac{\lambda^2\beta^2}{2}\left(v + \frac{1}{2}\right) , \tag{7.175}$$

where $v' = 0, 1, \ldots$.

Finally, the approximate energy reads

$$\mathcal{E} \approx \left(1 + \frac{\lambda^2\beta^4}{2}\right)\left(v + \frac{1}{2}\right) + \beta^2\left(v' + \frac{1}{2}\right)\sqrt{(1 - \lambda^2)(1 - \lambda^2\beta^4)} . \tag{7.176}$$

According to our discussion above this approximate expression should be accurate to order six. The radius of convergence of the Taylor expansion of equation (7.176) about $\beta = 0$ is $1/\sqrt{|\lambda|} > 1$ (remember that $|\lambda| < 1$ in order to have bound states). Through order six we have

$$\mathcal{E} \approx \left(v + \frac{1}{2}\right) + \left(v' + \frac{1}{2}\right)\sqrt{(1 - \lambda^2)}\beta^2$$
$$+ \left(v + \frac{1}{2}\right)\lambda^2\beta^4 - \frac{1}{2}\left(v' + \frac{1}{2}\right)\lambda^2\sqrt{(1 - \lambda^2)}\beta^6 + \mathcal{O}\left(\beta^{10}\right) . \tag{7.177}$$

The Taylor expansion of the exact eigenvalue (7.165) about $\beta = 0$ gives the same result through order six, except that the first neglected term is of order β^8. This difference is not surprising because the Born–Oppenheimer perturbation theory developed above is accurate through order six.

In Figure 7.10 we compare the exact ground-state energy ($v = v' = 0$) of the model (7.164) with the approximate expression (7.176) and the series (7.177) for two values of λ. The two approximate expressions are almost identical and their accuracy decreases as β increases as expected. Moreover, the accuracy of the approximate results also decreases with λ.

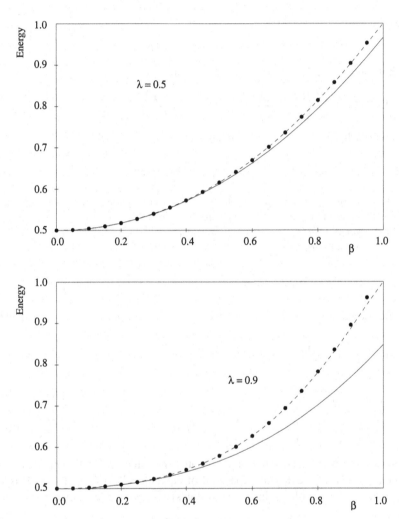

FIGURE 7.10
Ground-state energy of model (7.164) for two values of λ. The exact result, the approximate
expression (7.176), and the series (7.177) are, respectively, given by a continuous line, a broken
line, and points.

Chapter 8

Perturbation Theory for Scattering States in One Dimension

8.1 Introduction

In this chapter we briefly discuss the straightforward application of perturbation theory to scattering states of one-dimensional quantum-mechanical models. It is not our purpose to give an account of the methods commonly used to treat realistic problems but simply to show that some of the approaches developed for bound states can also be applied to scattering states. With that purpose in mind we consider simple one-dimensional models in which the interactions are nonzero only in finite regions of space. The reason for this restriction is that the calculation is much simpler.

8.2 On the Solutions of Second-Order Differential Equations

Two functions $u(x)$ and $v(x)$ are said to be linear dependent when there exists two nonzero constants c_1 and c_2 such that

$$c_1 u(x) + c_2 v(x) = 0 \tag{8.1}$$

for all values of x. Differentiating equation (8.1) with respect to x we obtain

$$c_1 u'(x) + c_2 v'(x) = 0 . \tag{8.2}$$

We can view equations (8.1) and (8.2) as a system of two linear equations with two unknowns c_1 and c_2. The determinant of such a system is the Wronskian

$$W(u, v) = uv' - u'v . \tag{8.3}$$

If $W(u, v)$ is nonzero for some value of x, then the only solution of the system of homogeneous linear equations is $c_1 = c_2 = 0$, and the functions are linearly independent. If, on the other hand, $W(u, v)$ is zero for all values of x, then there are solutions for nonzero values of c_1 and c_2, and the functions are linearly dependent.

Suppose that $u(x)$ and $v(x)$ are two solutions of the second order differential equation

$$Y''(x) = F(x)Y(x) . \tag{8.4}$$

If we multiply $v''(x) = F(x)v(x)$ by $u(x)$ and $u''(x) = F(x)u(x)$ by $v(x)$ and subtract, we conclude that the Wronskian is independent of x:

$$\left(uv' - u'v\right)' = \frac{d}{dx}W(u, v) = 0 . \tag{8.5}$$

Therefore if $W(u, v)$ is zero for a given value of x, then it is zero everywhere, and the solutions $u(x)$ and $v(x)$ of the differential equation (8.4) are linearly dependent.

If $W(u, v) \neq 0$ then

$$Y(x) = C_1 u(x) + C_2 v(x) \tag{8.6}$$

is a general solution of the differential equation (8.4). Taking into account that $W(u, u) = W(v, v) = 0$ we easily prove that C_1 and C_2 are given by

$$C_1 = \frac{W(v, Y)}{W(v, u)}, \quad C_2 = \frac{W(u, Y)}{W(u, v)} . \tag{8.7}$$

Notice that all the Wronskians in this equation are independent of x by virtue of the argument given above.

The practical value of the Wronskians in quantum mechanics was already pointed out some time ago [171].

8.3   The One-Dimensional Schrödinger Equation with a Finite Interaction Region

In what follows we consider a particle of mass m moving in a one-dimensional space under the effect of a force which is nonzero only in a finite region (x_L, x_R). The time-independent Schrödinger equation reads

$$-\frac{\hbar^2}{2m}\Psi''(x) + V(x)\Psi(x) = E\Psi(x) , \tag{8.8}$$

where

$$V(x) = \begin{cases} 0 & \text{if} \quad x < x_L \\ V_C(x) & \text{if} \quad x_L < x < x_R \\ 0 & \text{if} \quad x > x_R \end{cases} . \tag{8.9}$$

If $V_C(x)$ is continuous in (x_L, x_R) then $|V_C(x)|$ has a maximum V_0 there.

As we did so many times in preceding chapters, we first define dimensionless coordinate $q = (x - x_L)/L$, potential-energy function $\mathcal{V}(q) = V(Lq + x_L)/V_0$, and energy $\epsilon = E/V_0$, where $L = x_R - x_L$ and V_0 play the role of units of length and energy, respectively. Notice that $0 < q < 1$ and $|\mathcal{V}(q)| \leq 1$. In this way the problem reduces to solving the dimensionless Schrödinger equation

$$\Phi''(q) = 2a^2[\mathcal{V}(q) - \epsilon]\Phi(q) , \tag{8.10}$$

where the dimensionless parameter

$$a = \sqrt{\frac{mL^2V_0}{\hbar^2}} \tag{8.11}$$

is a measure of the potential strength.

The solution to the dimensionless Schrödinger equation for positive energy is

$$\Phi(x) = \begin{cases} \Phi_L(q) = A_L u_L(q) + B_L v_L(q) & \text{if} \quad q < 0 \\ \Phi_C(q) = A u(q) + B v(q) & \text{if} \quad 0 < q < 1 \\ \Phi_R(q) = A_R u_R(q) + B_R v_R(q) & \text{if} \quad q > 1 \end{cases} , \qquad (8.12)$$

where

$$u_L = \exp(ik_L q), v_L = \exp(-ik_L q), \ u_R = \exp(ik_R q), v_R = \exp(-ik_R q) , \qquad (8.13)$$

and $u(q)$ and $v(q)$ are two linearly independent solutions of the Schrödinger equation with the potential-energy function $\mathcal{V}_C(q) = V_C(Lq + x_L)/V_0$. In the present case

$$k_L = k_R = a\sqrt{2\epsilon} \qquad (8.14)$$

but we keep the notation sufficiently general so that the results apply to other situations as well.

We obtain four of the six coefficients A_L, B_L, A, B, A_R, and B_R from the equations given by the continuity conditions at $q = 0$ and $q = 1$, where the potential-energy function is discontinuous:

$$\begin{aligned} \Phi_L\left(0^-\right) &= \Phi_C\left(0^+\right), \ \Phi_L'\left(0^-\right) = \Phi_C'\left(0^+\right) , \\ \Phi_C\left(1^-\right) &= \Phi_R\left(1^+\right), \ \Phi_C'\left(1^-\right) = \Phi_R'\left(1^+\right) . \end{aligned} \qquad (8.15)$$

In order to solve the system of linear equations (8.15) for the chosen coefficients we systematically apply the Wronskian as in equation (8.7). For example, we obtain A_L and B_L in terms of A and B as follows:

$$A_L = \frac{W(v_L, u)_0}{W(v_L, u_L)_0} A + \frac{W(v_L, v)_0}{W(v_L, u_L)_0} B, \ B_L = \frac{W(u_L, u)_0}{W(u_L, v_L)_0} A + \frac{W(u_L, v)_0}{W(u_L, v_L)_0} B , \qquad (8.16)$$

where the subscript indicates that the functions are calculated at the point $q = 0$. Analogously, at $q = 1$ we write

$$A = \frac{W(v, u_R)_1}{W(v, u)_1} A_R + \frac{W(v, v_R)_1}{W(v, u)_1} B_R, \ B = \frac{W(u, u_R)_1}{W(u, v)_1} A_R + \frac{W(u, v_R)_1}{W(u, v)_1} B_R . \qquad (8.17)$$

Notice that equations (8.16) and (8.17) apply to any problem that we separate in three different spatial regions. The only requirement is that the pairs of functions (u_L, v_L), (u, v), and (u_R, v_R) are linearly independent solutions in each region.

Since $\mathcal{V}(q)$ and ϵ are real then $\Phi(q)^*$ is also a solution to the Schrödinger equation (8.10), and the Wronskian

$$W\left(\Phi, \Phi^*\right) = 2i\Im\left(\Phi^*\Phi'\right) , \qquad (8.18)$$

which is proportional to the current density, is independent of x. It follows from this equation that

$$|A_L|^2 - |B_L|^2 = |A_R|^2 - |B_R|^2 . \qquad (8.19)$$

If the flux of particles moves from left to right, then $B_R = 0$ and the transmission and reflection coefficients $T = |A_R/A_L|^2$ and $R = |B_L/A_L|^2$, respectively, satisfy the flux conservation condition $T + R = 1$.

On substituting equations (8.13) into equations (8.16) and (8.17) we obtain A_L and B_L in terms of A_R; in particular

$$\frac{A_R}{A_L} = \frac{2ik_L \exp(-ik_L) W(u, v)_1}{D},$$

$$D = \left[u'(0) + ik_L u(0)\right]\left[v'(1) - ik_L v(1)\right]$$
$$- \left[v'(0) + ik_L v(0)\right]\left[u'(1) - ik_L u(1)\right]. \tag{8.20}$$

In order to calculate the transmission coefficient, we have to solve the Schrödinger equation with the potential-energy function $\mathcal{V}_C(q)$ and obtain two linearly independent solutions $u(q)$ and $v(q)$. If we require that $u(q_0) = 1$, $u'(q_0) = 0$, $v(q_0) = 0$, and $v'(q_0) = 1$, then $W(u, v) = 1$ and the functions u and v are linearly independent. In such a case the transmission coefficient results to be $T = 4k_L^2/|D|^2$. However, when we solve the Schrödinger equation approximately $W(u, v)$ is close to but not exactly equal to unity and we therefore prefer to estimate T as

$$T = \left|\frac{A_R}{A_L}\right|^2 = \left|\frac{2k_L W(u, v)_1}{D}\right|^2. \tag{8.21}$$

Resonance states are a particular class of scattering states leading to outgoing waves in all channels [172]. In the case of the one-dimensional models treated here, the boundary conditions are $A_L = B_R = 0$ which determine complex values of the energy with precise physical interpretation. The real part of such energies are the resonance positions and their imaginary parts the resonance widths: $E = E_R - i\Gamma/(2\hbar)$.

In this chapter we show how to apply perturbation theory to the Schrödinger equation in the interaction region.

8.4 The Born Approximation

It is not difficult to derive the Born series for one-dimensional scattering models from the integration formulas of Appendix B. For simplicity and concreteness consider the Schrödinger equation (8.8), where the potential-energy function vanishes as $|x| \to \infty$. As usual, we define dimensionless coordinate $q = x/L$, energy $\mathcal{E} = mL^2 E/\hbar^2$, and potential-energy function $\lambda w(q) = mL^2 V(Lq)/\hbar^2$, in terms of an arbitrary unit of length L. We have purposely introduced a perturbation parameter λ because we are going to consider weak interactions.

We can view the dimensionless Schrödinger equation

$$\Phi''(q) + 2\mathcal{E}\Phi(q) = 2\lambda w(q)\Phi(q) \tag{8.22}$$

as a particular case of those discussed in Appendix B with $f(q) = 2\lambda w(q)\Phi(q)$; therefore, the general solution reads

$$\Phi(q) = C_1 \exp(-ikq) + C_2 \exp(ikq) - \frac{i\lambda}{k}\int_{q_i}^{q} \exp\left[ik(q - q')\right] w(q') \Phi(q') \, dq'$$
$$+ \frac{i\lambda}{k}\int_{q_i}^{q} \exp\left[ik(q' - q)\right] w(q') \Phi(q') \, dq', \quad k = \sqrt{2\mathcal{E}}, \tag{8.23}$$

where q_i is an arbitrary coordinate point. This solution satisfies the boundary conditions

$$\Phi(q) \to \exp(ikq) + B_L \exp(-ikq), \quad q \to -\infty; \quad \Phi(q) \to A_R \exp(ikq), \quad q \to \infty \tag{8.24}$$

provided that

$$C_1 + \frac{i\lambda}{k} \int_{q_i}^{-\infty} \exp\left(ikq'\right) w\left(q'\right) \Phi\left(q'\right) dq' = B_L, \tag{8.25}$$

$$C_2 - \frac{i\lambda}{k} \int_{q_i}^{-\infty} \exp\left(-ikq'\right) w\left(q'\right) \Phi\left(q'\right) dq' = 1, \tag{8.26}$$

$$C_1 + \frac{i\lambda}{k} \int_{q_i}^{\infty} \exp\left(ikq'\right) w\left(q'\right) \Phi\left(q'\right) dq' = 0, \tag{8.27}$$

$$C_2 - \frac{i\lambda}{k} \int_{q_i}^{\infty} \exp\left(-ikq'\right) w\left(q'\right) \Phi\left(q'\right) dq' = A_R. \tag{8.28}$$

Taking into account equations (8.26) and (8.27) we easily rearrange equation (8.23) in a more compact form:

$$\Phi(q) = \exp(ikq) - \frac{i\lambda}{k} \int_{-\infty}^{\infty} \exp\left(ik\left|q - q'\right|\right) w\left(q'\right) \Phi\left(q'\right) dq'. \tag{8.29}$$

Moreover, it follows from equations (8.25)–(8.28) that

$$B_L = -\frac{i\lambda}{k} \int_{-\infty}^{\infty} \exp\left(ikq'\right) w\left(q'\right) \Phi\left(q'\right) dq', \tag{8.30}$$

$$A_R = 1 - \frac{i\lambda}{k} \int_{-\infty}^{\infty} \exp\left(-ikq'\right) w\left(q'\right) \Phi\left(q'\right) dq'. \tag{8.31}$$

In order to compare the results of this section with those of the preceding one, it should be taken into consideration that here we have arbitrarily chosen $A_L = 1$ to derive the customary integral equations for the application of the Born approximation [173].

If the interaction is weak, we try an approximate solution in the form of a perturbation series

$$\Phi(q) = \sum_{j=0}^{\infty} \Phi_j(q)\lambda^j \tag{8.32}$$

which we substitute into equation (8.29) to derive a recurrence relation for the coefficients:

$$\Phi_j(q) = \delta_{j0} \exp(ikq) - \frac{i}{k} \int_{-\infty}^{\infty} \exp\left(ik\left|q - q'\right|\right) w\left(q'\right) \Phi_{j-1}\left(q'\right) dq' \tag{8.33}$$

that leads to the well-known Born approximation [173, 174].

In the first iteration we have $\Phi_0(q) = \exp(ikq)$ and

$$B_L \approx -\frac{i\lambda}{k} \int_{-\infty}^{\infty} \exp\left(2ikq'\right) w\left(q'\right) dq', \tag{8.34}$$

$$A_R \approx 1 - \frac{i\lambda}{k} \int_{-\infty}^{\infty} w\left(q'\right) dq'. \tag{8.35}$$

Notice that this approach is not restricted to a potential-energy function that vanishes outside a finite region, but it is necessary that $w(q)$ be integrable over all the range of values of q.

8.5   An Exactly Solvable Model: The Square Barrier

The square barrier is one of the simplest models commonly chosen to illustrate the tunnel effect in most textbooks on quantum mechanics [175]. We easily solve the Schrödinger equation because $V_C(x) = V_0 > 0$ is independent of x. It is customary to consider the two cases $E < V_0$ ($\epsilon < 1$) and $E > V_0$ ($\epsilon > 1$) separately. In the former we have $u(q) = \cosh(kq)$ and $v(q) = \sinh(kq)/k$, where $k = a\sqrt{2(1 - \epsilon)}$, whereas $u(q) = \cos(k'q)$, $v(q) = \sin(k'q)/k'$, and $k' = a\sqrt{2(\epsilon - 1)}$ hold for the latter. The transmission coefficient is therefore given by

$$T = \frac{4\epsilon(1 - \epsilon)}{4\epsilon(1 - \epsilon) + \sinh(k)^2}, \ \epsilon < 1, \tag{8.36}$$

$$T = \frac{4\epsilon(\epsilon - 1)}{4\epsilon(\epsilon - 1) + \sin(k')^2}, \ \epsilon > 1. \tag{8.37}$$

It is not difficult to verify that both expressions give exactly the same result when $\epsilon = 1$ ($E = V_0$):

$$T = \frac{2}{2 + a^2}. \tag{8.38}$$

In principle it is not necessary to use two expressions for the transmission coefficient because any of them applies to both situations. For example, when $\epsilon > 1$ we write $k = ik'$ and equation (8.36) becomes equation (8.37). However, we have written the transmission coefficient in terms of two formulas for clarity.

In what follows we apply two different perturbation expansions to the exact result (8.36), (8.37). When $E >> V_0$ we expand the transmission coefficient given by equation (8.37) in a V_0-power series. A straightforward calculation shows that

$$T = 1 - \frac{\sin(\alpha\sqrt{2E})^2}{4E^2}V_0^2 + \left[\frac{\alpha\sqrt{2}\sin(2\alpha\sqrt{2E})}{8E^{5/2}} - \frac{\sin(\alpha\sqrt{2E})^2}{4E^3}\right]V_0^3 + \cdots, \tag{8.39}$$

where $\alpha = \sqrt{mL^2/\hbar^2} = a/\sqrt{V_0}$. In this case the unperturbed model is the free particle and the whole interaction is chosen to be the perturbation.

In the first iteration, the Born approximation developed in the preceding section gives us

$$T = |A_R|^2 = 1 + \frac{\alpha^2 V_0^2}{2E} + \cdots, \tag{8.40}$$

which does not agree with equation (8.39). However, if we calculate the transmission coefficient as $T = 1 - |B_L|^2$, we obtain the first two terms of equation (8.39). Notice that $|A_R|^2 + |B_L|^2 > 1$ in the first iteration because the Born approximation violates unitarity or conservation of flux [173].

If we assume that a is a small parameter we can expand the transmission coefficient in a Taylor series about $a = 0$; both expressions (8.36), (8.37) give exactly the same perturbation series in powers of a^2:

$$T = 1 - \frac{a^2}{2\epsilon} + \frac{4\epsilon^2 - 4\epsilon + 3}{12\epsilon^2}a^4 - \frac{32\epsilon^4 - 64\epsilon^3 + 152\epsilon^2 - 120\epsilon + 45}{360\epsilon^3}a^6 + \cdots. \tag{8.41}$$

The first two terms of this perturbation series would resemble the Born approximation (8.40) if it were not for the sign of the second term. It is at first sight surprising that we obtain qualitatively different results from what appears to be the same expansion. In order to bring both results to agreement notice that we can rewrite the Born approximation $A_R \approx 1 - i\xi$, $\xi = a/k_L$, as $A_R \approx 1/(1 + i\xi)$ keeping the same order of accuracy. From this expression we already obtain the first two terms of the series (8.41): $T \approx 1/(1 + \xi^2) \approx 1 - \xi^2$.

It follows from equation (8.38) that the radius of convergence of the series (8.41) for $\epsilon = 1$ is determined by a pair of conjugate imaginary poles $a_s = \pm i\sqrt{2}$. We have calculated the roots of the Taylor expansion of $1/T$ about $a = 0$ for sufficiently great order, and it seems that for other values of ϵ the zero closest to origin in the complex a^2 plane is always real and negative. This calculation is straightforward if one resorts to the Maple commands *taylor* and *fsolve/complex*. Therefore, we can easily obtain the radius of convergence of the perturbation expansion (8.41) in terms of ϵ as follows: substitute $b = a^2$ in the inverse of any of the exact expressions (8.36) or (8.37), find the real roots of the resulting equation for given value of ϵ (using Maple command *fsolve*, for example), and select the one closest to origin $b_s = a_s^2$. Figure 8.1 shows the radius of convergence $r_c = |a_s|^2$ as a function of ϵ for the square barrier and for a parabolic barrier to be discussed later on in this chapter. It is worth noticing that we easily obtain the radius of convergence of the perturbation series for T from the roots of the perturbation series for $1/T$ (or from the zeros of the denominator D in equation (8.20)). In this way we can determine the radius of convergence of perturbation series for models that are not exactly solvable.

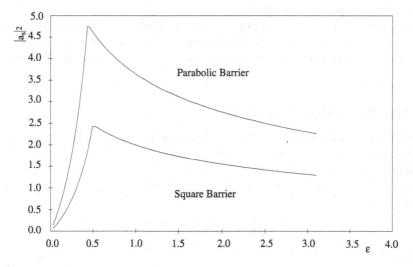

FIGURE 8.1
Radius of convergence $|a_s|^2$ of the perturbation series in powers of a^2 for the transmission coefficient for the square and parabolic barriers.

8.6 Nontrivial Simple Models

8.6.1 Accurate Nonperturbative Calculation

In order to test the results of perturbation theory we compare them with those provided by a simple accurate nonperturbative calculation based on a power series approach.

The dimensionless Schrödinger equation (8.10) for $0 < q < 1$ is $Y''(q) = F(q)Y(q)$, where $F(q) = 2a^2[V(q) - \epsilon]$. If the Taylor series about a point $0 < q_0 < 1$,

$$F(q) = \sum_{j=0}^{\infty} F_j (q - q_0)^j \, , \tag{8.42}$$

converges for all $0 \le q \le 1$, then we try a similar Taylor expansion for any solution $Y(q)$:

$$Y(q) = \sum_{j=0}^{\infty} Y_j (q - q_0)^j \, . \tag{8.43}$$

The recurrence relation

$$Y_{n+2} = \frac{1}{(n+1)(n+2)} \sum_{j=0}^{n} F_{n-j} Y_j, \ n = 0, 1, \ldots \tag{8.44}$$

yields the coefficients of the series (8.43) in terms of Y_0 and Y_1. The choices $Y_0 = 1$, $Y_1 = 0$ and $Y_0 = 0$, $Y_1 = 1$ give us two independent solutions $u(q)$ and $v(q)$, respectively. Since such series converge for $q = 0$ and $q = 1$ we obtain the transmission coefficient from equation (8.21) and use this result to test the perturbation calculation.

An alternative approach is to make use of the Maple command *dsolve* with the option *type = numeric* to obtain the linearly independent solutions and their derivatives at $q = 0$ and $q = 1$, and substitute them into equation (8.21). We show a simple procedure for this calculation in the program section.

8.6.2   First Perturbation Method

In what follows we discuss the application of perturbation theory to the Schrödinger equation in the interaction region. In order to simplify the notation we will omit the subscript in $\Phi_C(q)$ and $V_C(q)$ hoping that it may not be confusing.

The simplest perturbation approach consists of choosing the whole interaction as a perturbation. Consequently, we write

$$\Phi''(q) = 2a^2[\lambda V(q) - \epsilon]\Phi(q) \tag{8.45}$$

and expand

$$\Phi(q) = \sum_{j=0}^{\infty} \Phi_j(q)\lambda^j \, . \tag{8.46}$$

Substituting this expansion into equation (8.45) and collecting powers of λ, we obtain the perturbation equations

$$\Phi_j''(q) + k_L^2 \Phi_j(q) = 2a^2 V(q)\Phi_{j-1}(q), \ j = 1, 2, \ldots \tag{8.47}$$

that are inhomogeneous ordinary differential equations with constant coefficients that we easily solve hierarchically by means of the method developed in Appendix B. At each perturbation order we have to calculate

$$\Phi_j(q) = \frac{2a^2}{k_L} \int_0^q \sin[k_L(q - t)]V(t)\Phi_{j-1}(t) \, dt, \ j = 1, 2, \ldots \, . \tag{8.48}$$

Notice that we have chosen the lower limit of the integral so that $\Phi_j(0) = \Phi'_j(0) = 0$, which results in $\Phi(0) = \Phi_0(0)$ and $\Phi'(0) = \Phi'_0(0)$. At the end of the calculation we set the dummy perturbation parameter λ equal to unity.

We substitute $u_j(q)$ and $v_j(q)$ for $\Phi_j(q)$ in the equations above and obtain the perturbation series for the functions $u(q)$ and $v(q)$, respectively, that we substitute into equation (8.21) for the transmission coefficient T. If we choose the functions $u(q)$ and $v(q)$ such that

$$u(0) = 1, \; u'(0) = 0, \; v(0) = 0, \; v'(0) = 1 \,, \tag{8.49}$$

then

$$u_0(q) = \cos(k_L q), \; v_0(q) = \frac{\sin(k_L q)}{k_L} \,. \tag{8.50}$$

This perturbation method applies only when $\epsilon > 1 > |\mathcal{V}(q)|$ because we have assumed that the interaction is a small perturbation.

In order to test our perturbation equations we apply them to the square barrier and compare the results with the exact ones. In this case $\mathcal{V}(q) = 1$ in the interaction region. A straightforward calculation yields

$$u(q) = \cos(\theta) + \lambda \frac{aq \sin(\theta)}{\sqrt{2\epsilon}} + \lambda^2 \left[\frac{aq \sin(\theta)}{4\epsilon\sqrt{2\epsilon}} - \frac{(aq)^2 \cos(\theta)}{4\epsilon} \right] + \cdots \tag{8.51}$$

$$v(q) = \frac{\sin(\theta)}{a\sqrt{2\epsilon}} + \lambda \left[\frac{\sin(\theta)}{2a\epsilon\sqrt{2\epsilon}} - \frac{q \cos(\theta)}{2\epsilon} \right] + \cdots \,, \tag{8.52}$$

where $\theta = a\sqrt{2\epsilon}q$. We do not show terms of higher order because they become increasingly complicated. The calculation is straightforward and Maple greatly facilitates it. In order to verify equations (8.51) and (8.52) we compare them with the Taylor expansions of the exact functions

$$u(q) = \cos\left[a\sqrt{2(\epsilon - \lambda)}q\right], \; v(q) = \frac{\sin[a\sqrt{2(\epsilon - \lambda)}q]}{a\sqrt{2(\epsilon - \lambda)}} \tag{8.53}$$

about $\lambda = 0$. We do not discuss this perturbation method any further because it is not practical.

8.6.3   Second Perturbation Method

In order to develop a perturbation method that applies to all values of ϵ we choose the square barrier as the unperturbed model and write

$$\mathcal{V}(q, \lambda) = 1 + \lambda \Delta \mathcal{V}(q), \; \Delta \mathcal{V}(q) = \mathcal{V}(q) - 1 \,, \tag{8.54}$$

so that $\mathcal{V}(q, 0) = 1$ and $\mathcal{V}(q, 1) = \mathcal{V}(q)$ in the interaction region.

For simplicity we treat the cases $\epsilon < 1$ and $\epsilon > 1$ separately. The perturbation equations for the former case are

$$\Phi''_j(q) - k^2 \Phi_j(q) = 2a^2 \Delta \mathcal{V}(q)\Phi_{j-1}(q), \; j = 1, 2, \ldots, \; k = a\sqrt{2(1 - \epsilon)} \,, \tag{8.55}$$

and according to the results in Appendix B we obtain the corrections to $u(q)$ and $v(q)$ as follows:

$$\Phi_j(q) = \frac{2a^2}{k} \int_0^q \sinh[k(q - t)]\Delta \mathcal{V}(t)\Phi_{j-1}(t)\, dt \,, \tag{8.56}$$

where

$$u_0(q) = \cosh(kq), \quad v_0(q) = \frac{\sinh(kq)}{k} \tag{8.57}$$

satisfy the boundary conditions (8.49).

The solutions for $\epsilon > 1$ read

$$\Phi_j(q) = \frac{2a^2}{k} \int_0^q \sin[k(q-t)]\Delta\mathcal{V}(t)\Phi_{j-1}(t)\,dt \,, \tag{8.58}$$

where

$$u_0(q) = \cos(kq), \quad v_0(q) = \frac{\sin(kq)}{k}, \quad k = a\sqrt{2(\epsilon-1)} \,. \tag{8.59}$$

Taking the limit $k \to 0$ in any of the cases above we obtain the equations for $\epsilon = 1$:

$$\Phi_j(q) = 2a^2 \int_0^q (q-t)\Delta\mathcal{V}(t)\Phi_{j-1}(t)\,dt \,, \tag{8.60}$$

where

$$u_0(q) = 1, \quad v_0(q) = q \tag{8.61}$$

satisfy the boundary conditions (8.49).

As a simple nontrivial example we choose the parabolic barrier

$$V(x) = \begin{cases} 0 & \text{if } |x| > L \\ V_0(1 - 4x^2/L^2) & \text{if } |x| < L \end{cases} \tag{8.62}$$

that leads to the dimensionless potential-energy function

$$\mathcal{V}(q) = \begin{cases} 0 & \text{if } q < 0 \text{ or } q > 1 \\ 4q(1-q) & \text{if } 0 < q < 1 \end{cases} \,. \tag{8.63}$$

We have calculated several perturbation corrections using Maple but only the lowest order contributions are simple enough to be shown here. We have

$$\begin{aligned} u(q) &= \cosh(kq) + \lambda a^2 \left\{ \frac{2q(q-1)\cosh(kq)}{k^2} \right. \\ &\left. - \left[\frac{q(3-6q+4q^2)}{3k} + \frac{2(q-1)}{k^3} \right] \sinh(kq) \right\} + \cdots \end{aligned} \tag{8.64}$$

$$\begin{aligned} v(q) &= \frac{\sinh(kq)}{k} + \lambda a^2 \left\{ \left[\frac{1-2q+2q^2}{k^3} + \frac{2}{k^5} \right] \sinh(kq) \right. \\ &\left. - \left[\frac{q(3-6q+4q^2)}{3k^2} + \frac{2q}{k^4} \right] \cosh(kq) \right\} + \cdots \end{aligned} \tag{8.65}$$

for $\epsilon < 1$,

$$u(q) = 1 - \lambda a^2 \frac{q^2(3-4q+2q^2)}{3} + \cdots \tag{8.66}$$

$$v(q) = q - \lambda a^2 \frac{q^3(5-10q+6q^2)}{15} + \cdots \,, \tag{8.67}$$

for $\epsilon = 1$, and

$$
\begin{aligned}
u(q) &= \cos(kq) + \lambda a^2 \left\{ \frac{2q(1-q)\cos(kq)}{k^2} \right. \\
&\quad \left. - \left[\frac{q(3 - 6q + 4q^2)}{3k} + \frac{2(1-q)}{k^3} \right] \sin(kq) \right\} + \cdots
\end{aligned}
\tag{8.68}
$$

$$
\begin{aligned}
v(q) &= \frac{\sin(kq)}{k} + \lambda a^2 \left\{ \left[\frac{2}{k^5} - \frac{1 - 2q + 2q^2}{k^3} \right] \sin(kq) \right. \\
&\quad \left. + \left[\frac{q(3 - 6q + 4q^2)}{3k^2} - \frac{2q}{k^4} \right] \cos(kq) \right\} + \cdots
\end{aligned}
\tag{8.69}
$$

for $\epsilon > 1$. Notice that we obtain the case $\epsilon > 1$ by substituting ik for k in the case $\epsilon < 1$ as argued earlier for the square barrier.

It is instructive to discuss the form of the transmission coefficient for the simplest case $\epsilon = 1$. Through terms of second order we have

$$
\begin{aligned}
T &= \frac{8a^2}{4a^2(a^2 + 2) - 8a^4(a^2 + 5)\lambda/15 + 4a^4(6a^2 + 5)(2a^2 + 35)\lambda^2/1575 + \cdots} \\
&= \frac{2}{2 + a^2} + \frac{4a^2(a^2 + 5)}{(a^2 + 2)^2}\lambda + \frac{2a^2(-350 + 85a^2 + 36a^4 + 16a^6)}{1575(a^2 + 2)^3}\lambda^2 + \cdots,
\end{aligned}
\tag{8.70}
$$

where, as expected, the first term of the λ-power series is the transmission coefficient for the square barrier. It must be kept in mind that the rational approximation always gives better results than the power series because the convergence radius of the latter is determined by (usually complex) zeros of the denominator that do not affect the accuracy of the former.

We do not discuss this approach any further because we think that the third perturbation method developed below deserves more attention than the two just outlined.

8.6.4 Third Perturbation Method

Analytical calculation of perturbation corrections of sufficiently great order based on Maple procedures for the methods in the preceding sections is time and memory consuming. The remarkably simple form of the Taylor expansion about $a = 0$ of the transmission coefficient for the square barrier equation (8.41) suggests that it is worth trying a perturbation approach based on such a series. Fortunately, it is not difficult to develop such a method; we simply choose $\beta = a^2$ as perturbation parameter in the dimensionless Schrödinger equation (8.10) or, equivalently, we introduce a dummy perturbation parameter λ, as in the preceding methods, and write

$$
\Phi''(q) = 2\lambda a^2 [V(q) - \epsilon]\Phi(q) .
\tag{8.71}
$$

According to Appendix B the solutions to the perturbation equations

$$
\Phi_j''(q) = 2a^2 [V(q) - \epsilon]\Phi_{j-1}(q), \quad j = 1, 2, \ldots
\tag{8.72}
$$

are given by the simple formula

$$
\Phi_j(q) = 2a^2 \int_0^q (q - t)[V(t) - \epsilon]\Phi_{j-1}(t) \, dt, \quad j = 1, 2, \ldots ,
\tag{8.73}
$$

where $u_0(q) = 1$ and $v_0(q) = q$. The calculation is therefore as simple as the particular case $\epsilon = 1$ in the second perturbation method above.

It is not difficult to prove that the perturbation series for the eigenfunction $\Phi(q)$ converges uniformly in $q \in (0, 1)$ for all values of a and ϵ. It is sufficient to prove that the series of positive terms $|\Phi_j(q)|$ converges in that interval [96]. We first notice that $2a^2|\mathcal{V}(q) - \epsilon| \leq k^2 = 2a^2|1 - \epsilon|$, and that

$$\left|\Phi_j(q)\right| \leq k^2 \int_0^q (q - t) \left|\Phi_{j-1}(t)\right| dt . \tag{8.74}$$

By straightforward application of this recursive inequality to $\Phi_0(q) = u_0(q) = 1$ we conclude that $|u_j(q)| \leq (kq)^{2j}/(2j)!$. Analogously, if follows from $\Phi_0(q) = v_0(q) = q$ that $|v_j(q)| \leq k^{2j} q^{2j+1}/(2j + 1)!$. Therefore the series with terms $|u_j(q)|$ and $|v_j(q)|$ converge for all q and

$$\sum_{j=0}^{\infty} |u_j(q)| \leq \cosh(kq), \quad \sum_{j=0}^{\infty} |v_j(q)| \leq \frac{\sinh(kq)}{k} , \tag{8.75}$$

where the equality holds for the square barrier. Since the perturbation series for $u(q)$ and $v(q)$ converge for all a, ϵ, and q, then the present perturbation theory gives the transmission coefficient as accurately as desired by means of equation (8.21).

To test our perturbation equations and Maple program we have verified that they already yield the perturbation series in equation (8.41) for the square barrier. In what follows we show results for the parabolic barrier introduced above; that is to say, $\mathcal{V}(q) = 4q(1 - q)$ for $0 < q < 1$ and zero elsewhere.

The calculation is straightforward, and the simple Maple procedure shown in the program section allows one to obtain perturbation coefficients of sufficiently great order. The first terms of the linearly independent solutions are

$$
\begin{aligned}
u(q) &= 1 - a^2 q^2 \left(\epsilon - \frac{4q}{3} + \frac{2q^2}{3} \right) + a^4 q^4 \left[\frac{\epsilon^2}{6} - \frac{8\epsilon q}{15} \right. \\
&\quad \left. + \left(\frac{16}{45} + \frac{14\epsilon}{45} \right) q^2 - \frac{8q^3}{21} + \frac{2q^4}{21} \right] + \cdots
\end{aligned}
\tag{8.76}
$$

$$
\begin{aligned}
v(q) &= q - a^2 q^3 \left(\frac{\epsilon}{3} - \frac{2q}{3} + \frac{2q^2}{5} \right) + a^4 q^5 \left[\frac{\epsilon^2}{30} - \frac{2\epsilon q}{15} \right. \\
&\quad \left. + \left(\frac{8}{63} + \frac{26\epsilon}{315} \right) q^2 - \frac{16q^3}{105} + \frac{2q^4}{45} \right] + \cdots .
\end{aligned}
\tag{8.77}
$$

These functions are exact at $q = 0$, and the perturbation series converge at $q = 1$ for all values of a and ϵ. Substituting them into equation (8.21), and keeping terms of second order in λ (fourth order in a), we obtain the transmission coefficient

$$
\begin{aligned}
T &= \frac{\epsilon}{\epsilon + 2a^2/9 + (8/105 - 4\epsilon/45)\, a^4 + \cdots} \\
&= 1 - \frac{2a^2}{9\epsilon} + \frac{4(35 - 54\epsilon + 63\epsilon^2)a^4}{2835\epsilon^2} + \cdots .
\end{aligned}
\tag{8.78}
$$

The rational approximation is valid for all values of a and ϵ because the perturbation series for the functions $u(q)$ and $v(q)$ converge as proved above. On the other hand, the radius of convergence of the power series is limited by the zeros of the denominator (poles of T). Notice, for example, that the former gives the correct result $T = 0$ when $\epsilon = 0$ but the latter is singular for this value of ϵ, failure that shares with the Born approximation [173]. We have explicitly shown the power series in equation (8.78) for a purely pedagogical reason, but in order to obtain reasonable numerical results we directly substitute the perturbation series for the functions $u(q)$ and $v(q)$ into equation (8.21) and calculate T by means of the resulting rational expression.

In Figure 8.1 we show the radius of convergence $r_c = |a_s|^2$ of the perturbation series (8.78) obtained from the zero of $1/T$ closest to the origin of the a^2 complex plane. The curves $r_c(\epsilon)$ for the parabolic and square barrier (both of the same height V_0) are similar except that the former is greater.

Although the perturbation series for the linearly independent solutions converge for all values of a and ϵ, the rate of convergence depends on the values of these parameters. We calculate the rate of convergence of the rational perturbation approximation to T as $\log(|T^{exact} - T^{PT}|/T^{exact})$, where T^{exact} is given by any of the accurate nonperturbative methods discussed above in Section 8.6.1. In Figure 8.2 we appreciate that the rate of convergence decreases as either a or ϵ increases, and that the effect of the former is more noticeable as it is the actual perturbation parameter.

For completeness in Figure 8.3 we show the transmission coefficient for the parabolic barrier for three values of a. The perturbative and nonperturbative methods give exactly the same results but the latter are faster for great values of a.

An interesting application of the perturbation method is the calculation of resonance energies as roots of the denominator $|D|^2$ of the transmission coefficient T. Perturbation theory provides an analytical expression for $|D|^2$ in terms of a and ϵ. For a given value of a we solve $|D|^2 = 0$ for ϵ for increasing values of the perturbation order till convergence. Figure 8.4 shows the real and imaginary parts of the complex root for the square and parabolic barriers. Of the two complex conjugate roots we arbitrarily choose the one with $\Im(\epsilon) < 0$. Only energies with positive real parts should be interpreted as resonances, which appear for sufficiently large values of a. Although one of the barriers is smooth and the other exhibits sharp edges, the behavior of the root as a function of a is quite similar.

In principle, one can apply this perturbation approach to one-dimensional models supporting bound states. We do not discuss such a case here but leave it as an interesting exercise for the reader.

8.7 Perturbation Theory for Resonance Tunneling

In what follows we briefly illustrate a straightforward application of perturbation theory to a particle penetrating a barrier that is not restricted to a finite coordinate interval as in the preceding examples. More precisely, we assume that the potential-energy function $V(x)$ is positive definite, vanishes as $|x| \to \infty$, and exhibits a single maximum at $x = x_0$. In particular we are interested in resonance states $\Psi(x)$ that satisfy the boundary conditions

$$\Psi(x) \to \begin{cases} A_L \exp(-iKx) & \text{if} \quad x \to -\infty \\ A_R \exp(iKx) & \text{if} \quad x \to \infty \end{cases}, \tag{8.79}$$

where $K = \sqrt{2mE/\hbar^2}$ [172]. That is to say, we have outgoing waves in both channels. If we choose only incoming waves we obtain exactly the same resonance energies. Because the set of complex values of the energy that satisfy such boundary conditions is discrete, their calculation resembles that of bound-state energies.

As usual, we define dimensionless coordinates $q = x/\gamma$, energy $\epsilon = m\gamma^2 E/\hbar^2$, and potential-energy function $v(q) = m\gamma^2 V(\gamma q)/\hbar^2$ in terms of an appropriate length unit γ, so that we are left

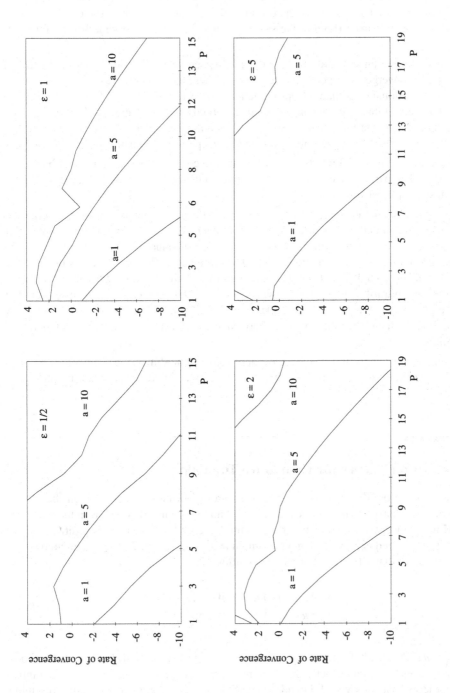

FIGURE 8.2

Rate of convergence $\log(|T^{exact} - T^{PT}|/T^{exact})$ of the rational perturbation approach to the transmission coefficient for a parabolic barrier. P is the perturbation order.

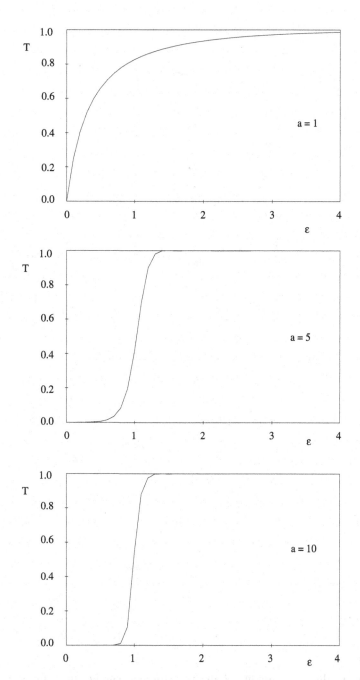

FIGURE 8.3
Transmission coefficient for the parabolic barrier for three values of a.

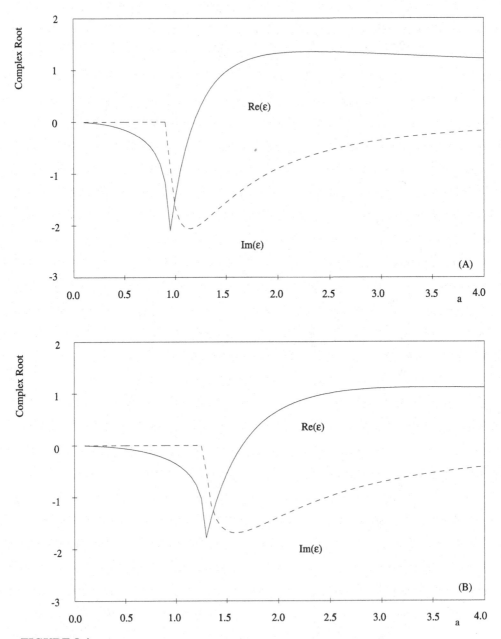

FIGURE 8.4

Complex root of the denominator of the transmission coefficient T for the square (A) and parabolic (B) barriers.

with the dimensionless Hamiltonian operator

$$\hat{\mathcal{H}} = \frac{m\gamma^2}{\hbar^2}\hat{H} = -\frac{1}{2}\frac{d^2}{dq^2} + v(q) . \tag{8.80}$$

In order to apply the polynomial approximation developed in Section 7.2.1 we expand

$$v(q) = \sum_{j=0}^{\infty} v_j (q - q_0)^j , \quad v_j = \frac{1}{j!}\frac{d^j v}{dq^j}\bigg|_{q=q_0} \tag{8.81}$$

about the maximum $q_0 = x_0/\gamma$ of $v(q)$ (notice that in this case $v_2 < 0$). We then introduce a perturbation parameter β by means of a second change of variable $z = (q - q_0)/\beta$, and obtain the dimensionless Hamiltonian operator

$$\hat{h} = \beta^2\left(\hat{\mathcal{H}} - v_0\right) = -\frac{1}{2}\frac{d^2}{dz^2} + \frac{1}{2}z^2 + \sum_{j=1}^{\infty} b_j\beta^j z^{j+2} , \tag{8.82}$$

where $\beta = (2v_2)^{-1/4}$ is a complex number, and the coefficients $b_j = v_{j+2}/(2v_2)$ are real.

We apply perturbation theory to the dimensionless Schrödinger equation $\hat{h}\Phi = e\Phi$ in the usual way and obtain

$$\Phi = \sum_{j=0}^{\infty} \Phi_j\beta^j, \quad e = \sum_{j=0}^{\infty} e_{2j}\beta^{2j}, \quad e_0 = n + \frac{1}{2}, \quad n = 0, 1, \ldots , \tag{8.83}$$

from which it follows that

$$\epsilon = v_0 + \sum_{j=0}^{\infty} e_{2j}\beta^{2j-2} \tag{8.84}$$

as discussed in Section 7.2.1. Notice that the energy corrections e_{2j} are real and that β^2 is purely imaginary; consequently, the terms with j odd are real and those with j even are imaginary.

As in the case of bound states considered in Section 7.2.1, we treat symmetric potential-energy functions $v(-q) = v(q)$ in a slightly different and more convenient way. In this case the maximum occurs at $q = 0$ and the Taylor expansion of the barrier reads

$$v(q) = \sum_{j=0}^{\infty} v_j q^{2j}, \quad v_j = \frac{1}{(2j)!}\frac{d^{2j}v}{dq^{2j}}\bigg|_{q=0} . \tag{8.85}$$

By straightforward application of perturbation theory to $\hat{h}\Phi = e\Phi$, where

$$\begin{aligned}
\hat{h} &= \beta\left(\hat{\mathcal{H}} - v_0\right) = -\frac{1}{2}\frac{d^2}{dz^2} + \frac{1}{2}z^2 + \sum_{j=1}^{\infty} b_j\beta^j z^{2j+2} , \\
z &= \frac{q}{\sqrt{\beta}}, \quad b_j = \frac{v_j}{2v_2}, \quad \beta = \frac{1}{\sqrt{2v_1}} ,
\end{aligned} \tag{8.86}$$

we obtain a perturbation series of the form

$$\epsilon = v_0 + \sum_{j=0}^{\infty} e_j\beta^{j-1} . \tag{8.87}$$

The energy coefficients e_j are real because the potential parameters b_j are real, and the value of the perturbation parameter β is purely imaginary because $v_1 < 0$. Again the terms with j odd are real and those with j even are imaginary.

Any of the methods developed in Chapters 2 and 3 is suitable for obtaining the perturbation series (8.83) or (8.87). In particular, the methods of Dalgarno and Stewart and Fernández and Castro clearly show that every perturbation correction to the eigenfunction Φ_j is a polynomial function times the Gaussian function $\exp(-z^2/2)$, and therefore square integrable in $-\infty < z < \infty$. The application of such methods is straightforward as the reader may easily verify; on the other hand, it is not obvious at first sight that the method of Swenson and Danforth developed in Chapter 3 is suitable for this problem because the Hamiltonian operator $\hat{h}$ is not hermitian (remember that β is complex). For this reason we discuss the latter approach with some detail in what follows.

In order to apply the method of Swenson and Danforth we introduce the c-product

$$(f|g) = \int_{-\infty}^{\infty} f(z)g(z)\,dz \,, \tag{8.88}$$

which is a complex number with finite modulus if the complex-valued functions $f(z)$ and $g(z)$ are square integrable [176]. Notice that $(f|f)$ cannot play the role of a norm because it is a complex number; however we can normalize $f(z)$ so that $(f|f) = 1$. The advantage of the c-product is that the hypervirial and Hellmann–Feynman theorems

$$\left(\Phi \left| \left[\hat{h}, \hat{W}\right] \right| \Phi \right) = 0, \quad \frac{\partial e}{\partial \beta}(\Phi|\Phi) = \left(\Phi \left| \frac{\partial \hat{h}}{\partial \beta} \right| \Phi \right) , \tag{8.89}$$

respectively, are valid provided that $\hat{W}$ is a linear operator. The reader may easily prove them by simply repeating the arguments given in Section 3.2. Consequently, it makes sense to define the complex moments $Z_k = (\Phi|z^k|\Phi)$ normalized as $Z_0 = 1$, and expand them in powers of β:

$$Z_k = \sum_{j=0}^{\infty} Z_{k,j}\beta^j \,. \tag{8.90}$$

The application of the method of Swenson and Danforth is therefore straightforward following the lines indicated in Section 3.3.1 [177].

It is always instructive to apply approximate methods to exactly solvable models. In this case we choose the Eckart barrier [178]

$$v(q) = \frac{A\exp(q+q_m)}{1 + \exp(q+q_m)} + \frac{B\exp(q+q_m)}{[1 + \exp(q+q_m)]^2}, \quad q_m = \ln\left(\frac{A+B}{B-A}\right) , \tag{8.91}$$

where $B > |A|$. This potential-energy function exhibits a maximum $v_0 = (A+B)^2/(4B)$ at $q = 0$, and is symmetric about $q = q_m = 0$ when $A = 0$. Figure 8.5 shows the Eckart barrier $v(q)$ for three values of A.

It is not difficult to solve the Schrödinger equation with the Eckart potential in terms of hypergeometric functions and obtain the transmission and reflection coefficients in terms of gamma functions [178]. Setting the appropriate coefficients of the exact solution equal to zero, in order to have only outgoing or incoming waves, we obtain the resonance energies [177]

$$\epsilon^{\pm} = \frac{\left\{\left[\sqrt{8B-1} \pm (2n+1)i\right]^2 + 8A\right\}^2}{32\left[\sqrt{8B-1} \pm (2n+1)i\right]^2} , \tag{8.92}$$

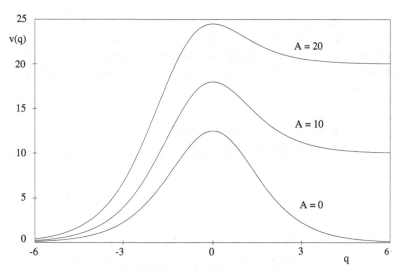

FIGURE 8.5
Dimensionless Eckart barrier for $B = 50$ and three values of A.

where $n = 0, 1, \ldots$.

The polynomial approximation yields the perturbation series (8.84), where

$$\beta = \left[-\frac{8B^3}{(B^2 - A^2)^2} \right]^{1/4} . \tag{8.93}$$

However, it is easier to study the convergence properties of the closely related expansion in powers of $1/\sqrt{B}$, which we easily derive from equation (8.92) either by hand or, much faster and more easily, by means of the Maple command *series*:

$$\epsilon = \frac{B}{4} \mp ie_0 \sqrt{\frac{B}{8}} - \frac{4e_0^2 + 1}{32} + \frac{A}{2} \pm \frac{ie_0}{64} \sqrt{\frac{2}{B}} + \frac{A^2}{4B} \pm \frac{i\sqrt{2}e_0 A^2}{4B^{3/2}} \pm \frac{i\sqrt{2}e_0}{2048 B^{3/2}} + \cdots . \tag{8.94}$$

When $A = 0$ all the resonance energies (8.92) exhibit the same branch point at $B_s = 1/8$ and the series (8.94) converge for all $B > 1/8$. When $A \neq 0$ there are two singular points $B_{1s} = 1/8$ and $B_{2s} = -n(n + 1)/2$, so that the series (8.94) converge for all $B > 1/8$ when $n = 0$ and for all $B > n(n + 1)/2$ otherwise. As it usually happens with the polynomial approximation, the radius of convergence decreases with the quantum number n. A straightforward way of testing the perturbation method described above is to expand the β-perturbation series in powers of $1/\sqrt{B}$ and compare the result with equation (8.94). To this end the Maple command *series* proves extremely useful. In particular, notice that this further expansion is not necessary when $A = 0$ because $\beta^2 = \pm i(8/B)^{1/2}$, from which we conclude that the perturbation series (8.87) converges for all $|\beta| < \sqrt{8}$.

As a simple nontrivial example we choose the Gaussian barrier

$$V(x) = V_0 \exp \left[-(x/\gamma)^2 \right] , \tag{8.95}$$

where $V_0 > 0$ and $\gamma > 0$ determine the strength and range of the interaction, respectively. The dimensionless Hamiltonian operator reads

$$\hat{\mathcal{H}} = -\frac{1}{2} \frac{d^2}{dq^2} + A \exp \left(-q^2 \right) , \quad A = \frac{m\gamma^2 V_0}{\hbar^2} . \tag{8.96}$$

Table 8.1 shows the first coefficients of the perturbation series (8.87), where $v_0 = A$ and $\beta = 1/\sqrt{-2A}$. The Maple programs for the calculation of tunnel resonances and for the application of the deep-well approximation are identical, and one obtains them by straightforward modification of the Maple procedures in the program section that illustrate the implementation of the method of Swenson and Danforth.

Table 8.1   Coefficients of the Perturbation Series for the Resonances of the Gaussian Barrier

$$e_1 = -\frac{3}{32} - \frac{3}{8} e_0{}^2$$

$$e_2 = -\frac{1}{768} e_0 - \frac{11}{192} e_0{}^3$$

$$e_3 = \frac{141}{16384} - \frac{1}{6144} e_0{}^2 - \frac{85}{3072} e_0{}^4$$

$$e_4 = \frac{300383}{11796480} e_0 + \frac{91}{294912} e_0{}^3 - \frac{4351}{245760} e_0{}^5$$

$$e_5 = -\frac{9}{2097152} + \frac{265567}{5242880} e_0{}^2 + \frac{343}{589824} e_0{}^4 - \frac{38633}{2949120} e_0{}^6$$

Figure 8.6 shows that the perturbation series for the Gaussian barrier is divergent. However, for sufficiently large values of A we obtain reasonable results by means of the truncation criteria adopted in Section 6.4. In order to obtain the *exact* resonances necessary for the calculation of the logarithmic error displayed in Figure 8.6, we resorted to the practical Riccati–Padé method [162]. In principle one can improve the accuracy of the perturbation series for resonances by means of the methods discussed in Section 6.5 for bound states.

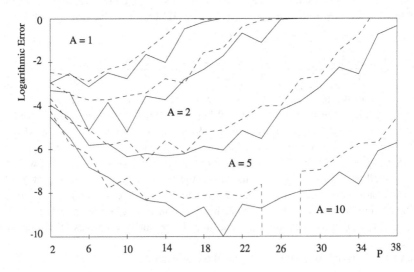

FIGURE 8.6

Logarithmic errors $\log |[Re(\epsilon_{exact}) - Re(\epsilon_{approx})]/Re(\epsilon_{exact})|$ **(continuous line) and** $\log |[Im(\epsilon_{exact}) - Im(\epsilon_{approx})]/Im(\epsilon_{exact})|$ **(broken line) for the resonance of the Gaussian barrier with** $e_0 = 1/2$ **in terms of the perturbation order** P.

Chapter 9

Perturbation Theory in Classical Mechanics

9.1 Introduction

Although the title of this book refers only to quantum mechanics, in this chapter we outline the application of perturbation theory in classical mechanics. It is not our purpose to give a thorough account of the subject that may be found in other books [179, 180], but simply to show that some of the approaches developed in preceding chapters for quantum systems are suitable for classical ones with just slight modifications. We hope that this fact may facilitate a unified teaching of perturbation methods in undergraduate and graduate courses. We first consider the simplest perturbation expansion that applies when the amplitude of the motion is sufficiently small. This approach is based on the Taylor expansion of the nonlinear force (or the anharmonic potential-energy function) around the origin and resembles the polynomial approximation in quantum mechanics discussed in Chapter 7. We obtain the perturbation series for the trajectory of a particle moving in a one-dimensional space under the effect of an arbitrary nonlinear conservative force, and the perturbation series for the period of the motion. Straightforward integration of the perturbation equations gives rise to secular terms that one easily removes by appropriately scaling the frequency. We choose the simple pendulum as an illustrative example. Most of that discussion is based on an appropriate modification and adaptation of a recent pedagogical article on the subject [181].

Later, we concentrate on Hamilton's equations of motion that allow the development of perturbation theory in operator form that is reminiscent of the interaction picture in quantum mechanics, already discussed in Section 1.3.1. We explicitly consider secular and canonical perturbation theories using one-dimensional anharmonic oscillators as illustrative examples.

Finally, we show that it is easier to obtain the canonical perturbation series for separable models by means of a simple approach based on the hypervirial and Hellmann–Feynman theorems that closely resemble the method of Swenson and Danforth discussed in Section 3.3.

9.2 Dimensionless Classical Equations

Throughout this book we have transformed physical equations into dimensionless mathematical equations before solving them either exactly or approximately. In this chapter we proceed exactly in the same way with the classical equations of motion. For simplicity we consider a one-dimensional motion and point out that the treatment of more degrees of freedom follows exactly the same lines.

For concreteness consider a particle of mass m moving along a one-dimensional trajectory $x(t)$ with velocity $\dot{x}(t) = dx(t)/dt$, under the effect of a conservative force $F(x) = -dV(x)/dx$. From the Lagrangian

$$\mathcal{L}(\dot{x}, x) = \frac{m}{2}\dot{x}^2 - V(x) \tag{9.1}$$

we obtain the conjugate linear moment $p_x = \partial \mathcal{L}/\partial \dot{x} = m\dot{x}$ and the Hamiltonian function

$$H(p_x, x) = p_x \dot{x} - \mathcal{L}(\dot{x}, x) = \frac{m}{2}p_x^2 + V(x). \tag{9.2}$$

In order to obtain dimensionless equations of motion we define dimensionless coordinate $q = x/L$ and time $s = \omega t + \phi$, where L, ω, and ϕ are arbitrary length, frequency, and phase, respectively. The dimensionless Lagrangian and Hamiltonian read

$$\tilde{\mathcal{L}}(\dot{q}, q) = \frac{\mathcal{L}(\dot{x}, x)}{m\omega^2 L^2} = \frac{\dot{q}^2}{2} - v(q), \tag{9.3}$$

$$\tilde{H}(p, q) = \frac{p^2}{2} + v(q), \tag{9.4}$$

respectively, where $\dot{q} = dq(s)/ds$, $v(q) = V(Lq)/(m\omega^2 L^2)$, and $p = \partial \tilde{\mathcal{L}}/\partial \dot{q} = \dot{q} = p_x/(m\omega L)$. The equations of motion

$$m\ddot{x} = F(x) = -\frac{dV(x)}{dx} \tag{9.5}$$

become

$$\ddot{q} = f(q) = -\frac{dv(q)}{dq} \tag{9.6}$$

with the initial conditions

$$Lq(\phi) = x(0), \quad \omega L\dot{q}(\phi) = \dot{x}(0). \tag{9.7}$$

9.3   Polynomial Approximation

Consider Newton's second law of motion (9.5) with initial conditions $x(0)$ and $\dot{x}(0)$ at $t = 0$, and assume a bounded trajectory: $|x(t)| < \infty$ for all t. Without loss of generality we place the coordinate origin at the stable equilibrium position, so that $F(0) = 0$ and $F'(0) < 0$. If $F(x)$ is analytic at $x = 0$ we expand it in a Taylor series:

$$F(x) = \sum_{j=1}^{\infty} F_j x^j, \quad F_j = \frac{1}{j!}\frac{d^j F}{dx^j}\bigg|_{x=0}. \tag{9.8}$$

Choosing $\omega = \sqrt{-F_1/m}$ we can write the dimensionless equation of motion (9.6) as

$$\ddot{q}(s) + q(s) = \sum_{j=1}^{\infty} a_j L^j q(s)^{j+1}, \quad a_j = -\frac{F_{j+1}}{F_1}. \tag{9.9}$$

If the amplitude of the motion is sufficiently small, then it is reasonable to look for a solution in the form of a perturbation series

$$q(s) = \sum_{j=0}^{\infty} q_j(s) L^j \,, \tag{9.10}$$

sometimes called the straightforward expansion [182]. In order to facilitate the application of perturbation theory to equation (9.9) we rewrite the sum as follows

$$\sum_{j=1}^{\infty} a_j L^j q(s)^{j+1} = \sum_{j=1}^{\infty} G_j(s) L^j \,, \tag{9.11}$$

where the coefficients $G_j(s)$ are nonlinear functions of the coefficients $q_j(s)$. We obtain the former systematically by means of the following equations:

$$G_n = \sum_{j=1}^{n} a_j q_{j+1,n-j}, \ n = 1, 2, \ldots, \tag{9.12}$$

where $q_{j+1,k}$ denotes one of the coefficients of the series,

$$q^{j+1} = \sum_{k=0}^{\infty} q_{j+1,k} L^k \,, \tag{9.13}$$

that satisfy the recurrence relation

$$q_{j+1,k} = \sum_{i=0}^{k} q_i q_{j,k-i} \,. \tag{9.14}$$

Notice that $q_{1,k} = q_k$, and $q_{0,k} = \delta_{0k}$. For example, the first two coefficients $G_j(s)$ are

$$G_1(s) = a_1 q_0(s)^2, \ G_2(s) = 2a_1 q_0(s) q_1(s) + a_2 q_0(s)^3 \,. \tag{9.15}$$

It follows from the equation of motion (9.9) and from the expansions (9.10) and (9.11) that the coefficients $q_j(s)$ are solutions to the differential equations

$$\ddot{q}_j + q_j = G_j, \ j = 0, 1, \ldots, \tag{9.16}$$

where $G_0 = 0$. If we arbitrarily choose $q_0(0) = 1$, $\dot{q}_0(0) = 0$, $q_j(0) = 0$, and $\dot{q}_j(0) = 0$ for all $j > 0$ (that is to say $q(0) = 1$ and $\dot{q}(0) = 0$), then the solutions to the perturbation equations take a particularly simple form. According to the integration formulas developed in Appendix B we have $q_0(s) = \cos(s)$, and

$$q_j(s) = \int_0^s \sin\left(s - s'\right) G_j\left(s'\right) ds' \tag{9.17}$$

for all $j > 0$. This choice may not be the most convenient way of taking into account arbitrary initial conditions $x(0)$ and $\dot{x}(0)$, because once we have a sufficiently great number N of perturbation coefficients $q_j(s)$ we should obtain the undetermined parameters L and ϕ from the rather complicated equations

$$\sum_{j=0}^{N} q_j(\phi) L^{j+1} = x(0), \ \omega \sum_{j=0}^{N} \dot{q}_j(\phi) L^{j+1} = \dot{x}(0) \,. \tag{9.18}$$

If, for example, we choose $q_j(s)$ exactly as above for $j > 0$ (so that equation (9.17) remains unchanged) but $q_0(s) = A\cos(s) + B\sin(s)$, and $\phi = 0$, then we simply have $A = x(0)/L$ and $B = \dot{x}(0)/(\omega L)$. However, we prefer the simpler form of $q(s)$ for the discussion below, which does not require the initial conditions explicitly.

In the program section we show a set of simple Maple procedures for the application of the polynomial approximation just described with any form of $q_0(s)$. When $q_0(s) = \cos(s)$ the first two perturbation corrections to the trajectory are

$$
\begin{aligned}
q_1(s) &= a_1\left[\frac{1}{2} - \frac{1}{3}\cos(s) - \frac{1}{6}\cos(2s)\right] \\
q_2(s) &= -\frac{a_1^2}{3} + \frac{58a_1^2 + 9a_2}{288}\cos(s) + \frac{a_1^2}{9}\cos(2s) \\
&\quad + \frac{2a_1^2 - 3a_2}{96}\cos(3s) + \frac{10a_1^2 + 9a_2}{24}s\sin(s)\,.
\end{aligned}
\tag{9.19}
$$

The unbounded term $s\sin(s)$ in the correction of second order is incompatible with a bounded motion. Such secular terms are well known in perturbation theory and are commonly removed by means of, for example, the Lindstedt–Poincaré technique [182] that we discuss later in this chapter.

9.3.1   Odd Force

The equations developed above apply to any nonlinear force $F(x)$ provided that $F_1 < 0$. However, if $F(x)$ is an odd function of x it is convenient to proceed in a different way in order to derive a more efficient algorithm. In such a case we write the Taylor expansion as

$$
F(x) = \sum_{j=1}^{\infty} F_j x^{2j-1}, \ \ F_j = \frac{1}{(2j-1)!}\left.\frac{d^{2j-1}F}{dx^{2j-1}}\right|_{x=0},
\tag{9.20}
$$

and realize that the resulting equation for $q(s)$

$$
\ddot{q}(s) + q(s) = \sum_{j=1}^{\infty} a_j L^{2j} q(s)^{2j+1}, \ \ a_j = -\frac{F_{j+1}}{F_1}
\tag{9.21}
$$

suggests the perturbation parameter $\lambda = L^2$. Writing

$$
q(s) = \sum_{j=0}^{\infty} q_j(s)\lambda^j\,,
\tag{9.22}
$$

and proceeding as before one obtains the same perturbation equations (9.16), except that the inhomogeneous terms G_j are different. The reader may easily derive the appropriate equations for their systematic calculation, and verify that the first two of them are

$$
G_1 = a_1 q_0^3, \ \ G_2 = 3a_1 q_0^2 q_1 + a_2 q_0^5\,.
\tag{9.23}
$$

Straightforward application of equation (9.17) shows that

$$q_1(s) = a_1 \left[\frac{1}{32} \cos(s) - \frac{1}{32} \cos(3s) + \frac{3}{8} s \sin(s) \right]$$

$$q_2(s) = \frac{69a_1^2 + 128a_2}{3072} \cos(s) - \frac{3a_1^2 + 5a_2}{128} \cos(3s)$$
$$+ \frac{3a_1^2 - 8a_2}{3072} \cos(5s) + \frac{3a_1^2 + 10a_2}{32} s \sin(s)$$
$$- \frac{9a_1^2}{256} s \sin(3s) - \frac{9a_1^2}{128} s^2 \cos(s) , \quad (9.24)$$

where we see secular terms in the contributions of first and second order.

9.3.2 Period of the Motion

From the perturbation expansion for $q(s)$ we easily obtain a perturbation expansion for the period τ. For all t we have $x(t + \tau) = x(t)$, so that $q(s + \tau') = q(s)$, where $\tau' = \omega\tau$, and, in particular, $q(\tau') = q(0) = 1$. The perturbation expansion for τ' is

$$\tau' = \sum_{j=0}^{\infty} \tau_j' \lambda^j, \ \tau_0' = 2\pi , \quad (9.25)$$

in terms of the perturbation parameter λ chosen to be $\lambda = L$ or $\lambda = L^2$ in either of the two cases discussed above (expansions (9.8) or (9.20), respectively). Straightforward differentiation of $q(\tau') = 1$ with respect to λ yields

$$\frac{\partial q}{\partial \lambda} (\tau') + \dot{q} (\tau') \frac{\partial \tau'}{\partial \lambda} = 0 \quad (9.26)$$

and

$$\frac{\partial^2 q}{\partial \lambda^2} (\tau') + 2 \frac{\partial \dot{q}}{\partial \lambda} (\tau') \frac{\partial \tau'}{\partial \lambda} + \ddot{q} (\tau') \left(\frac{\partial \tau'}{\partial \lambda} \right)^2 + \dot{q} (\tau') \frac{\partial^2 \tau'}{\partial \lambda^2} = 0 . \quad (9.27)$$

When $\lambda \to 0$ these two equations lead to

$$\tau_1' = \lim_{s \to 2\pi} \left[\frac{q_1(s)}{\dot{q}_0(s)} \right] , \quad (9.28)$$

and

$$\tau_2' = \lim_{s \to 2\pi} \left[\frac{2q_2(s) + 2\dot{q}_1(s)\tau_1' + \ddot{q}(s)\tau_1'^2}{2q_0'(s)} \right] , \quad (9.29)$$

respectively. Substituting the expressions of $q_j(s)$ derived earlier we obtain [181]

$$\tau_1' = 0, \ \tau_2' = \frac{\pi}{12} \left(10a_1^2 + 9a_2 \right) \quad (9.30)$$

for the general case, and

$$\tau_1' = \frac{3\pi}{4} a_1, \ \tau_2' = \frac{\pi}{128} \left(57a_1^2 + 80a_2 \right) \quad (9.31)$$

for an odd force.

Proceeding in the same way we obtain perturbation corrections of higher order; aided by Maple we have derived the results shown in Table 9.1.

Table 9.1   Perturbation Series for the Period

Arbitrary Force

$\tau_1 = 0$

$\tau_2 = \frac{3}{4}\pi a_2 + \frac{5}{6}\pi a_1{}^2$

$\tau_3 = -\frac{1}{2} a_1 \pi a_2 - \frac{5}{9} a_1{}^3 \pi$

$\tau_4 = \frac{275}{96}\pi a_1{}^2 a_2 + \frac{57}{128}\pi a_2{}^2 + \frac{385}{288}\pi a_1{}^4 + \frac{7}{4}\pi a_1 a_3 + \frac{5}{8}\pi a_4$

$\tau_5 = -\frac{35}{8}\pi a_2 a_1{}^3 - \frac{385}{216}\pi a_1{}^5 - \frac{8}{3}\pi a_3 a_1{}^2 - \frac{35}{32}\pi a_2{}^2 a_1 - \frac{3}{10}\pi a_3 a_2 - \frac{5}{6} a_1 \pi a_4$

$\tau_6 = \frac{63}{80} a_3{}^2 \pi + \frac{35}{64} a_6 \pi + \frac{10535}{1536}\pi a_1{}^2 a_2{}^2 + \frac{8435}{768}\pi a_1{}^4 a_2 + \frac{315}{1024}\pi a_2{}^3$

$\qquad + \frac{175}{32} a_1 a_2 \pi a_3 + \frac{49}{64}\pi a_2 a_4 + \frac{385}{96}\pi a_1{}^2 a_4 + \frac{371}{48} a_1{}^3 \pi a_3 + \frac{15}{8} a_1 a_5 \pi$

$\qquad + \frac{103565}{31104}\pi a_1{}^6$

$\tau_7 = -\frac{4445}{256} a_1{}^3 a_2{}^2 \pi - \frac{875}{512} a_1 a_2{}^3 \pi - \frac{21}{32}\pi a_3 a_2{}^2 - \frac{763}{48}\pi a_2 a_1{}^2 a_3 - \frac{275}{96} a_2 a_1 a_4 \pi$

$\qquad - \frac{385}{48} a_1{}^3 a_4 \pi - \frac{71995}{3456}\pi a_1{}^5 a_2 - \frac{539}{36} a_1{}^4 a_3 \pi - \frac{85085}{15552} a_1{}^7 \pi - \frac{119}{40} a_1 a_3{}^2 \pi$

$\qquad - \frac{1}{2} a_4 a_3 \pi - \frac{3}{14} a_2 a_5 \pi - \frac{335}{84} a_1{}^2 a_5 \pi - \frac{35}{32} a_6 a_1 \pi$

$\tau_8 = \frac{1425}{256} a_1 a_2 \pi a_5 + \frac{2550625}{49152}\pi a_1{}^4 a_2{}^2 + \frac{18865}{1024}\pi a_2 a_1{}^2 a_4 + \frac{28847}{512}\pi a_2 a_1{}^3 a_3$

$\qquad + \frac{24661}{2048}\pi a_2{}^2 a_1 a_3 + \frac{2683}{384}\pi a_1 a_3 a_4 + \frac{223265}{16384}\pi a_1{}^2 a_2{}^3 + \frac{30345}{131072}\pi a_2{}^4$

$\qquad + \frac{6551545}{663552} a_1{}^8 \pi + \frac{1197}{512}\pi a_2 a_3{}^2 + \frac{1401}{2048}\pi a_2 a_6 + \frac{52661}{3840}\pi a_1{}^2 a_3{}^2$

$\qquad + \frac{16615}{3072}\pi a_1{}^2 a_6 + \frac{4535}{384} a_1{}^3 \pi a_5 + \frac{182875}{9216}\pi a_1{}^4 a_4 + \frac{153769}{4608}\pi a_1{}^5 a_3$

$\qquad + \frac{4919915}{110592}\pi a_1{}^6 a_2 + \frac{515}{1536}\pi a_4{}^2 + \frac{3335}{4096}\pi a_2{}^2 a_4 + \frac{63}{128} a_8 \pi + \frac{385}{192} a_1 a_7 \pi$

$\qquad + \frac{99}{64} a_5 a_3 \pi$

Odd Force

$\tau_1 = \frac{3}{4}\pi a_1$

$\tau_2 = \frac{57}{128} a_1{}^2 \pi + \frac{5}{8}\pi a_2$

$\tau_3 = \frac{49}{64}\pi a_1 a_2 + \frac{315}{1024} a_1{}^3 \pi + \frac{35}{64} a_3 \pi$

$\tau_4 = \frac{1401}{2048}\pi a_1 a_3 + \frac{30345}{131072}\pi a_1{}^4 + \frac{515}{1536}\pi a_2{}^2 + \frac{3335}{4096} a_1{}^2 \pi a_2 + \frac{63}{128} a_4 \pi$

$\tau_5 = \frac{8965\,\pi\,a_1 a_2{}^2}{12288} + \frac{27335\,a_1{}^3\,\pi\,a_2}{32768} + \frac{24255\,a_1{}^2\,\pi\,a_3}{32768} + \frac{3201\,\pi\,a_1 a_4}{5120} + \frac{1243\,\pi\,a_2 a_3}{2048}$

$\qquad + \frac{231\,a_5\,\pi}{512} + \frac{193347\,a_1{}^5\,\pi}{1048576}$

$\tau_6 = \frac{1770195}{2097152} a_1{}^4 \pi a_2 + \frac{448805}{393216} a_1{}^2 \pi a_2{}^2 + \frac{403165}{524288} a_1{}^3 \pi a_3 + \frac{22449}{32768} a_1{}^2 \pi a_4$

$\qquad + \frac{4751}{8192}\pi a_1 a_5 + \frac{43935}{32768}\pi a_1 a_2 a_3 + \frac{18165}{65536} a_3{}^2 \pi + \frac{5127969}{33554432} a_1{}^6 \pi + \frac{16285}{73728} a_2{}^3 \pi$

$\qquad + \frac{1147}{2048} a_4 \pi a_2 + \frac{429}{1024} a_6 \pi$

$\tau_7 = \frac{188475}{262144} a_1{}^3 \pi a_4 + \frac{84115}{131072} a_1{}^2 \pi a_5 + \frac{1112895}{524288} a_1{}^2 a_2 a_3 \pi + \frac{31179}{57344}\pi a_1 a_6$

$\qquad + \frac{137935}{196608}\pi a_1 a_2{}^3 + \frac{14263095}{16777216} a_1{}^5 \pi a_2 + \frac{1636985}{1048576} a_1{}^3 \pi a_2{}^2 + \frac{13198185}{16777216} a_1{}^4 \pi a_3$

$\qquad + \frac{20495}{16384}\pi a_1 a_4 a_2 + \frac{325425}{524288}\pi a_1 a_3{}^2 + \frac{8439}{16384} a_4 a_3 \pi + \frac{120835}{196608}\pi a_2{}^2 a_3$

$\qquad + \frac{35002539}{268435456}\pi a_1{}^7 + \frac{6435}{16384} a_7 \pi + \frac{12865}{24576} a_2 a_5 \pi$

9.3.3 Removal of Secular Terms

In what follows we briefly review the Lindstedt–Poincaré technique for removing unbound secular terms [182]. For concreteness we apply this straightforward procedure to the expansion (9.8) showing that equation (9.17) is also useful for that purpose.

If we define $q(s)$ as before, where $s = \sqrt{\gamma}\omega t + \phi$, the dimensionless equation of motion becomes

$$\gamma \ddot{q}(s) + q(s) = \sum_{j=1}^{\infty} a_j L^j q(s)^{j+1}, \quad a_j = -\frac{F_{j+1}}{F_1}. \tag{9.32}$$

Substituting the perturbation series (9.10), and

$$\gamma = \sum_{j=0}^{\infty} \gamma_j L^j, \tag{9.33}$$

we obtain

$$\ddot{q}_0 + q_0 = 0, \quad \ddot{q}_j + q_j = G_j - \sum_{m=1}^{j} \gamma_m \ddot{q}_{j-m}. \tag{9.34}$$

The Lindstedt–Poincaré technique consists of choosing the coefficients γ_j in such a way that the right-hand side of equation (9.34) is free from terms that, being solutions of the homogeneous equation $\ddot{y}(s) + y(s) = 0$, give rise to unbound terms after integration [182]. Since G_1 does not contain such terms we set $\gamma_1 = 0$. At second order we choose

$$\gamma_2 = -\left(\frac{5a_1^2}{6} + \frac{3a_2}{4}\right) \tag{9.35}$$

to remove the term proportional to cos(s) from

$$G_2 = \frac{1}{12}\left(3a_2 - 2a_1^2\right)\cos(3s) - \frac{a_1^2}{3}\cos(2s) + \left(\frac{5a_1^2}{6} + \frac{3a_2}{4}\right)\cos(s) - \frac{a_1^2}{3}. \tag{9.36}$$

The resulting correction of second order is periodic:

$$\begin{aligned} q_2(s) &= -\frac{a_1^2}{3} + \frac{58a_1^2 + 9a_2}{288}\cos(s) + \frac{a_1^2}{9}\cos(2s) \\ &\quad + \frac{2a_1^2 - 3a_2}{96}\cos(3s). \end{aligned} \tag{9.37}$$

9.3.4 Simple Pendulum

As an illustrative example consider the equation of motion for the simple pendulum [183]

$$\ddot{\theta} = F(\theta) = -\frac{g}{l}\sin(\theta), \tag{9.38}$$

where l is the pendulum length, θ is the angle subtended with respect to the equilibrium position, and g is the gravitational acceleration. If we apply the method developed above with $x = \theta$ and $m = 1$ we have an expansion like (9.20) because the force is an odd function of θ. Since $F_1 = -g/l$

and $F_2 = g/(6l)$, we have $a_1 = 1/6$, and $a_2 = -1/120$ that we substitute into equation (9.31) and obtain

$$\tau_1' = \frac{\pi}{8}, \quad \tau_2' = \frac{11\pi}{1536} \tag{9.39}$$

which agree with the first terms of the expansion of the integral representation of the period of the pendulum [183].

9.4   Canonical Transformations in Operator Form

9.4.1   Hamilton's Equations of Motion

We consider a classical dynamical system described by a set of generalized coordinates $\mathbf{q} = (q_1, q_2, \ldots, q_n)$ and conjugate momenta $\mathbf{p} = (p_1, p_2, \ldots, p_n)$. The trajectory in phase space is given by Hamilton's equations of motion

$$\dot{q}_j = \frac{\partial H}{\partial p_j}, \quad \dot{p}_j = -\frac{\partial H}{\partial q_j}, \quad j = 1, 2, \ldots n, \tag{9.40}$$

where $H = H(\mathbf{q}, \mathbf{p}, t)$ is the Hamiltonian [184]. Wherever necessary we explicitly indicate the dependence of $\mathbf{q}$ and $\mathbf{p}$ on the initial conditions $\mathbf{q}_0$ and $\mathbf{p}_0$ at a given time (say $t = 0$) as $\mathbf{q} = \mathbf{q}(\mathbf{q}_0, \mathbf{p}_0, t)$ and $\mathbf{p} = \mathbf{p}(\mathbf{q}_0, \mathbf{p}_0, t)$, respectively.

The velocity of change of a general function $F(\mathbf{q}, \mathbf{p})$ is given by

$$\frac{\partial F}{\partial t} = \sum_{j=1}^{n} \left(\frac{\partial F}{\partial q_j} \dot{q}_j + \frac{\partial F}{\partial p_j} \dot{p}_j \right) = \sum_{j=1}^{n} \left(\frac{\partial F}{\partial q_j} \frac{\partial H}{\partial p_j} - \frac{\partial F}{\partial p_j} \frac{\partial H}{\partial q_j} \right) = \{F, H\}_{[q,p]}, \tag{9.41}$$

where $\{F, H\}_{[q,p]}$ denotes the well-known Poisson bracket [185]. In general, for any two functions $F(\mathbf{q}, \mathbf{p})$ and $G(\mathbf{q}, \mathbf{p})$ we write

$$\{F, G\}_{[q,p]} = \sum_{j=1}^{n} \left(\frac{\partial F}{\partial q_j} \frac{\partial G}{\partial p_j} - \frac{\partial F}{\partial p_j} \frac{\partial G}{\partial q_j} \right). \tag{9.42}$$

In particular, the coordinates and momenta satisfy

$$\{q_i, p_j\}_{[q,p]} = \delta_{ij}. \tag{9.43}$$

9.4.2   General Poisson Brackets

It is convenient for our purposes to generalize the Poisson brackets for an arbitrary set of $2n$ variables $\mathbf{x} = \{x_1, x_2, \ldots, x_n, x_{n+1}, \ldots, x_{2n}\}$ as

$$\{F, G\}_x = \sum_{j=1}^{n} \left(\frac{\partial F}{\partial x_j} \frac{\partial G}{\partial x_{n+j}} - \frac{\partial F}{\partial x_{n+j}} \frac{\partial G}{\partial x_j} \right). \tag{9.44}$$

We can rewrite this equation in a more compact form in terms of the antisymmetric matrix

$$\mathbf{J} = \begin{pmatrix} 0 & 1 \\ -1 & 0 \end{pmatrix}, \tag{9.45}$$

where $\mathbf{0}$ and $\mathbf{1}$ are the $n \times n$ zero and unit matrices, respectively [185]. One easily verifies that

$$\{F, G\}_x = \sum_{i=1}^{2n} \sum_{j=1}^{2n} J_{ij} \frac{\partial F}{\partial x_i} \frac{\partial G}{\partial x_j} \,, \tag{9.46}$$

where J_{ij} denotes the element in the ith row and jth column of $\mathbf{J}$. In particular,

$$\{x_i, x_j\}_x = J_{ij} \,. \tag{9.47}$$

Consider the set $\mathcal{V} = \{A, B, C, \dots\}$ of differentiable functions of $\mathbf{x}$. The Poisson brackets satisfy

$$\{A, B\} \in \mathcal{V} \,, \tag{9.48}$$
$$\{A, B + C\} = \{A, B\} + \{A, C\} \,, \tag{9.49}$$
$$\alpha\{A, B\} = \{\alpha A, B\} = \{A, \alpha B\} \,, \tag{9.50}$$
$$\{A, A\} = 0 \Rightarrow \{A, B\} = -\{B, A\} \,, \tag{9.51}$$
$$\{A, \{B, C\}\} + \{C, \{A, B\}\} + \{B, \{C, A\}\} = 0 \,, \tag{9.52}$$

where α is a complex number. These equations define a Lie algebra [48]. One easily proves the properties (9.48)–(9.51), but the proof of Jacobi's identity (9.52), although straightforward, is rather tedious [185]. However, using Maple we can easily verify Jacobi's identity for particular cases. In the program section there is a set of simple procedures for that purpose.

9.4.3 Canonical Transformations

A change of variables

$$y_j = y_j(\mathbf{x}), \quad j = 1, 2, \dots, 2n \tag{9.53}$$

is said to be canonical if $\{y_i, y_j\}_x = \{x_i, x_j\}_x = J_{ij}$. Expanding the Poisson bracket for any two functions $F(\mathbf{y})$ and $G(\mathbf{y})$, and using the chain rule we easily prove that

$$\{F, G\}_x = \sum_{i=1}^{2n} \sum_{j=1}^{2n} \frac{\partial F}{\partial y_i} \frac{\partial G}{\partial y_j} \{y_i, y_j\}_x \,. \tag{9.54}$$

This equation states that

$$\{F, G\}_x = \{F, G\}_y \,, \tag{9.55}$$

provided that the transformation (9.53) is canonical.

Given a differentiable function $F(\mathbf{x})$ we define an operator

$$\hat{F} = \sum_{i=1}^{2n} \sum_{j=1}^{2n} J_{ij} \frac{\partial F}{\partial x_i} \frac{\partial}{\partial x_j} \,, \tag{9.56}$$

such that $\hat{F}G = \{F, G\}_x$ for any differentiable function $G(\mathbf{x})$. In what follows we consider a change of variables $\mathbf{y} = \mathbf{y}(\mathbf{x}, \lambda)$ depending on a parameter λ in such a way that $\mathbf{y}(\mathbf{x}, 0) = \mathbf{x}$, and

$$\frac{\partial y_i}{\partial \lambda} = \hat{W} y_i \,, \tag{9.57}$$

where $\hat{W}$ is the operator generated by a differentiable function $W(\mathbf{x}, \lambda)$ according to equation (9.56). By means of equation (9.57) and the chain rule it is not difficult to show that

$$\frac{dF(\mathbf{y})}{d\lambda} = \{W, F\}_x = \hat{W}F \ . \tag{9.58}$$

Assuming that $(d/d\lambda)(\partial F/\partial x_i) = (\partial/\partial x_i)(dF/d\lambda)$ we easily prove that

$$\frac{d}{d\lambda}\{F(\mathbf{y}), G(\mathbf{y})\}_x = \{\{W, F\}, G\}_x + \{F, \{W, G\}\}_x = \{W, \{F, G\}\}_x \ , \tag{9.59}$$

where we have used Jacobi's identity in order to derive the second equality.

We denote the transformation $\mathbf{x} \rightarrow \mathbf{y}$ with generator $W(\mathbf{x}, \lambda)$ and parameter λ as

$$y_i(x, \lambda) = T_W(x_i, \lambda) = \hat{T}_W(\lambda)x_i \ , \tag{9.60}$$

where the transformation operator $\hat{T}_W$ is a solution to the differential equation

$$\frac{d}{d\lambda}\hat{T}_W = \hat{W}\hat{T}_W, \ \hat{T}_W(0) = \hat{1} \ . \tag{9.61}$$

Differentiating the equation $\hat{T}_W\hat{T}_W^{-1} = \hat{T}_W^{-1}\hat{T}_W = \hat{1}$, which defines the inverse $\hat{T}_W^{-1}$ of $\hat{T}_W$, with respect to λ we easily obtain

$$\frac{d}{d\lambda}\hat{T}_W^{-1} = -\hat{T}_W^{-1}\hat{W}, \ \hat{T}_W^{-1}(0) = \hat{1} \ . \tag{9.62}$$

It follows from equation (9.58) and the initial condition $\mathbf{y}(\mathbf{x}, 0) = \mathbf{x}$ that

$$F(\mathbf{y}) = F\left(\hat{T}_W\mathbf{x}\right) = \hat{T}_W F(\mathbf{x}) \ . \tag{9.63}$$

Analogously, from equation (9.59) we have

$$\{F(\mathbf{y}), G(\mathbf{y})\}_x = \hat{T}_W\{F(\mathbf{x}), G(\mathbf{x})\}_x \ , \tag{9.64}$$

which shows that the transformation (9.60) is canonical:

$$\{y_i, y_j\}_x = \hat{T}_W\{x_i, x_j\}_x = \hat{T}_W J_{ij} = J_{ij} \ . \tag{9.65}$$

For this reason we can write

$$\frac{dy_i}{d\lambda} = \{W, y_i\}_x = \{\hat{W}, y_i\}_y = \sum_{j=1}^{2n} J_{ji}\frac{\partial W}{\partial y_j} \ , \tag{9.66}$$

which is equivalent to

$$\frac{dy_i}{d\lambda} = -\frac{\partial W}{\partial y_{i+n}}, \ \frac{dy_{i+n}}{d\lambda} = \frac{\partial W}{\partial y_i}, \ i = 1, 2, \ldots, n \ . \tag{9.67}$$

These expressions give Hamilton's equations of motion as a particular case when $W = -H$, $\lambda = t$, $\mathbf{y} = (\mathbf{q}, \mathbf{p})$, and $\mathbf{x} = (\mathbf{q}_0, \mathbf{p}_0)$.

If W is independent of λ we have

$$\hat{T}_W(\lambda) = \exp\left(\lambda\hat{W}\right), \ \hat{T}_W(\lambda)^{-1} = \exp\left(-\lambda\hat{W}\right) \ . \tag{9.68}$$

It is left to the reader to prove that in this particular case the function $W(x)$ is invariant under the canonical transformation of variables $\mathbf{x} \to \mathbf{y}$: $W(\mathbf{y}) = W(\mathbf{x})$. The exponential operator (9.68) may be expressed as a Taylor series about $\lambda = 0$:

$$\hat{T}_W(\lambda) = \sum_{k=0}^{\infty} \frac{\lambda^k}{k!} \hat{W}^k \,. \tag{9.69}$$

The operator form of the canonical transformation just outlined provides a useful expression for the solution of the inhomogeneous differential equation

$$\frac{d}{d\lambda} Y(\mathbf{x}, \lambda) = \hat{W}(\mathbf{x}, \lambda) Y(\mathbf{x}, \lambda) + F(\mathbf{x}, \lambda) \tag{9.70}$$

with a given initial condition $Y(\mathbf{x}, 0)$. Choosing $Y(\mathbf{x}, \lambda) = \hat{T}_W K(\mathbf{x}, \lambda)$, we easily solve the resulting differential equation for $K(\mathbf{x}, \lambda)$ $dK/d\lambda = \hat{T}_W^{-1} F$ obtaining

$$Y(\mathbf{x}, \lambda) = \int_0^{\lambda} \hat{T}_W(\lambda) \hat{T}_W(\lambda')^{-1} F(\mathbf{x}, \lambda') \, d\lambda' + \hat{T}_W(\lambda) Y(\mathbf{x}, 0) \,. \tag{9.71}$$

The operator form of canonical transformations based on Lie algebras offers some advantages over the more familiar canonical transformation commonly used in classical mechanics [185], and has been extensively studied by several authors [186]–[191]. An interesting physical application would be as follows: suppose that $\mathbf{x} = (\mathbf{q}, \mathbf{p})$ is a vector in phase space, and that $\mathbf{y} = \mathbf{y}(\mathbf{x}, \lambda)$ is a new set of generalized coordinates and conjugate momenta. The equations of motion for the Hamiltonian in the new variables $\tilde{H}(\mathbf{y}) = H(\mathbf{x}(\mathbf{y})) = H(\mathbf{x})$ read

$$\dot{y}_i = \{y_i, H(\mathbf{x})\}_x = \left\{ y_i, \tilde{H}(\mathbf{y}) \right\}_y \,. \tag{9.72}$$

A convenient choice of the transformation may lead to simpler equations of motion.

9.5 The Evolution Operator

We can view the solution $\mathbf{x} = \{\mathbf{q}, \mathbf{p}\}$ of Hamilton's equations of motion (9.40) as a canonical transformation $\mathbf{x} = \mathbf{x}(\mathbf{x}_0, t)$ depending on the parameter $\lambda = t$ and write

$$\mathbf{x} = \hat{T}_H(t)\mathbf{x}_0, \quad \frac{d}{dt} \hat{T}_H(t) = -\hat{H}\hat{T}_H(t), \quad \hat{T}_H(0) = \hat{1} \,. \tag{9.73}$$

This expression of the trajectory in phase space is reminiscent of the evolution of the state vector in quantum mechanics discussed in Section 1.3.

If the Hamiltonian is independent of time we easily obtain an explicit formal expression for the evolution operator: $\hat{T}_H(t) = \exp(-t\hat{H})$, where $H(\mathbf{x}) = H(\mathbf{x}_0)$ [185]. However, even in this simpler case we are not able to derive exact analytical solutions to Hamilton's equations of motion, except for particular models. The substitution of the expansion $\hat{T}_H = \hat{1} - t\hat{H} + (t\hat{H})^2/2 + \dots$ for the exponential operator yields a Taylor series for $\mathbf{x}$ about $t = 0$.

9.5.1 Simple Examples

In what follows we consider Hamiltonian functions

$$H = \frac{p_x^2}{2m} + k_M x^M \tag{9.74}$$

that we easily transform into their dimensionless form

$$H = m\omega^2 L^2 \left(\frac{p^2}{2} + q^M \right) \tag{9.75}$$

by means of the change of variables already introduced above: $x = Lq$, $p_x = m\omega Lp$, $s = \omega t$, where the arbitrary length L and frequency ω satisfy

$$\frac{k_M L^{M-2}}{m\omega^2} = 1 \, . \tag{9.76}$$

In the case of the harmonic oscillator M = 2 it is preferable to choose $k_2/m\omega^2 = 1/2$ so that

$$H = \frac{m\omega^2}{2} \left(p^2 + q^2 \right) \, . \tag{9.77}$$

Later on, we apply perturbation theory to the anharmonic oscillator,

$$H = \frac{p_x^2}{2m} + \frac{m\omega^2 x^2}{2} + k_M x^M \, , \tag{9.78}$$

that we may rewrite as

$$H = m\omega^2 L^2 \left(\frac{p^2}{2} + \frac{q^2}{2} + \lambda q^M \right) , \quad \lambda = \frac{k_M L^{M-2}}{m\omega^2} \, . \tag{9.79}$$

Although we believe that one should always transform physical equations into dimensionless mathematical expressions, we will sacrifice our philosophy in this chapter to facilitate comparison of present results with those of other authors.

One of the simplest models is given by the Hamiltonian $H(q, p) = p^2/(2m) + fq$, where m and f are real numbers, which applies, for example, to a particle of mass m under the effect of a gravitational field ($f = mg$). In this case the expansion of the exponential operator $\hat{T}_H$ yields the exact result

$$\begin{aligned} q &= \exp\left[-t\hat{H}(\mathbf{x}_0) \right] q_0 = q_0 + \frac{p_0 t}{m} - \frac{f t^2}{2m} \\ p &= \exp\left[-t\hat{H}(\mathbf{x}_0) \right] p_0 = p_0 - ft \end{aligned} \tag{9.80}$$

because the series terminate when $\hat{H}^3 q_0 = 0$ and $\hat{H}^2 p_0 = 0$ [185].

Another interesting example is the harmonic oscillator

$$H(q, p) = \frac{p^2}{2m} + \frac{m\omega^2 q^2}{2} \, , \tag{9.81}$$

where m and ω are, respectively, the mass and frequency. The exact solution is

$$q = q_0 \cos(\omega t) + \frac{p_0}{m\omega} \sin(\omega t), \quad p = p_0 \cos(\omega t) - m\omega q_0 \sin(\omega t) \, . \tag{9.82}$$

As discussed above, the application of the expansion of the time evolution operator $\hat{T}_H = \exp[-t\hat{H}(\mathbf{x}_0)]$ to a given function of the initial coordinates and momenta $F(\mathbf{x}_0)$ leads to the Taylor series of $F(\mathbf{x})$ about $t = 0$. For example, in the case of the harmonic oscillator the expansion

$$\left(\hat{1} - t\hat{H}(\mathbf{x}_0) + \frac{t^2}{2}\hat{H}(x_0)^2 - \frac{t^3}{6}\hat{H}(x_0)^3 + \frac{t^4}{24}\hat{H}(x_0)^4 - \ldots\right)q_0$$

$$= q_0 + \frac{p_0 t}{m} - \frac{\omega^2 q_0 t^2}{2} - \frac{\omega^2 p_0 t^3}{6m} + \frac{\omega^4 q_0 t^4}{24} + \cdots \tag{9.83}$$

exactly agrees with the Taylor series of q in equation (9.82) about $t = 0$.

By means of canonical transformations we may obtain new dynamical variables that render the equations of motion simpler. For example, if a function $z(q, p)$ satisfies $\hat{H}z = \vartheta z$, where ϑ is a complex number, then without much thinking we realize that $z = z_0 \exp(-\vartheta t)$. Notice that what we are actually doing is seeking for eigenfunctions of the operator $\hat{H}$.

Taking into account that in the case of the harmonic oscillator $\hat{H}q = -p/m$ and $\hat{H}p = m\omega^2 q$, we conclude that the simplest eigenfunctions of $\hat{H}$ are linear combinations of the coordinate and momentum: $z = c_1 q + c_2 p$. A straightforward calculation shows that there are two solutions given by

$$a = c_1\left(q + \frac{ip}{m\omega}\right), \quad \vartheta = i\omega; \quad b = \frac{1}{2c_1}(im\omega q + p), \quad \vartheta = -i\omega, \tag{9.84}$$

where c_1 is an arbitrary complex number. Notice that the transformation $(q, p) \to (a, b)$ is canonical because $\{a, b\}_{[q,p]} = 1$. We arbitrarily choose $c_1 = \sqrt{m\omega/2}$ so that

$$a = \sqrt{\frac{m\omega}{2}}\left(q + \frac{ip}{m\omega}\right), \quad b = i\sqrt{\frac{m\omega}{2}}\left(q - \frac{ip}{m\omega}\right) = ia^*. \tag{9.85}$$

The inverse transformation is

$$q = \frac{1}{\sqrt{2m\omega}}(a - ib) = \sqrt{\frac{2}{m\omega}}\Re(a), \quad p = \sqrt{\frac{m\omega}{2}}(b - ia) = \sqrt{2m\omega}\Im(a). \tag{9.86}$$

In terms of these new variables the equations of motion have the simpler form $\dot{a} = -i\omega a$ and $\dot{b} = i\omega b$ so that one easily obtains the solutions

$$a = a_0 \exp(-i\omega t), \quad b = b_0 \exp(i\omega t). \tag{9.87}$$

The pairs of alternative initial conditions (a_0, b_0) and (q_0, p_0) relate each other exactly in the same way as (a, b) and (q, p) do, namely, through equations (9.85) and (9.86). Taking this fact into account, one easily recovers the solution (9.82) from equation (9.87). In terms of the new variables the Hamiltonian becomes

$$H(q(a, b), p(a, b)) = -i\omega ab = \omega|a|^2 = \omega|a_0|^2. \tag{9.88}$$

In order to express the canonical transformation just discussed in operator form, we consider the functions

$$K_0(a, b) = \frac{ab}{2}, \quad K_+(a, b) = \frac{a^2}{4}, \quad K_-(a, b) = \frac{b^2}{2} \tag{9.89}$$

and their corresponding operators

$$\hat{K}_0 = \frac{1}{2}\left(b\frac{\partial}{\partial b} - a\frac{\partial}{\partial a}\right), \quad \hat{K}_+ = \frac{a}{2}\frac{\partial}{\partial b}, \quad \hat{K}_- = -b\frac{\partial}{\partial a}. \tag{9.90}$$

We expect the canonical transformation to be given by the product of exponential operators $\hat{U} = \exp(\beta \hat{K}_-) \exp(\alpha \hat{K}_+) \exp(\theta \hat{K}_0)$. Taking into account that

$$\hat{K}_0 a = -\frac{a}{2}, \quad \hat{K}_+ a = 0, \quad \hat{K}_- a = -b,$$

$$\hat{K}_0 b = \frac{b}{2}, \quad \hat{K}_+ b = \frac{a}{2}, \quad \hat{K}_- b = 0 \tag{9.91}$$

it is not difficult to verify that

$$\hat{U}a = \exp(-\theta/2)(a - \beta b), \quad \hat{U}b = \exp(\theta/2)\left[(1 - \alpha\beta/2)b + \alpha a/2\right]. \tag{9.92}$$

The reader may find detailed discussions of the application of exponential operators in most books on Lie algebras [4, 7]. If we require that $q = \hat{U}a$ and $p = \hat{U}b$, then it follows from equations (9.92) and (9.86) that $\alpha = -i$, $\beta = i$, and $\theta = \ln(2m\omega)$; that is to say:

$$\hat{U} = \exp\left(i\hat{K}_-\right) \exp\left(-i\hat{K}_+\right) \exp\left[\ln(2m\omega)\hat{K}_0\right]. \tag{9.93}$$

The inverse transformation is given by the operator

$$\hat{U}^{-1} = \exp\left[-\ln(2m\omega)\hat{K}_0\right] \exp\left(i\hat{K}_+\right) \exp\left(-i\hat{K}_-\right), \tag{9.94}$$

where $\hat{K}_0$, $\hat{K}_+$, and $\hat{K}_-$ have exactly the same form (9.90) except that a and b are replaced with q and p, respectively, so that $a = \hat{U}^{-1}q$, and $b = \hat{U}^{-1}p$.

9.6 Secular Perturbation Theory

If we cannot solve the equations of motion for the Hamiltonian

$$H(\mathbf{x}, t) = H_0(\mathbf{x}) + \lambda H'(\mathbf{x}, t), \quad \mathbf{x} = (\mathbf{q}, \mathbf{p}) \tag{9.95}$$

exactly, but we expect $\lambda H'$ to be just a small correction to the known dynamics of H_0, then we may resort to perturbation theory. Here we present this approximate method in a way that closely resembles perturbation theory in the interaction picture of quantum mechanics outlined in Section 1.3.1. Expanding the trajectory in phase space as

$$\mathbf{x} = \sum_{j=0}^{\infty} \mathbf{x}^{(j)} \lambda^j, \tag{9.96}$$

and taking into account the equations of motion for the Hamiltonian (9.95), we obtain the following differential equation for the coefficients $\mathbf{x}^{(j)}$

$$\dot{\mathbf{x}}^{(j)} = -\hat{H}_0 \mathbf{x}^{(j)} - \hat{H}' \mathbf{x}^{(j-1)}, \quad j = 0, 1, \ldots, \tag{9.97}$$

where $\mathbf{x}^{(-1)} = \mathbf{0}$. Because each of these equations is a particular case of equation (9.70), the solutions are

$$\mathbf{x}^{(0)}(\mathbf{x}_0, t) = \hat{T}_{H_0}(t)\mathbf{x}_0$$

$$\mathbf{x}^{(j)}(\mathbf{x}_0, t) = \int_0^t \hat{T}_{H_0}\left(t - t'\right) \hat{\mathbf{x}}^{(j-1)}\left(\mathbf{x}_0, t'\right) H'\left(\mathbf{x}_0, t'\right) dt', \quad j > 0, \tag{9.98}$$

where $\hat{T}_{H_0}(t) = \exp[-t\hat{H}_0(\mathbf{x}_0)]$. Notice that the approximate solution satisfies the initial conditions at every perturbation order because $\mathbf{x}^{(0)}(\mathbf{x}_0, 0) = \mathbf{x}_0$ and $\mathbf{x}^{(j)}(\mathbf{x}_0, 0) = \mathbf{0}$ for all $j > 0$. We calculate the corrections recursively for $j = 1, 2, \ldots$, taking into account that $\hat{T}_{H_0}(t)F(\mathbf{x}_0, t) = F(\hat{T}_{H_0}(t)\mathbf{x}_0, t) = F(\mathbf{x}^{(0)}(\mathbf{x}_0, t), t)$, where $\mathbf{x}^{(0)}(\mathbf{x}_0, t)$ is the known unperturbed trajectory. The problem reduces to the calculation of the integral in equation (9.98).

This straightforward perturbation theory is called secular because the corrections $\mathbf{x}^{(j)}$ commonly exhibit unbounded (secular) terms in the case of bounded motion [189]. We have already discussed this point earlier in this chapter for the polynomial approximation.

9.6.1 Simple Examples

In what follows we consider weak perturbations of the well-known dynamics of the harmonic oscillator

$$H_0(q, p) = \frac{p^2}{2m} + \frac{m\omega^2 q^2}{2} \tag{9.99}$$

that gives rise to the trajectory in phase space given by equation (9.82). In the program section we show a set of simple Maple procedures for the application of perturbation theory according to equation (9.98). As a particular example we consider a cubic perturbation $H'(q, p) = q^3$ that makes the resulting potential-energy function a well with a barrier of height $V(q_m) = m^3\omega^6/(54\lambda^2)$ at $q_m = -m\omega^2/(3\lambda)$. Figure 9.1 (a) shows this potential-energy function and $V(q_m)$ for arbitrary values of m, ω, and $\lambda < 0$. In order to shorten the size of the results we substitute the particular initial condition $p_0 = 0$. For example, the first two corrections to the trajectory are

$$q^{(1)} = \frac{q_0^2}{2m\omega^2}[\cos(2\omega t) + 2\cos(\omega t) - 3]$$

$$q^{(2)} = \frac{q_0^3}{16m^2\omega^4}[3\cos(3\omega t) + 16\cos(2\omega t) + 29\cos(\omega t) + 60\omega t \sin(\omega t) - 48] . \tag{9.100}$$

Notice the secular term $t\sin(\omega t)$ that makes $|q^{(2)}|$ grow unboundedly even for initial conditions leading to periodic motion.

The polynomial approximation discussed in Section 9.3 and the secular perturbation theory in operator form give exactly the same result. To obtain the trajectory for the cubic oscillator from the former approach we set $F_1 = -m\omega^2$, $F_2 = -3\lambda$, and $F_j = 0$ for all $j > 2$ in equation (9.8), so that $a_1 = -F_2/F_1 = -3\lambda/(m\omega^2)$, and $a_j = 0$ for all $j > 1$ in equation (9.9). By straightforward inspection of equations (9.19) and (9.100), one easily verifies that $L^{j+1}q_j(\omega t)$ and $\lambda^j q^{(j)}(t)$ agree for $j = 1, 2$ provided that $L = q_0$.

It is instructive (and also a suitable test for the equations and programs) to consider the perturbation $H'(q, p) = m\omega^2 q^2/2$, because it enables us to compare the perturbation series with the Taylor expansion of the exact solution obtained by substitution of $\omega\sqrt{1 + \lambda}$ for ω in equation (9.82).

Present secular perturbation theory in operator form closely resembles the application of perturbation theory to the Heisenberg equations of motion in quantum mechanics discussed in Section 1.3. It is instructive to compare classical and quantum-mechanical results. For that purpose we consider dimensionless anharmonic oscillators of the form

$$H = \frac{p^2}{2} + \frac{q^2}{2} + \lambda q^M, M = 3, 4, \ldots . \tag{9.101}$$

Table 9.2 shows $q^{(1)}$ and $q^{(2)}$ for $M = 3$ and $q^{(1)}$ for $M = 4$, obtained by means of the Maple program mentioned above with $m = \omega = 1$. Comparing them with the corresponding quantum-mechanical expressions in Table 1.2 we realize that the former are formally identical with the real

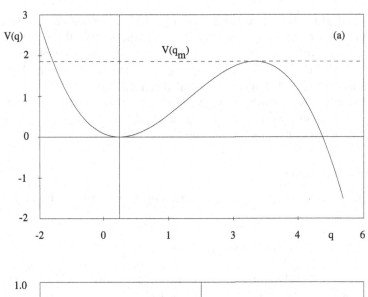

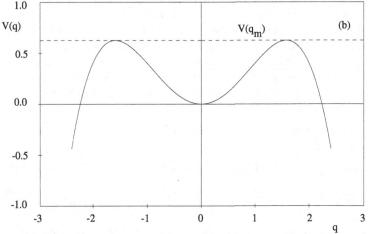

FIGURE 9.1
Potential-energy functions $V(q) = m\omega^2 q^2/2 - |\lambda|q^3$ **(a) and** $V(q) = m\omega^2 q^2/2 - |\lambda|q^4$ **(b), and their barrier heights** $V(q_m)$.

parts of the latter. The imaginary parts of the quantum-mechanical results arise in the process of ordering the noncommutative coordinate and momentum operators according to the rule $[\hat{q}, \hat{p}] = i$ and have no classical counterpart.

9.6.2   Construction of Invariants by Perturbation Theory

The total rate of change of a function $F(\mathbf{q}, \mathbf{p}, t)$ with time is

$$\frac{dF}{dt} = \left(\frac{\partial F}{\partial t}\right)_{q,p} + \{F, H\}_{[q,p]} \,. \tag{9.102}$$

Table 9.2 Secular Perturbation Theory in Operator Form for the

Dimensionless Anharmonic Oscillators $H = \dfrac{p^2}{2} + \dfrac{q^2}{2} + \lambda q^M$

$$M = 3$$

$$q^{(1)} = q_0\,p_0\,\sin(2\,t) - 2\,q_0\,p_0\,\sin(t) + \left(-\frac{p_0^2}{2} + \frac{q_0^2}{2}\right)\cos(2\,t)$$

$$+\left(2\,p_0^2 + q_0^2\right)\cos(t) - \frac{3\,p_0^2}{2} - \frac{3\,q_0^2}{2}$$

$$q^{(2)} = \left(2\,p_0^3 - p_0\,q_0^2\right)\sin(2\,t) + \left(\frac{15\,t\,q_0^3}{4} + \frac{15\,t\,q_0\,p_0^2}{4} + \frac{5\,p_0^3}{16} + \frac{65\,p_0\,q_0^2}{16}\right)\sin(t)$$

$$+\left(\frac{9\,p_0\,q_0^2}{16} - \frac{3\,p_0^3}{16}\right)\sin(3\,t) + \left(q_0^3 + 4\,q_0\,p_0^2\right)\cos(2\,t)$$

$$+\left(-\frac{15\,t\,p_0^3}{4} - \frac{55\,q_0\,p_0^2}{16} - \frac{15\,t\,p_0\,q_0^2}{4} + \frac{29\,q_0^3}{16}\right)\cos(t)$$

$$+\left(-\frac{9\,q_0\,p_0^2}{16} + \frac{3\,q_0^3}{16}\right)\cos(3\,t) - 3\,q_0^3$$

$$M = 4$$

$$q^{(1)} = \left(\frac{3\,q_0^2\,p_0}{8} - \frac{p_0^3}{8}\right)\sin(3\,t) + \left(-\frac{21\,q_0^2\,p_0}{8} - \frac{9\,p_0^3}{8} - \frac{3\,t\,q_0^3}{2} - \frac{3\,t\,q_0\,p_0^2}{2}\right)\sin(t)$$

$$+\left(-\frac{3\,q_0\,p_0^2}{8} + \frac{q_0^3}{8}\right)\cos(3\,t) + \left(\frac{3\,t\,q_0^2\,p_0}{2} + \frac{3\,t\,p_0^3}{2} - \frac{q_0^3}{8} + \frac{3\,q_0\,p_0^2}{8}\right)\cos(t)$$

A function $I(\mathbf{q}, \mathbf{p}, t)$ is called an invariant if $dI/dt = 0$; that is to say, if

$$\left(\frac{\partial I}{\partial t}\right)_{q,p} = \{H, I\}_{[q,p]} = \hat{H}\,I\,. \tag{9.103}$$

The reader may easily verify that with just a change of sign in the main equation, the method outlined above proves suitable for constructing invariants by means of perturbation theory according to equation (9.103) [192].

9.7 Canonical Perturbation Theory

In what follows we outline a classical perturbation theory free from secular terms which we present in a way that closely resembles the perturbation theory for operators in quantum mechanics discussed in Section 3.5. Canonical perturbation theory has been known for a long time and widely applied to numerous problems in classical mechanics. Here we just mention some references [186]–[189], [193]–[195] where the interested reader may look up others. In order to illustrate the main ideas underlying canonical perturbation theory, we choose a simple anharmonic oscillator in one dimension.

Consider the Hamiltonian $H(q, p) = H_0(q, p) + \lambda H'(q, p)$, where H_0 corresponds to a harmonic oscillator and the perturbation H' is a polynomial function of q and p. If the degree of the polynomial H' is greater than two, then the dynamical problem is nonlinear.

The complex variables a and b defined in equation (9.85) considerably simplify the application of canonical perturbation theory exactly as the creation and annihilation operators are suitable for the application of perturbation theory for operators in quantum mechanics. We explicitly indicate this change of dynamical variables as

$$\tilde{H}(a, b) = \hat{U} H(a, b) = H(q(a, b), p(a, b)) = \tilde{H}_0(a, b) + \lambda \tilde{H}'(a, b), \tag{9.104}$$

where $\tilde{H}_0(a, b) = -i\omega ab$ as in equation (9.88).

The aim of the approach is to find a change of variables

$$a(A, B) = \hat{T} A, \quad b(A, B) = \hat{T} B \tag{9.105}$$

with inverse

$$A(a, b) = \hat{T}^{-1} a, \quad B(a, b) = \hat{T}^{-1} b \tag{9.106}$$

such that the transformed Hamiltonian depends on A and B only through $\tilde{H}_0(A, B) = -i\omega AB$; that is to say:

$$\hat{T} \tilde{H}(A, B) = \tilde{H}\left(\hat{T} A, \hat{T} B\right) = K\left(\tilde{H}_0(A, B), \lambda\right) . \tag{9.107}$$

Notice that $\tilde{H}_0(A, B)$ is a constant of the motion:

$$\frac{d}{dt} \tilde{H}_0(A, B) = \left\{\tilde{H}_0, K\right\} = 0 . \tag{9.108}$$

The equations of motion for the new dynamical variables A and B are

$$\dot{A} = \frac{\partial K}{\partial B} = -i\Omega A, \quad \dot{B} = -\frac{\partial K}{\partial A} = i\Omega B , \tag{9.109}$$

where $\Omega = \omega(\partial K/\partial \tilde{H}_0)$ is a constant of the motion because it is a function of $\tilde{H}_0$. Therefore, the solutions are given by

$$A = A_0 \exp(-i\Omega t), \quad B = B_0 \exp(i\Omega t) , \tag{9.110}$$

where A_0 and B_0 are appropriate initial conditions at $t = 0$.

Later on we will show that the Hamiltonian for the harmonic oscillator in terms of the action J is $\tilde{H}_0 = J\omega$, so that $\Omega = \omega(\partial K/\partial \tilde{H}_0) = \partial K/\partial J$ is one of Hamilton's equations for the action-angle variables [196].

In order to express the canonical transformation in a more familiar way we write q and p in terms of a new coordinate Q and conjugate momentum P instead of the complex variables A and B. Taking into account equation (9.92) we write

$$Q = \hat{U} A, \quad P = \hat{U} B . \tag{9.111}$$

It is not difficult to verify that we have carried out the following transformations of the Hamiltonian

$$K\left(\tilde{H}_0, \lambda\right) = \hat{T} \tilde{H} = \hat{T}\hat{U} H = K\left(\hat{U} H_0, \lambda\right) = \hat{U} K\left(H_0, \lambda\right) , \tag{9.112}$$

from which it follows that

$$K\left(H_0(Q, P), \lambda\right) = H(q(Q, P), p(Q, P)) = \hat{U}^{-1}\hat{T}\hat{U} H(Q, P) . \tag{9.113}$$

Accordingly, the canonical transformation of the coordinate and momentum is

$$\mathbf{x} = \hat{U}^{-1} \hat{T} \hat{U} \mathbf{y}, \ \ \mathbf{y} = \hat{U}^{-1} \hat{T}^{-1} \hat{U} \mathbf{x}, \tag{9.114}$$

where $\mathbf{x} = (q, p)$ and $\mathbf{y} = (Q, P)$.

The transformed Hamiltonian K is commonly said to be in (Birkhoff–Gustavson) normal form [186, 188, 189], [193]–[195]. In practice it is not possible to obtain the normal form exactly (except for some trivial problems), and one resorts to an approximation. In what follows we show how to obtain the normal form as a perturbation series:

$$K\left(\tilde{H}_0, \lambda\right) = \sum_{j=0}^{\infty} K_j\left(\tilde{H}_0\right) \lambda^j, \ \ K_0 = \tilde{H}_0 . \tag{9.115}$$

We choose a transformation operator $\hat{T}$ given by equation (9.61) with an operator $\hat{W}$ to be determined by perturbation theory. By straightforward substitution of the power series

$$\hat{T} = \sum_{j=0}^{\infty} \lambda^j \hat{T}_j, \ \hat{T}_0 = \hat{1} , \tag{9.116}$$

$$\hat{W} = \sum_{j=0}^{\infty} \lambda^j \hat{W}_j , \tag{9.117}$$

into equation (9.61), we obtain a recurrence relation for the operator coefficients $\hat{T}_n$ in terms of the operator coefficients $\hat{W}_j$:

$$\hat{T}_n = \frac{1}{n} \sum_{j=0}^{n-1} \hat{W}_j \hat{T}_{n-j-1} . \tag{9.118}$$

Exactly in the same way we derive a recurrence relation for the coefficients of the expansion of the inverse operator

$$\hat{T}^{-1} = \sum_{j=0}^{\infty} \lambda^j \left(\hat{T}^{-1}\right)_j , \ \left(\hat{T}^{-1}\right)_0 = \hat{1} \tag{9.119}$$

from equation (9.62) as follows:

$$\left(\hat{T}^{-1}\right)_n = -\frac{1}{n} \sum_{j=0}^{n-1} \left(\hat{T}^{-1}\right)_{n-j-1} \hat{W}_j . \tag{9.120}$$

The expansion of equation (9.107) in a perturbation series leads to

$$\hat{T}_n \tilde{H}_0 + \hat{T}_{n-1} \tilde{H}' = K_n \tag{9.121}$$

that we rewrite as follows:

$$\widehat{\tilde{H}_0 W_n} = \sum_{j=0}^{n-1} \hat{W}_j \hat{T}_{n-j} \tilde{H}_0 + (n+1)\hat{T}_n \tilde{H}' - (n+1)K_{n+1} = F_n - (n+1)K_{n+1} . \tag{9.122}$$

Notice that at the *nth* step the unknowns are W_n and K_{n+1} because

$$F_n = \sum_{j=0}^{n-1} \hat{W}_j \hat{T}_{n-j} \tilde{H}_0 + (n+1) \hat{T}_n \tilde{H}' \tag{9.123}$$

contains terms already calculated previously.

In order to obtain those unknowns we simply take into account that $\widehat{\tilde{H}}_0 a^j b^k = i(j-k)\omega a^j b^k$. On choosing K_{n+1} so that $F_n - (n+1)K_{n+1}$ is free from diagonal terms $a^j b^j$ we can apply

$$\widehat{\tilde{H}}_0^{-1} a^j b^k = \frac{a^j b^k}{(j-k)\omega i}, \ j \neq k \tag{9.124}$$

to obtain

$$W_n = \widehat{\tilde{H}}_0^{-1} \left[F_n - (n+1)K_{n+1} \right] . \tag{9.125}$$

An alternative way is to write

$$\left(\widehat{\tilde{H}}_0 + \alpha \right)^{-1} = \int_{-\infty}^{0} \exp\left[t \left(\widehat{\tilde{H}}_0 + \alpha \right) \right] dt, \ \alpha > 0 , \tag{9.126}$$

and to take into account that $(\widehat{\tilde{H}}_0 + \alpha)^{-1} K_{n+1} = K_{n+1}/\alpha$ because $\widehat{\tilde{H}}_0 K_{n+1} = 0$. Therefore, if we define

$$W_n(\alpha) = \int_{-\infty}^{0} \exp\left[t \left(\widehat{\tilde{H}}_0 + \alpha \right) \right] F_n \, dt , \tag{9.127}$$

we have

$$K_{n+1} = \frac{1}{n+1} \lim_{\alpha \to 0} \alpha W_n(\alpha) , \tag{9.128}$$

and

$$W_n = \lim_{\alpha \to 0} \left[W_n(\alpha) - \frac{(n+1)K_{n+1}}{\alpha} \right] . \tag{9.129}$$

These expressions are suitable for the systematic application of computer algebra because we can easily program the effect of $\exp(t\widehat{\tilde{H}}_0)$ on any function of a and b as $\exp(t\widehat{\tilde{H}}_0)G(a,b) = G(a\exp(i\omega t), b\exp(-i\omega t))$.

It follows from inspection of equation (9.114) that we can express the operators $\hat{U}$ and $\hat{T}$ either in terms of q and p or in terms of Q and P, omitting explicit reference to the intermediate complex variables a, b, A and B. Proceeding in this way greatly facilitates programming the equations for the transformation. It is worth noting that the canonical transformations in operator form (9.114) give both the direct and inverse change of variables, $\mathbf{y(x)}$ and $\mathbf{x(y)}$, explicitly. On the other hand, the traditional canonical transformations known in classical mechanics since long ago [185], and widely applied to many problems of current interest [197]–[199], give implicit expressions. In fact, the traditional canonical transformation for the simple anharmonic oscillator discussed here is given by a generating function $F_2(q, P) = qP + S(q, P)$ as

$$Q = q + \frac{\partial S(q, P)}{\partial P} , \tag{9.130}$$

$$P = p - \frac{\partial S(q, P)}{\partial q} . \tag{9.131}$$

Consequently, to obtain $Q(q, p)$ and $P(q, p)$ explicitly, we should first solve equation (9.131) for $P(q, p)$ and substitute it into (9.130) to derive $Q(q, p)$. This approach is considerably more cumbersome than the operator form of the canonical transformations, and in the end both yield the new coordinate and momentum as perturbation series:

$$Q = q + \sum_{j=1}^{\infty} Q_j(q, p)\lambda^j, \quad P = p + \sum_{j=1}^{\infty} P_j(q, p)\lambda^j. \tag{9.132}$$

In the program section we show a set of simple Maple procedures for the systematic application of canonical perturbation theory according to the equations above. Choosing the cubic and quartic perturbations $H' = q^3$ and $H' = q^4$, respectively, as illustrative examples we obtain the results in Table 9.3. If we substitute $\lambda = f/6$ and $\lambda = f/24$ in the cubic and quartic cases, respectively, our results through second order agree with the expressions derived recently by means of the traditional canonical transformations [199]. It is straightforward to generalize the equations given above in order to treat anharmonic oscillators in more than one dimension [194].

In closing this section we mention that we can entirely omit the intermediate transformation to the complex variables a and b, because equations (9.127)–(9.129) are valid for any set of variables. If, for example, we choose q and p, this approach offers no difficulty because we know the effect of $\exp(t\widehat{H}_0)$ on them. Although in this way we bypass the transformations $\hat{U}$ and $\hat{U}^{-1}$ the resulting procedure is slower because the integrals are more complicated.

9.8 The Hypervirial Hellmann–Feynman Method (HHFM)

The simplest and most efficient way of obtaining the canonical perturbation series for separable classical models is based on the hypervirial and Hellmann–Feynman theorems [200]. This approach is reminiscent of the method of Swenson and Danforth discussed in Section 3.3 that facilitates the application of perturbation theory in quantum mechanics. In order to apply the HHFM we have first to present the classical problem in a way that closely resembles a quantum-mechanical one, and develop the classical counterparts of the hypervirial and Hellmann–Feynman theorems given in Section 3.2.

For simplicity and concreteness, we consider the periodic motion of a one-dimensional model with Hamiltonian

$$H(q, p) = \frac{p^2}{2m} + V(q). \tag{9.133}$$

The time average of a function $F(q, p)$ is given by

$$\overline{F} = \frac{1}{\tau} \int_0^{\tau} F(q(t), p(t)) \, dt, \tag{9.134}$$

where τ is the period of the motion. In particular,

$$\overline{\frac{dG}{dt}} = \overline{\{G, H\}} = \frac{1}{\tau}[G(\tau) - G(0)] = 0 \tag{9.135}$$

is the classical hypervirial theorem. Choosing $G(q, p) = f(q)p$ we obtain

$$2E\overline{f'} - 2\overline{f'V} - \overline{fV'} = 0, \tag{9.136}$$

Table 9.3 Canonical Perturbation Theory in Operator Form for

$$H = \frac{p^2}{2m} + \frac{m\,\omega^2 q^2}{2} + \lambda\,q^M \ \text{(Continued)}.$$

M = 3

$K_1 = 0$

$K_2 = -\frac{15}{4}\,\frac{H_0{}^2}{\omega^6 m^3}$

$K_3 = 0$

$K_4 = -\frac{705}{16}\,\frac{H_0{}^3}{\omega^{12} m^6}$

$K_5 = 0$

$K_6 = -\frac{115755}{128}\,\frac{H_0{}^4}{\omega^{18} m^9}$

$Q_1 = \frac{q^2}{m\,\omega^2} + 2\,\frac{p^2}{m^3\,\omega^4}$

$Q_2 = \frac{7}{16}\,\frac{q^3}{m^2\,\omega^4} - \frac{77}{16}\,\frac{q\,p^2}{m^4\,\omega^6}$

$Q_3 = \frac{53}{16}\,\frac{q^4}{m^3\,\omega^6} + \frac{465}{16}\,\frac{q^2\,p^2}{m^5\,\omega^8} + \frac{115}{8}\,\frac{p^4}{m^7\,\omega^{10}}$

$Q_4 = -\frac{28869}{512}\,\frac{q\,p^4}{m^8\,\omega^{12}} + \frac{4795}{512}\,\frac{q^5}{m^4\,\omega^8} - \frac{12869}{256}\,\frac{q^3\,p^2}{m^6\,\omega^{10}}$

$Q_5 = \frac{317095}{512}\,\frac{q^2\,p^4}{m^9\,\omega^{14}} + \frac{25571}{64}\,\frac{p^2\,q^4}{m^7\,\omega^{12}} + \frac{51131}{256}\,\frac{p^6}{m^{11}\,\omega^{16}} + \frac{22183}{512}\,\frac{q^6}{m^5\,\omega^{10}}$

$Q_6 = -\frac{12574859}{8192}\,\frac{q^3\,p^4}{m^{10}\,\omega^{16}} - \frac{7575957}{8192}\,\frac{q\,p^6}{m^{12}\,\omega^{18}} - \frac{1496055}{8192}\,\frac{q^5\,p^2}{m^8\,\omega^{14}}$
$\qquad + \frac{1474623}{8192}\,\frac{q^7}{m^6\,\omega^{12}}$

$P_1 = -\frac{2q\,p}{m\,\omega^2}$

$P_2 = \frac{43}{16}\,\frac{q^2\,p}{m^2\,\omega^4} - \frac{17}{16}\,\frac{p^3}{m^4\,\omega^6}$

$P_3 = -\frac{55}{8}\,\frac{q\,p^3}{m^5\,\omega^8} - 16\,\frac{q^3\,p}{m^3\,\omega^6}$

$P_4 = -\frac{2949}{256}\,\frac{q^2\,p^3}{m^6\,\omega^{10}} + \frac{4171}{512}\,\frac{q^4\,p}{m^4\,\omega^8} - \frac{10389}{512}\,\frac{p^5}{m^8\,\omega^{12}}$

$P_5 = -\frac{50485}{256}\,\frac{q^5\,p}{m^5\,\omega^{10}} - \frac{16811}{256}\,\frac{q\,p^5}{m^9\,\omega^{14}} - \frac{1185}{4}\,\frac{q^3\,p^3}{m^7\,\omega^{12}}$

$P_6 = -\frac{3274377}{8192}\,\frac{p^7}{m^{12}\,\omega^{18}} - \frac{2213997}{8192}\,\frac{q^6\,p}{m^6\,\omega^{12}} - \frac{3978499}{8192}\,\frac{q^4\,p^3}{m^8\,\omega^{14}}$
$\qquad - \frac{6691551}{8192}\,\frac{q^2\,p^5}{m^{10}\,\omega^{16}}$

$W_0 = -\frac{1}{12}\,\frac{\sqrt{2}\,(9q^2\,p + p^3 + i\,(9q\,p^2 + q^3))}{(m\,\omega)^{(3/2)}\,\omega}$

$W_1 = \frac{3}{16}\,\frac{8\,(q^3\,p + p^3\,q) + i\,(p^4 - q^4)}{m^3\,\omega^5}$

Table 9.3 *(Cont.)* Canonical Perturbation Theory in Operator Form for

$$H = \frac{p^2}{2\,m} + \frac{m\,\omega^2\,q^2}{2} + \lambda\,q^M$$

$$M = 4$$

$$K_1 = \tfrac{3}{2}\,\frac{H_0{}^2}{m^2\,\omega^4}$$

$$K_2 = -\tfrac{17}{4}\,\frac{H_0{}^3}{m^4\,\omega^8}$$

$$K_3 = \tfrac{375}{16}\,\frac{H_0{}^4}{m^6\,\omega^{12}}$$

$$K_4 = -\tfrac{10689}{64}\,\frac{H_0{}^5}{m^8\,\omega^{16}}$$

$$K_5 = \tfrac{87549}{64}\,\frac{H_0{}^6}{m^{10}\,\omega^{20}}$$

$$K_6 = -\tfrac{3132399}{256}\,\frac{H_0{}^7}{m^{12}\,\omega^{24}}$$

$$Q_1 = \frac{5\,q^3}{8\,m\,\omega^2} + \frac{9\,q\,p^2}{8\,m^3\,\omega^4}$$

$$Q_2 = -\tfrac{77}{128}\,\frac{q^5}{m^2\,\omega^4} - \tfrac{461}{64}\,\frac{q^3\,p^2}{m^4\,\omega^6} - \tfrac{493}{128}\,\frac{q\,p^4}{m^6\,\omega^8}$$

$$Q_3 = \tfrac{949}{1024}\,\frac{q^7}{m^3\,\omega^6} + \tfrac{56423}{1024}\,\frac{q^3\,p^4}{m^7\,\omega^{10}} + \tfrac{37731}{1024}\,\frac{q^5\,p^2}{m^5\,\omega^8} + \tfrac{19257}{1024}\,\frac{q\,p^6}{m^9\,\omega^{12}}$$

$$Q_4 = -\tfrac{3471229}{32768}\,\frac{q\,p^8}{m^{12}\,\omega^{16}} - \tfrac{8237815}{16384}\,\frac{q^5\,p^4}{m^8\,\omega^{12}} - \tfrac{1422157}{8192}\,\frac{q^7\,p^2}{m^6\,\omega^{10}}$$
$$- \tfrac{3434637}{8192}\,\frac{q^3\,p^6}{m^{10}\,\omega^{14}} - \tfrac{56829}{32768}\,\frac{q^9}{m^4\,\omega^8}$$

$$Q_5 = \tfrac{957363}{262144}\,\frac{q^{11}}{m^5\,\omega^{10}} + \tfrac{721934803}{131072}\,\frac{q^5\,p^6}{m^{11}\,\omega^{16}} + \tfrac{841618031}{262144}\,\frac{q^3\,p^8}{m^{13}\,\omega^{18}}$$
$$+ \tfrac{169181199}{262144}\,\frac{q\,p^{10}}{m^{15}\,\omega^{20}} + \tfrac{488103927}{131072}\,\frac{q^7\,p^4}{m^9\,\omega^{14}} + \tfrac{205785691}{262144}\,\frac{q^9\,p^2}{m^7\,\omega^{12}}$$

$$Q_6 = -\tfrac{35117705}{4194304}\,\frac{q^{13}}{m^6\,\omega^{12}} - \tfrac{230944788423}{4194304}\,\frac{q^5\,p^8}{m^{14}\,\omega^{20}} - \tfrac{102328046375}{4194304}\,\frac{q^9\,p^4}{m^{10}\,\omega^{16}}$$
$$- \tfrac{7267119483}{2097152}\,\frac{q^{11}\,p^2}{m^8\,\omega^{14}} - \tfrac{58263714797}{1048576}\,\frac{q^7\,p^6}{m^{12}\,\omega^{18}} - \tfrac{51873744571}{2097152}\,\frac{q^3\,p^{10}}{m^{16}\,\omega^{22}}$$
$$- \tfrac{17331036841}{4194304}\,\frac{q\,p^{12}}{m^{18}\,\omega^{24}}$$

$$P_1 = -\frac{15\,q^2\,p}{8\,m\,\omega^2} - \frac{3\,p^3}{8\,m^3\,\omega^4}$$

$$P_2 = \tfrac{835}{128}\,\frac{q^4\,p}{m^2\,\omega^4} + \tfrac{371}{64}\,\frac{q^2\,p^3}{m^4\,\omega^6} + \tfrac{131}{128}\,\frac{p^5}{m^6\,\omega^8}$$

$$P_3 = -\tfrac{4227}{1024}\,\frac{p^7}{m^9\,\omega^{12}} - \tfrac{24943}{1024}\,\frac{q^6\,p}{m^3\,\omega^6} - \tfrac{29109}{1024}\,\frac{q^2\,p^5}{m^7\,\omega^{10}} - \tfrac{53025}{1024}\,\frac{q^4\,p^3}{m^5\,\omega^8}$$

$$P_4 = \tfrac{1335299}{8192}\,\frac{q^2\,p^7}{m^{10}\,\omega^{14}} + \tfrac{3000483}{8192}\,\frac{q^6\,p^3}{m^6\,\omega^{10}} + \tfrac{6642697}{16384}\,\frac{q^4\,p^5}{m^8\,\omega^{12}} + \tfrac{3049571}{32768}\,\frac{q^8\,p}{m^4\,\omega^8}$$
$$+ \tfrac{652835}{32768}\,\frac{p^9}{m^{12}\,\omega^{16}}$$

$$P_5 = -\tfrac{27973029}{262144}\,\frac{p^{11}}{m^{15}\,\omega^{20}} - \tfrac{94246393}{262144}\,\frac{q^{10}\,p}{m^5\,\omega^{10}} - \tfrac{596711177}{262144}\,\frac{q^8\,p^3}{m^7\,\omega^{12}}$$
$$- \tfrac{551855349}{131072}\,\frac{q^6\,p^5}{m^9\,\omega^{14}} - \tfrac{416637985}{131072}\,\frac{q^4\,p^7}{m^{11}\,\omega^{16}} - \tfrac{265471245}{262144}\,\frac{q^2\,p^9}{m^{13}\,\omega^{18}}$$

$$P_6 = \tfrac{151826257737}{4194304}\,\frac{q^8\,p^5}{m^{10}\,\omega^{16}} + \tfrac{27329970085}{2097152}\,\frac{q^{10}\,p^3}{m^8\,\omega^{14}} + \tfrac{13856811429}{2097152}\,\frac{q^2\,p^{11}}{m^{16}\,\omega^{22}}$$
$$+ \tfrac{45562683987}{1048576}\,\frac{q^6\,p^7}{m^{12}\,\omega^{18}} + \tfrac{104791019465}{4194304}\,\frac{q^4\,p^9}{m^{14}\,\omega^{20}} + \tfrac{5863791911}{4194304}\,\frac{q^{12}\,p}{m^6\,\omega^{12}}$$
$$+ \tfrac{2566494119}{4194304}\,\frac{p^{13}}{m^{18}\,\omega^{24}}$$

$$W_0 = -\frac{8\,p^3\,q + 8\,q^3\,p + i\,(q^4 - p^4)}{16\,m^2\,\omega^3}$$

$$W_1 = -\frac{9\,p^5\,q - 9\,q^5\,p + i\,(99\,q^4\,p^2 + 99\,q^2\,p^4 + q^6 + p^6)}{24\,m^4\,\omega^6}$$

where $E = H$ is the energy of the motion. In addition to the different meaning of the expectation values, this expression differs from the quantum-mechanical one in equation (3.16) in a term proportional to $\hbar^2$ that arises from the noncommuting properties of the coordinate and momentum operators.

In order to obtain the classical Hellmann–Feynman theorem, we suppose that the potential-energy function depends on a parameter λ: $V = V(q, \lambda)$. The action is [196]

$$J = \frac{1}{2\pi} \oint p \, dq = \frac{1}{\pi} \int_{q_1}^{q_2} \sqrt{2m(E - V)} \, dq \, , \qquad (9.137)$$

where the first integral is taken along the periodic trajectory, and the turning points q_1 and q_2 in the second integral are roots of $V(q_1, \lambda) = V(q_2, \lambda) = E$. Equation (9.137) gives the energy as a function of λ and J that we keep constant. Differentiating equation (9.137) with respect to λ, and taking into account that $p(q_1) = p(q_2) = 0$ and $\partial J / \partial \lambda = 0$, we obtain

$$p(q_2) \frac{\partial q_2}{\partial \lambda} - p(q_1) \frac{\partial q_1}{\partial \lambda} + \int_{q_1}^{q_2} \frac{\partial p}{\partial \lambda} \, dq = \int_{q_1}^{q_2} \left(\frac{\partial E}{\partial \lambda} - \frac{\partial V}{\partial \lambda} \right) \dot{q}^{-1} \, dq$$

$$= \frac{1}{2} \oint \left(\frac{\partial E}{\partial \lambda} - \frac{\partial V}{\partial \lambda} \right) \dot{q}^{-1} \, dq = \frac{1}{2} \int_0^\tau \left(\frac{\partial E}{\partial \lambda} - \frac{\partial V}{\partial \lambda} \right) \, dt = 0 \, , \qquad (9.138)$$

which becomes the classical Hellmann–Feynman theorem

$$\frac{\partial E}{\partial \lambda} = \overline{\frac{\partial V}{\partial \lambda}} \, . \qquad (9.139)$$

This expression is identical to the quantum-mechanical Hellmann–Feynman theorem in equation (3.7) except for the different meaning of the expectation value.

9.8.1   One-Dimensional Models with Polynomial Potential-Energy Functions

For simplicity, in what follows we consider anharmonic oscillators with polynomial potentials

$$V(q, \lambda) = \frac{m\omega^2}{2} q^2 + \lambda q^M, \ M = 3, 4, \dots \qquad (9.140)$$

as simple nontrivial illustrative examples. We will obtain the perturbation series for the energy

$$E(J, \lambda) = \sum_{i=0}^{\infty} E_i(J) \lambda^i \qquad (9.141)$$

in terms of a fixed value of J that resembles the quantum number of the quantum-mechanical models. The coefficient $E_0(J)$ is the energy of the harmonic oscillator for the same value of the action. In this case the calculation of the integral in equation (9.137) is straightforward giving

$$J = \frac{E_0}{\omega} \, . \qquad (9.142)$$

Because the potential-energy function of the anharmonic oscillator is a polynomial function of q, it is convenient to choose

$$f(q) = q^{N-1}, \ N = 1, 2, \dots \qquad (9.143)$$

so that equation (9.136) becomes

$$2(N-1)EQ_{N-2} - m\omega^2 N Q_N - \lambda(2N+M-2)Q_{N+M-2} = 0, \tag{9.144}$$

where

$$Q_N = \overline{q^N}. \tag{9.145}$$

Substituting the perturbation expansion

$$Q_N = \sum_{i=0}^{\infty} Q_{N,i}\lambda^i \tag{9.146}$$

and the energy series (9.141) into the hypervirial equation (9.144), we obtain a recurrence relation for the perturbation coefficients:

$$Q_{N,p} = \frac{1}{mN\omega^2}\left[2(N-1)\sum_{j=0}^{p}E_j Q_{N-2,p-j} - (2N+M-2)Q_{N+M-2,p-1}\right]. \tag{9.147}$$

The Hellmann–Feynman theorem

$$\frac{\partial E}{\partial \lambda} = Q_M \tag{9.148}$$

provides a necessary relationship between the coefficients of the series (9.141) and (9.146):

$$E_j = \frac{1}{j}Q_{M,j-1}, \; j = 1, 2, \ldots. \tag{9.149}$$

We can obtain all the perturbation corrections to the energy and averages Q_N from equations (9.147) and (9.149). It follows from $Q_0 = 1$ that the initial condition is $Q_{0,p} = \delta_{p0}$. In order to obtain E_p we have to calculate $Q_{N,j}$ for all $j = 0, 1, \ldots, p-1$, and $N = 1, 2, \ldots, (p - j)(M-2)+2$. Notice that equations (9.147) and (9.149) are so similar to their quantum-mechanical counterparts in Section 3.3.1 that one easily writes a set of Maple procedures for the classical models by slight modification of the quantum-mechanical ones already given in the program section.

Aided by Maple we calculate as many analytical perturbation corrections as desired. Table 9.4 shows results for the cubic ($M = 3$) and quartic ($M = 4$) perturbations as illustrative examples. Writing the Hamiltonian of the anharmonic oscillators in dimensionless form the reader may easily verify that the actual perturbation parameter is $\beta = \lambda m^{-M/2}\omega^{-(M+2)/2}$, and that the energy coefficients satisfy $E_j(m, \omega) = \omega E_j(1, 1)m^{-Mj/2}\omega^{-(M+2)j/2}$. Notice that if M is odd, then $E_{2j+1} = 0$ and $Q_{N,i} = 0$ if $N+i$ is odd as already discussed in preceding chapters for the quantum-mechanical counterpart.

The HHFM and the operator method discussed earlier give exactly the same canonical perturbation series for the energy because $E_0 = H_0$. However, the HHFM is much faster and is therefore preferable when one is interested in the calculation of perturbation corrections to the energy of sufficiently large order. The main disadvantage of the HHFM is that it only applies to separable models and that it does not give the new coordinate and momentum explicitly.

9.8.2 Radius of Convergence of the Canonical Perturbation Series

The singularity of $E(J, \lambda)$ closest to the origin in the complex λ plane determines the radius of convergence of the canonical perturbation series. Apparently it is possible to locate this singularity

Table 9.4   Canonical Perturbation Theory for the Anharmonic

Oscillators $H = \dfrac{p^2}{2\,m} + \dfrac{m\,\omega^2\,q^2}{2} + \lambda\,q^M$ by Means of the

Hypervirial Hellmann–Feynman Method *(Continued)*

$$M = 3$$

$$E_0 = \omega J$$
$$E_1 = 0$$
$$E_2 = -\tfrac{15}{4}\,\frac{E_0{}^2}{m^3\,\omega^6}$$
$$E_3 = 0$$
$$E_4 = -\tfrac{705}{16}\,\frac{E_0{}^3}{m^6\,\omega^{12}}$$
$$E_5 = 0$$
$$E_6 = -\tfrac{115755}{128}\,\frac{E_0{}^4}{m^9\,\omega^{18}}$$

$$Q_{1,0} = 0$$
$$Q_{1,1} = -3\,\frac{E_0}{m^2\,\omega^4}$$
$$Q_{1,2} = 0$$
$$Q_{1,3} = -45\,\frac{E_0{}^2}{m^5\,\omega^{10}}$$
$$Q_{1,4} = 0$$
$$Q_{1,5} = -\tfrac{19035}{16}\,\frac{E_0{}^3}{m^8\,\omega^{16}}$$

$$Q_{2,0} = \frac{E_0}{m\,\omega^2}$$
$$Q_{2,1} = 0$$
$$Q_{2,2} = 15\,\frac{E_0{}^2}{m^4\,\omega^8}$$
$$Q_{2,3} = 0$$
$$Q_{2,4} = \tfrac{6345}{16}\,\frac{E_0{}^3}{m^7\,\omega^{14}}$$
$$Q_{2,5} = 0$$

if one simply assumes that the series converges as long as the potential-energy function supports the action J [200]. For simplicity we illustrate this argument by means of the simple oscillators with polynomial potentials $V(q) = m\omega^2 q^2/2 + \lambda q^M$, $M = 2, 3, 4, \ldots$.

The exact expression of the energy of the harmonic oscillator $M = 2$

$$E(J, \lambda) = \omega J \sqrt{1 + \frac{2\lambda}{m\omega^2}} \tag{9.150}$$

exhibits a singular point at $\lambda_s = -m\omega^2/2$, and we realize that there is no bounded motion when $\lambda < \lambda_s$; therefore the argument applies to this trivial example.

In the case of an even anharmonic perturbation $M = 4, 6, \ldots$, the action J is supported for all $\lambda > 0$ because the potential is an infinite well. If, on the other hand, λ is negative, then the potential-energy function exhibits two symmetrical maxima $V(q_m)$ at $q = \pm q_m$ and there will be no periodic motion for $E > V(q_m)$ (see Figure 9.1 b for an example). Therefore, there is a critical negative value $\lambda = \lambda_c$ for which $E(J, \lambda_c) = V(q_m)$ as depicted in Figure 9.1 b.

We can obtain an exact expression of λ_c for the quartic case that we conveniently rewrite as

Table 9.4 *(Cont.)* Canonical Perturbation Theory for the
Anharmonic Oscillators $H = \dfrac{p^2}{2\,m} + \dfrac{m\,\omega^2\,q^2}{2} + \lambda\,q^M$ by Means of
the Hypervirial Hellmann–Feynman Method

$$M = 4$$

$$E_1 = \tfrac{3}{2}\,\frac{E_0{}^2}{m^2\,\omega^4}$$

$$E_2 = -\tfrac{17}{4}\,\frac{E_0{}^3}{m^4\,\omega^8}$$

$$E_3 = \tfrac{375}{16}\,\frac{E_0{}^4}{m^6\,\omega^{12}}$$

$$E_4 = -\tfrac{10689}{64}\,\frac{E_0{}^5}{m^8\,\omega^{16}}$$

$$E_5 = \tfrac{87549}{64}\,\frac{E_0{}^6}{m^{10}\,\omega^{20}}$$

$$E_6 = -\tfrac{3132399}{256}\,\frac{E_0{}^7}{m^{12}\,\omega^{24}}$$

$$Q_{2,0} = \frac{E_0}{m\,\omega^2}$$

$$Q_{2,1} = -3\,\frac{E_0{}^2}{m^3\,\omega^6}$$

$$Q_{2,2} = \tfrac{85}{4}\,\frac{E_0{}^3}{m^5\,\omega^{10}}$$

$$Q_{2,3} = -\tfrac{375}{2}\,\frac{E_0{}^4}{m^7\,\omega^{14}}$$

$$Q_{2,4} = \tfrac{117579}{64}\,\frac{E_0{}^5}{m^9\,\omega^{18}}$$

$$Q_{2,5} = -\tfrac{612843}{32}\,\frac{E_0{}^6}{m^{11}\,\omega^{22}}$$

$$Q_{6,0} = \frac{5\,E_0{}^3}{2\,m^3\,\omega^6}$$

$$Q_{6,1} = -\tfrac{165}{8}\,\frac{E_0{}^4}{m^5\,\omega^{10}}$$

$$Q_{6,2} = \tfrac{3129}{16}\,\frac{E_0{}^5}{m^7\,\omega^{14}}$$

$$Q_{6,3} = -\tfrac{31983}{16}\,\frac{E_0{}^6}{m^9\,\omega^{18}}$$

$$Q_{6,4} = \tfrac{2742687}{128}\,\frac{E_0{}^7}{m^{11}\,\omega^{22}}$$

$V(q) = m\omega^2 q^2/2 - |\lambda| q^4$. We have

$$q_m = \frac{\omega}{2}\sqrt{\frac{m}{|\lambda|}}\,. \tag{9.151}$$

From the critical-point condition

$$E = V(q_m) = |\lambda_c|\,q_m^4 \tag{9.152}$$

we obtain

$$E - V(q) = |\lambda_c|\left(q^2 - q_m^2\right)^2 \tag{9.153}$$

and

$$J = \frac{\sqrt{2m|\lambda_c|}}{\pi} \int_{-q_m}^{q_m} \left(q_m^2 - q^2 \right) dq = \frac{\sqrt{2}m^2\omega^3}{6\pi|\lambda_c|} \tag{9.154}$$

that suggests the radius of convergence

$$|\lambda_c| = \frac{\sqrt{2}m^2\omega^3}{6\pi J} . \tag{9.155}$$

The potential for an odd anharmonic perturbation $M = 3, 5, \ldots$, exhibits a barrier at $q_m > 0$ if $\lambda < 0$ or at $q_m < 0$ if $\lambda > 0$. Therefore, for a given value of the action J there is a critical value λ_c of λ such that E equals the top of the barrier $V(q_m)$ indicated in Figure 9.1 a. The simplest illustrative example is $V(q) = m\omega^2 q^2/2 - |\lambda|q^3$ because we can obtain λ_c exactly. In this case

$$q_m = \frac{m\omega^2}{3|\lambda|}, \ V(q_m) = \frac{|\lambda|q_m^3}{2} , \tag{9.156}$$

and the critical condition $E = V(q_m)$ leads to

$$E - V(q) = |\lambda_c|(q - q_m)^2 \left(q + \frac{q_m}{2} \right) , \tag{9.157}$$

and

$$J = \frac{\sqrt{2m|\lambda_c|}}{\pi} \int_{-\frac{q_m}{2}}^{q_m} (q_m - q)\sqrt{q + \frac{q_m}{2}} \, dq = \frac{m^3\omega^5}{15\pi|\lambda_c|^2} , \tag{9.158}$$

which suggests the radius of convergence

$$|\lambda_c| = \sqrt{\frac{m^3\omega^5}{15\pi J}} . \tag{9.159}$$

Strictly speaking, the argument given above tells us that the perturbation series diverges when $|\lambda| > |\lambda_c|$, but in principle there could be a singular point λ_s closer to the origin that would make the convergence radius of the perturbation series to be smaller than $|\lambda_c|$. Only for the harmonic oscillator we are certain that $\lambda_c = \lambda_s$ as follows from equation (9.150). In order to verify if there were singular points satisfying $|\lambda_s| < |\lambda_c|$ for the anharmonic oscillators (9.140) we resorted to the numerical method developed in Section 6.2.1 and obtained both λ_s and the exponent a. From the energy coefficients through order $P = 200$ for dimensionless anharmonic oscillators ($m = \omega = 1$) with $M = 3, 4, 5$, and 6 we constructed the sequences for the location and exponent of the singular point assuming an algebraic singularity as in the harmonic case. Although the sequences did not appear to converge, they yielded results close to the ones predicted by the argument above.

In any case the radius of convergence of the classical perturbation series is nonzero in contrast with the quantum-mechanical counterpart. This fact has been taken into account in recent discussions and applications of methods for improving the convergence properties of the quantum-mechanical perturbation series [121].

9.8.3 Nonpolynomial Potential-Energy Function

If the potential $V(x)$ is not a polynomial function of the coordinate x we apply perturbation theory by means of a polynomial approximation similar to that discussed above in Section 9.3. We suppose that $V(x)$ has a minimum at $x = x_e$ and choose the energy origin such that $V(x_e) = 0$. We therefore have $V'(x_e) = 0$ and $V''(x_e) = k > 0$, where k is the force constant that we write in terms of the mass m and frequency $\omega = \sqrt{k/m}$. Expanding $V(x)$ about its minimum we have

$$H = \frac{p^2}{2m} + \frac{m\omega^2 q^2}{2} + \sum_{j=1}^{\infty} V_{j+2} \lambda^j q^{j+2} , \tag{9.160}$$

where $q = x - x_e$, $p = m\dot{x} = m\dot{q}$, $V_j = (1/j!)(d^j V/dx^j)_{x=x_e}$, and λ is a dummy perturbation parameter that we set equal to unity at the end of the calculation.

We easily apply the HHFM to the perturbed harmonic oscillator in equation (9.160), and expand both the energy and time averages $Q_N = \overline{q^N}$ in the λ-power series. Proceeding exactly as in the case of the polynomial interactions discussed above we obtain the following recurrence relation for the perturbation corrections to the time averages:

$$Q_{N,i} = \frac{1}{m\omega^2 N} \left[2(N-1) \sum_{j=0}^{i} E_j Q_{N-2,i-j} - \sum_{j=1}^{i} (2N+j) V_{j+2} Q_{N+j,i-j} \right] . \tag{9.161}$$

In addition to it, the Hellmann–Feynman theorem gives us an expression for the perturbation corrections to the energy:

$$E_i = \frac{1}{i} \sum_{j=1}^{i} j V_{j+2} Q_{j+2,i-j} . \tag{9.162}$$

From these two expressions and the initial condition $Q_{0,j} = \delta_{0j}$, we obtain the perturbation corrections E_{i+1}, $Q_{N,i}$ for all $i = 0, 1, \ldots, p-1$, $N = 1, 2, \ldots, p-i+2$. One easily writes a Maple program for this problem by slight modification of the one in the program section for the method of Swenson and Danforth.

As an interesting illustrative example, we choose the Morse oscillator given by the anharmonic potential

$$V(x) = D \left[1 - \exp\left(-\frac{x - x_e}{\gamma} \right) \right]^2 , \tag{9.163}$$

where D is the depth of the potential well, x_e is the equilibrium coordinate, and γ is a length parameter that determines the range of the interaction. Substituting the Taylor coefficients

$$V_j = D \left[\delta_{j0} + \frac{(-1)^j}{j! \gamma^j} (2^j - 2) \right] \tag{9.164}$$

into equations (9.161) and (9.162) we obtain the perturbation corrections shown in Table 9.5. Notice that exactly as in its quantum-mechanical counterpart discussed in Section 7.2.1, the perturbation series for the energy reduces to just two terms:

$$E = E_0 - \frac{E_0^2}{4D} = J\omega - \frac{J^2 \omega^2}{4D} , \tag{9.165}$$

where $\omega = \sqrt{2D/(m\gamma^2)}$. Solving for J we obtain

$$J(E) = \frac{2D}{\omega} \left(1 + \sqrt{1 - \frac{E}{D}} \right) = \sqrt{2mD\gamma^2} \left(1 + \sqrt{1 - \frac{E}{D}} \right) . \tag{9.166}$$

Table 9.5 Hypervirial Hellmann–Feynman Method for the
Classical Morse Oscillator

$$E_0 = \omega J$$

$$E_1 = 0$$

$$E_2 = -\frac{1}{4} \frac{E_0{}^2}{D}$$

$$E_3 = 0$$

$$E_4 = 0$$

$$E_5 = 0$$

$$E_6 = 0$$

$$Q_{1,1} = \frac{3 L E_0}{4 D}$$

$$Q_{1,2} = 0$$

$$Q_{1,3} = \frac{7}{32} \frac{E_0{}^2 L}{D^2}$$

$$Q_{1,4} = 0$$

$$Q_{1,5} = \frac{5}{64} \frac{E_0{}^3 L}{D^3}$$

$$Q_{1,6} = 0$$

$$Q_{1,7} = \frac{31}{1024} \frac{E_0{}^4 L}{D^4}$$

$$Q_{1,8} = 0$$

$$Q_{1,9} = \frac{63}{5120} \frac{E_0{}^5 L}{D^5}$$

$$Q_{2,1} = 0$$

$$Q_{2,2} = \frac{23}{32} \frac{E_0{}^2 L^2}{D^2}$$

$$Q_{2,3} = 0$$

$$Q_{2,4} = \frac{109}{288} \frac{E_0{}^3 L^2}{D^3}$$

$$Q_{2,5} = 0$$

$$Q_{2,6} = \frac{1117}{6144} \frac{E_0{}^4 L^2}{D^4}$$

$$Q_{2,7} = 0$$

$$Q_{2,8} = \frac{1639}{19200} \frac{E_0{}^5 L^2}{D^5}$$

$$Q_{2,9} = 0$$

The other root of equation (9.165) leads to an unphysical negative frequency, whereas from equation (9.166) we obtain the well-known result [196]:

$$\Omega(E) = \frac{\partial H}{\partial J} = \frac{\partial E}{\partial J} = \omega \sqrt{1 - \frac{E}{D}} = \sqrt{2mD\gamma^2} \sqrt{1 - \frac{E}{D}} \; . \tag{9.167}$$

Surprisingly, perturbation theory provides a simple way of obtaining the exact action and frequency for the Morse oscillator. Exactly as discussed in Section 7.2.1 for the quantum-mechanical case, the energy series has a finite number of terms in contrast with the perturbation expansions for the averages Q_N which do not terminate.

Table 9.5 shows that $E_{2j+1} = 0$ and $Q_{N,i} = 0$ if $N + i$ is odd, which follows from the fact that λ appears in the Hamiltonian only in terms of the form $\lambda^j q^{j+2}$, exactly as in the application of the polynomial approximation in quantum mechanics discussed in Section 7.2.1.

9.9 Central Forces

As an example of application of the HHFM to classical separable models, in what follows we consider a system of two particles that interact by means of central forces. The potential $V(r)$ depends on the distance r between the particles, and the Lagrangian in spherical coordinates for the relative motion reads

$$\mathcal{L} = \frac{m}{2}\left[\dot{r} + r^2\dot{\theta}^2 + r^2\dot{\phi}^2\sin(\theta)^2\right] - V(r) \, , \tag{9.168}$$

where m is the reduced mass. The Hamiltonian function is given by

$$H = \frac{1}{2m}\left[p_r^2 + \frac{p_\theta^2}{r^2} + \frac{p_\phi^2}{r^2\sin(\theta)^2}\right] + V(r) \tag{9.169}$$

in terms of the momenta

$$p_r = m\dot{r}, \quad p_\theta = mr^2\dot{\theta}, \quad p_\phi = mr^2\dot{\phi}\sin(\theta)^2 \, . \tag{9.170}$$

It follows from the equations of motion that p_ϕ and $p_\theta^2 + p_\phi^2/\sin(\theta)^2$ are independent of time.

It is well known that in the case of central forces the angular momentum $\mathbf{L} = \mathbf{r} \times \mathbf{p}$ is a constant of the motion, and one easily verifies that $L^2 = r^2p^2 - (\mathbf{r} \cdot \mathbf{p})^2$, where $L = |\mathbf{L}|$, $r = |\mathbf{r}|$, and $p = |\mathbf{p}|$. Taking into account that $\mathbf{r} \cdot \mathbf{p} = rp_r$ we conclude that

$$H = \frac{1}{2m}\left(p_r^2 + \frac{L^2}{r^2}\right) + V(r) \, . \tag{9.171}$$

Comparing equations (9.169) and (9.171) we realize that $L^2 = p_\theta^2 + p_\phi^2/\sin(\theta)^2$.

The equation of motion for the variable r is

$$\dot{p}_r = \frac{L^2}{mr^3} - V'(r) \, . \tag{9.172}$$

Taking into account equations (9.171) and (9.172) we easily obtain

$$\frac{d}{dt}r^N p_r = 2NEr^{N-1} - \frac{(N-1)L^2}{m}r^{N-3} - 2NVr^{N-1} - r^N V' \, , \tag{9.173}$$

where $E = H$ is the energy, which is also a constant of the motion. It is clear that we can apply the HHFM exactly as in the one-dimensional case if we simply substitute the effective potential $L^2/(2mr^2) + V(r)$ for V in the equations above.

9.9.1 Perturbed Kepler Problem

For concreteness we concentrate on the potential

$$V(r) = -\frac{A}{r} + \lambda r^K , \qquad (9.174)$$

which we view as a perturbed Kepler problem when $A > 0$.

Substituting equation (9.174) into equation (9.173), and arguing as in the one-dimensional case discussed earlier, we easily derive the hypervirial relation

$$2NER_{N-1} - \frac{(N-1)L^2}{m}R_{N-3} + (2N-1)AR_{N-2} - (2N+K)\lambda R_{N+K-1} = 0 , \qquad (9.175)$$

where $R_N = \overline{r^N}$. As before, we expand the energy and time averages in the λ-power series

$$R_N = \sum_{j=0}^{\infty} R_{N,j}\lambda^j, \; E = \sum_{j=0}^{\infty} E_j\lambda^j , \qquad (9.176)$$

the Hellmann–Feynman theorem $\partial E/\partial \lambda = R_K$ leads to

$$E_s = \frac{1}{s}R_{K,s-1}, \; s = 1, 2, \dots , \qquad (9.177)$$

and the initial condition $R_{0,s} = \delta_{0s}$ follows from $R_0 = 1$.

When $K > 0$ we substitute $N + 1$ for N in equation (9.175), and expand it in a λ-power series obtaining the following recurrence relation for the perturbation corrections

$$R_{N,s} = \frac{1}{2(N+1)E_0}\left[\frac{NL^2}{m}R_{N-2,s} - (2N+1)AR_{N-1,s}\right.$$
$$\left. -2(N+1)\sum_{j=1}^{s} E_j R_{N,s-j} + (2N+K+2)R_{N+K,s-1}\right] \qquad (9.178)$$

valid for $N \neq -1$. We obtain $R_{-1,s}$ from the expansion of equation (9.175) with $N = 1$:

$$R_{-1,s} = -\frac{2}{A}E_s + (K+2)R_{K,s-1} . \qquad (9.179)$$

In order to calculate E_p we have to obtain $R_{N,s}$ for $s = 0, 1, \dots, p-1$, $N = 1, 2, \dots, (p-s)K$ by means of equation (9.178), and $R_{-1,s}$ from equation (9.179), taking into account the initial condition.

When $K < -2$ we substitute $N + 3$ for N in equation (9.175) and expand it in a λ-power series to obtain

$$R_{N,s} = \frac{m}{(N+2)L^2}\left[(2N+5)AR_{N+1,s} + 2(N+3)\sum_{j=0}^{s} E_j R_{N+2,s-j}\right.$$
$$\left. -(2N+K+6)R_{N+K+2,s-1}\right] \qquad (9.180)$$

valid when $N \neq -2$. In order to obtain $R_{-2,s}$ we make use of the Hellmann–Feynman theorem as $\partial E/\partial L = LR_{-2}/m$, which leads to

$$R_{-2,s} = \frac{m}{L}\frac{\partial E_s}{\partial L} . \qquad (9.181)$$

The calculation of E_p requires $R_{N,s}$ for all $s = 0, 1, \ldots, p-1, N = -3, -4, \ldots, (p-s)(K+2)-2$ that follow from equation (9.180), and $R_{-2,s}$ from equation (9.181).

In order to round off the discussion above we only need an expression of the unperturbed energy E_0 in terms of the actions

$$J_r = \frac{1}{2\pi} \oint p_r \, dr, \ J_\theta = \frac{1}{2\pi} \oint p_\theta \, d\theta, \ J_\phi = \frac{1}{2\pi} \oint p_\phi \, d\phi = p_\phi \,. \tag{9.182}$$

It is well known that

$$E_0 = -\frac{mA^2}{2J^2} \,, \tag{9.183}$$

where $J = J_r + J_\theta + J_\phi = L + J_r$ [200, 201].

The calculation for positive and negative values of K is straightforward even by hand; however, if one needs results of high perturbation order it is advisable to resort to computer algebra. It is not difficult to write a set of simple Maple procedures for the application of the HHFM following the lines indicated in Section 3.3.2 for the quantum-mechanical counterpart. In Table 9.6 we show results for $K = 1$ and $K = -3$ which were obtained earlier by the same method [200].

The HHFM is most probably the simplest and most efficient method for the calculation of canonical perturbation series for separable classical systems [200], and is a straightforward transcription of the quantum-mechanical method of Swenson and Danforth discussed in Section 3.3.

Table 9.6   Hypervirial and Hellmann–Feynman Method for Classical Models with Central Forces: $H = \dfrac{p^2}{2\,m} - \dfrac{A}{r} + \lambda\, r^K$

(Continued).

$$K = 1$$

$$E_0 = -\frac{m\,A^2}{2\,J^2}$$

$$E_1 = \frac{-L^2 + 3\,J^2}{2\,A\,m}$$

$$E_2 = \frac{1}{8}\,\frac{3\,L^4\,J^2 - 7\,J^6}{m^3\,A^4}$$

$$E_3 = \frac{1}{16}\,\frac{-10\,L^6\,J^4 - 7\,L^4\,J^6 + 33\,J^{10}}{m^5\,A^7}$$

$$E_4 = \frac{1}{64}\,\frac{84\,L^8\,J^6 + 90\,L^6\,J^8 + 99\,L^4\,J^{10} - 465\,J^{14}}{m^7\,A^{10}}$$

$$E_5 = \frac{1}{64}\,\frac{-198\,L^{10}\,J^8 - 264\,L^8\,J^{10} - 364\,L^6\,J^{12} - 465\,L^4\,J^{14} + 1995\,J^{18}}{m^9\,A^{13}}$$

$$R_{1,0} = \frac{1}{2}\,\frac{-L^2 + 3\,J^2}{A\,m}$$

$$R_{1,1} = \frac{1}{4}\,\frac{3\,L^4\,J^2 - 7\,J^6}{m^3\,A^4}$$

$$R_{1,2} = \frac{1}{16}\,\frac{-30\,L^6\,J^4 - 21\,L^4\,J^6 + 99\,J^{10}}{m^5\,A^7}$$

$$R_{1,3} = \frac{1}{16}\,\frac{84\,L^8\,J^6 + 90\,L^6\,J^8 + 99\,L^4\,J^{10} - 465\,J^{14}}{m^7\,A^{10}}$$

$$R_{1,4} = \frac{1}{64}\,\frac{-990\,L^{10}\,J^8 - 1320\,L^8\,J^{10} - 1820\,L^6\,J^{12} - 2325\,L^4\,J^{14} + 9975\,J^{18}}{m^9\,A^{13}}$$

$$R_{-1,0} = \frac{A\,m}{J^2}$$

$$R_{-1,1} = \frac{1}{2}\,\frac{-L^2 + 3\,J^2}{A^2\,m}$$

$$R_{-1,2} = \frac{1}{2}\,\frac{3\,L^4\,J^2 - 7\,J^6}{A^5\,m^3}$$

$$R_{-1,3} = \frac{1}{16}\,\frac{-70\,L^6\,J^4 - 49\,L^4\,J^6 + 231\,J^{10}}{A^8\,m^5}$$

$$R_{-1,4} = \frac{1}{32}\,\frac{420\,L^8\,J^6 + 450\,L^6\,J^8 + 495\,L^4\,J^{10} - 2325\,J^{14}}{A^{11}\,m^7}$$

Table 9.6 *(Cont.)* Hypervirial and Hellmann–Feynman Method for Classical Models with Central Forces: $H = \dfrac{p^2}{2\,m} - \dfrac{A}{r} + \lambda\,r^K$

$$K = -3$$

$E_1 = \frac{m^3 A^3}{L^3 J^3}$

$E_2 = \frac{1}{4} \frac{3 A^4 m^5 L^2 - 6 A^4 m^5 L J - 15 A^4 m^5 J^2}{L^7 J^5}$

$E_3 = \frac{1}{4} \frac{-15 A^5 m^7 L^3 - 27 A^5 m^7 L^2 J + 45 A^5 m^7 L J^2 + 105 A^5 m^7 J^3}{L^{11} J^6}$

$E_4 = \frac{1}{64} \left(-126 A^6 m^9 L^5 + 405 A^6 m^9 L^4 J + 3540 A^6 m^9 L^3 J^2 + 5490 A^6 m^9 L^2 J^3 \right.$
$\left. -6390 A^6 m^9 L J^4 - 15015 A^6 m^9 J^5 \right) / \left(L^{15} J^8 \right)$

$E_5 = \frac{1}{64} \left(1008 A^7 m^{11} L^6 + 3465 A^7 m^{11} L^5 J - 4599 A^7 m^{11} L^4 J^2 - 49050 A^7 m^{11} L^3 J^3 \right.$
$\left. -74610 A^7 m^{11} L^2 J^4 + 63945 A^7 m^{11} L J^5 + 153153 A^7 m^{11} J^6 \right) / \left(L^{19} J^9 \right)$

$R_{1,0} = \frac{1}{2} \frac{-L^2 + 3 J^2}{A\,m}$

$R_{1,1} = \frac{1}{4} \frac{3 L^4 J^2 - 7 J^6}{m^3 A^4}$

$R_{1,2} = \frac{1}{16} \frac{-30 L^6 J^4 - 21 L^4 J^6 + 99 J^{10}}{m^5 A^7}$

$R_{1,3} = \frac{1}{16} \frac{84 L^8 J^6 + 90 L^6 J^8 + 99 L^4 J^{10} - 465 J^{14}}{m^7 A^{10}}$

$R_{1,4} = \frac{1}{64} \frac{-990 L^{10} J^8 - 1320 L^8 J^{10} - 1820 L^6 J^{12} - 2325 L^4 J^{14} + 9975 J^{18}}{m^9 A^{13}}$

$R_{-1,0} = \frac{A\,m}{J^2}$

$R_{-1,1} = -3 \frac{A^2 m^3}{L^3 J^3}$

$R_{-1,2} = \frac{-3 m^5 A^3 L^2 + 6 m^5 A^3 L J + 15 m^5 A^3 J^2}{L^7 J^5}$

$R_{-1,3} = \frac{1}{4} \frac{75 m^7 A^4 L^3 + 135 m^7 A^4 L^2 J - 225 m^7 A^4 L J^2 - 525 m^7 A^4 J^3}{L^{11} J^6}$

$R_{-1,4} = \frac{1}{32} \left(378 m^9 A^5 L^5 - 1215 m^9 A^5 L^4 J - 10620 m^9 A^5 L^3 J^2 - 16470 m^9 A^5 L^2 J^3 \right.$
$\left. +19170 m^9 A^5 L J^4 + 45045 m^9 A^5 J^5 \right) / \left(L^{15} J^8 \right)$

Maple Programs

Since we are not experts but simply naive Maple users, our programs may not be efficient or elegant. However, the reader may profit from the fact that our procedures are just straightforward translations of the equations developed in the different chapters of this book into simple Maple code. The procedures given below are not foolproof, but if they are executed as indicated, they certainly produce the results displayed in the tables and figures of this book.

Programs for Chapter 1

1) Calculation of perturbation corrections to the energies and stationary states of the anharmonic oscillator (1.54) with $M = 4$.
Kronecker delta function δ_{ij}:

```
delta:=proc(i,j) if i = j then 1 else 0 fi end:
```

Calculation of the matrix elements $Q(m, j, n) = \langle m|q^j|n\rangle$ in the harmonic oscillator basis set according to equation (1.55):

```
Q:=proc (m,j,n)  option remember;
if n=0 then delta(m,n) else
expand(sqrt(n/2)*Q(m,j−1,n−1) +sqrt((n+1)/2)*Q(m,j−1,n+1))
fi;
end:
```

Matrix elements of the dimensionless perturbation in the harmonic oscillator basis set $H1(m, n) = \langle m|\hat{\mathcal{H}}'|n\rangle$:

```
H1:=proc(m,n) option remember;
Q(m,4,n);
end:
```

Cutoff function used to force $C_{n+j,n,s} = 0$ if $|j| > 4s$:

```
cut:=proc(j,p)
if abs(j)<=4*p then 1 else 0 fi;
end:
```

Calculation of the perturbation corrections $E_{n,p}$ and $C_{mn,p}$ according to equations (1.10), (1.12), and (1.13). For convenience we write $c[j, n, s] = C_{n+j,n,s}$

```
PT:=proc(n,P)
local i,j,k,p,s; global e,c;
e[n,0]:=n+1/2;
for j from −4*P to 4*P do c[j,n,0]:=delta(j,0) od;
```

for p from 1 to P do
e[n,p]:=expand(sum('H1(n,n+k)
*c[k,n,p−1]*cut(k,p−1) ', 'k '=−4..4)*
*−sum('e[n,s]*c[0,n,p−s] ', 's '=1..p−1));*
for j from −4*p to 4*p do
if j=0 then
c[0,n,p]:=−1/2*sum('c[i,n,s] *c[i,n,p−s]*cut(i,s)*cut(i,p−s) ',
'i '=−4*p+4..4*p−4) ', 's '=1..p−1);
else
c[j,n,p]:=1/j*(sum('e[n,s]*c[j,n,p−s]*cut(j,p−s) ', 's '=1..p)
−sum('H1(n+j,n+k)*c[k,n,p−1]*cut(k,p−1) ', 'k '=j−4..j+4))
fi;
od;
od;
end:

Perturbation corrections to the matrix elements $\langle \Psi_m | \hat{q}^k | \Psi_n \rangle$ calculated according to equation (1.17).

Qpert:=proc(m,k,n,P)
local s,p,j:
PT(m,P):
PT(n,P):
for p from 0 to P do
simplify(sum('sum('sum('c[i,m,s] *c[j,n,p−s]*cut(i,s)*cut(j,p−s)
*Q(m+i,k,n+j) ', 'i '=−4*s..4*s) '
'j '=−4*(p−s)..4*(p−s)) ', 's '=0..p)):
od:
end:

To calculate the perturbation corrections $E_{n,s}$ for $s = 0, 1, \ldots, P$ simply execute $PT(n, P)$, where P must be a positive integer, and n may be either a positive integer indicating a particular state ($n = 0, 1, \ldots$) or just a generic variable name. The perturbation correction of order P for $\langle \Psi_n | \hat{q}^k | \Psi_{n+j} \rangle$ is given by $Qpert(n,k,j,P)$ where k, j, and P must be integers and n may by either an integer or a generic variable name.

Programs for Chapter 2

2) Method of Dalgarno and Stewart for the ground state of hydrogen in a uniform magnetic field. Construction of the factor functions $F_j(r, u)$ according to equation (2.23)

funF:=proc(j)
local i,k:global F,c,r,u:
if j<0 then F[j]:=0 elif j=0 then F[j]:=1 else
F[j]:=sum('sum('c[j,i,k]*r ^i ', 'i '=0..3*j)*u ^(2*k) ', 'k '=0..j):
fi:
end:

Substitution of the factor functions into equation (2.22) and calculation of their coefficients

eqj:=proc(j)
local i,i1,i2,ind,sol: global c,e,F,u,r,equ:
equ:=simplify(−(1/2)*diff(funF(j),r\$2)−(1/r)*diff(funF(j),r)

```
 +((u ^2−1)/(2*r ^2))*diff(funF(j),u$2)
 +(u/r^2)*diff(funF(j),u)+diff(funF(j),r)
 +r^2*(1−u^2)*funF(j−1)−sum('e[i]*funF(j−i) ', 'i '=1..j)):
  equ:=collect(equ,[u,r]):
  for i1 from 2*j by −2 to 0 do
  for i2 from −2 to 3*j do
  ind:=indets(coeff(coeff(equ,u,i1),r,i2)):
  if ind <>{} then
  sol:=solve({coeff(coeff(equ,u,i1),r,i2)},ind[1]):
  assign(sol):
  equ:=collect(equ,[u,r]):
  fi:
  od:
  od:
  e[j]:=e[j]:
  end:
  norma:=proc(j)
  local i,k,ind,nor,sol: global psi,F,c,r,u:
  psi[0]:=sqrt(2)*exp(−r):
  for k from 1 to j do
  psi[k]:=F[k]*exp(−r):
  nor:=int(int(sum('psi[i]*psi[k−i] ', 'i '=0..k)*r ^2 ,r=0..infinity),u=−1..1):
  ind:=indets(nor):
  if ind <>{} then
  sol:=solve({nor},ind):
  assign(sol):
  fi:
  F[j]:=collect(simplify(F[j]),[u,r]):
  od:
  end:
```

The procedure below calls the other ones to calculate the perturbation corrections to the energy and eigenfunction through order k. To this end simply execute $PT(k)$, where k is a positive integer. However, if one only needs perturbation corrections to the energy it is sufficient to call $eqj(1)$, $eqj(2)$, ..., $eqj(k)$.

```
  PT:=proc(k)
  local j:
  for j from 1 to k do
  eqj(j):
  norma(j):
  od:
  end:
```

3) Method of Fernández and Castro for the dimensionless anharmonic oscillator $\hat{H} = (-d^2/dq^2 + q^2)/2 + \lambda q^4$

Construction of the functions A_k and B_k according to equation (2.58)

```
  funciones:=proc(k)
  local j,i: global A,B,a,b,alpha,beta,q:
  alpha[0]:=0: beta[0]:=0:
  alpha[1]:=1: beta[1]:=1:
  for j from 2 to k do
```

```
alpha[j]:=beta[j-1]+3:
beta[j]:=alpha[j-1]+1:
od:
if k=0 then
A[k]:=1:B[k]:=0 else
A[k]:=sum( 'a[k,i]*q ^(2*i) ', 'i '=1..alpha[k]):
B[k]:=sum( 'b[k,i]*q ^(2*i+1) ', 'i '=0..beta[k]):
fi:
```

Construction of the perturbation equations of order $k = 1, 2, \ldots$ followed by calculation of the coefficients a_{ki} and b_{ki} starting with the coefficient of the largest coordinate power down to zero.

```
eqk:=proc(k)
local j,i,n1,n2,n,ind,ind1,sol:
global A,B,a,b,equ1,equ2,e,q:
for j from 0 to k do
funciones(j) :
od:
equ1:=diff(A[k],q$2)+2*(q ^2-2*e[0])*diff(B[k],q)+2*q*B[k]
+2*(e[1]-q ^4)*A[k-1]+2*sum( 'e[j]*A[k-j] ', 'j '=2..k):
equ1:=collect(simplify(equ1),q):
n1:=degree(equ1,q):
equ2:=diff(B[k],q$2)+2*diff(A[k],q)+2*(e[1]-q ^4)*B[k-1]
+2*sum( 'e[j]*B[k-j] ', 'j '=2..k):
equ2:=collect(simplify(equ2),q):
n2:=degree(equ2,q):
while equ1 <> 0 or equ2<>0 do
if n1>=n2 then
ind:=indets(coeff(equ1,q,n1)) minus {e[0]}:
if ind <>{} then
sol:=solve({coeff(equ1,q,n1)},ind1):
assign(sol):
equ1:=collect(simplify(equ1),q):
n1:=degree(equ1,q):
fi:
else
ind:=indets(coeff(equ2,q,n2)) minus{e[0]}:
if ind <>{} then
sol:=solve({coeff(equ2,q,n2)},ind1):
assign(sol):
equ2:=collect(simplify(equ2),q):
n2:=degree(equ2,q):
fi:
fi:
od:
A[k]:=collect(simplify(A[k]),q):
B[k]:=collect(simplify(B[k]),q):
end:
```

In order to obtain perturbation corrections through order P simply execute *eqk(1)*, *eqk(2)*, ..., *eqk(P)*.

Programs for Chapter 3

4) Method of Swenson and Danforth for the dimensionless anharmonic oscillator $\hat{H} = (-d^2/dx^2 + x^2)/2 + \lambda x^{2K}$.

delta:=proc(i,j) if $i = j$ then 1 else 0 fi end:
Modified Heaviside function
trunca:=proc(x) if $x < 0$ then 0 else 1 fi end:
Global variable K for the power of the perturbation; for example,
K:=2:
for the quartic perturbation.
Calculation of perturbation corrections through order P according to equations (3.24), (3.23), and (3.26). Simply execute *PT(P)*:

PT:=proc(P)
local i,j,m: global X,e,K:
for i from 0 to P−1 do
X[0,i]:=delta(0,i):
for j from 1 to $(P−i)*(K−1)+1$ do
X[j,i]:=simplify(1/(2*j)*((j−1)/2*(4*(j−1) ^2−1)*X[j−2,i]
+2*(2*j−1)*sum('e[m]*X[j−1,i−m] ', 'm '=0..i)
−trunca(i−1)*2*(2*j+K−1)*X[j+K−1,i−1])):
od:
e[i+1]:=1/(i+1)*X[K,i]:
od:
end:

5) Moment method for the lowest energies of two-dimensional anharmonic oscillators $\hat{H} = -\nabla^2/2 + (x^2 + y^2)/2 + \lambda(ax^4 + by^4 + 2cx^2y^2)$.

trunca:=proc(i) if $i<0$ then 0 else 1: fi: end:
delta:=proc(i,j) if $i=j$ then 1 else 0: fi: end:
Calculation of perturbation corrections to the energy and moments of the ground state according to equation (3.133), (3.131), and (3.137):

PT0:=proc(p)
local i,j,m,k: global F,e,a,b,c:
for m from 0 to p−1 do
for i from 0 by 2 to $4*(p−m)$ do
for j from 0 by 2 to $4*(p−m)$ do
if i=0 and j=0 then F[i,j,m]:=delta(m,0) else
F[i,j,m]:=simplify(1/(i+j)*(i*(i−1)/2*F[i−2,j,m]+j*(j−1)/2*F[i,j−2,m]
+trunca(m−1)*sum('e[k]*F[i,j,m−k] ', 'k '=1..m)
trunca(m−1)*(a*F[i+4,j,m−1]+b*F[i,j+4,m−1]+2*c*F[i+2,j+2,m−1]))):
fi:
od:
od:
e[m+1]:=simplify(a*F[4,0,m]+b*F[0,4,m]+2*c*F[2,2,m]):
od:
end:

We do not show the procedures for the two excited states with $N = 1$ and for the state $N = 2$ (o, o) because they are similar to the one just given. On the other hand, the procedure for the coupling

degenerate states $N = 2$ (e, e) is noticeably different from the one above, and for that reason it is
already shown below:

```
PT2ee:=proc(p)
local i,j,m,k,sol: global F,e,a,b,c,R:
for m from 0 to p-1 do
for i from 0 by 2 to 4*(p-m)+2 do
for j from 0 by 2 to 4*(p-m)+2 do
if i=0 and j=0 then
F[0,0,m]:=1/2*trunca(m-1)*(a*F[4,0,m-1]+b*F[0,4,m-1]
+2*c*F[2,2,m-1]-sum( 'e[k]*F[0,0,m-k] ', 'k '=1..m)):
F[0,0,m]:=simplifica(F[0,0,m])
elif i=0 and j=2 then
F[0,2,m]:=delta(m,0)/2+1/2*F[0,0,m]:
elif i=2 and j=0 and m=0 then
F[2,0,0]:=1/(4*c)*(9*(a-b)+R):
elif i+j>2 then
F[i,j,m]:=1/(i+j-2)*(i*(i-1)/2*F[i-2,j,m]+j*(j-1)/2*F[i,j-2,m]
+trunca(m-1)*sum( 'e[k]*F[i,j,m-k] ', 'k '=1..m)
-trunca(m-1)*(a*F[i+4,j,m-1]+b*F[i,j+4,m-1]+2*c*F[i+2,j+2,m-1])):
fi:
F[i,j,m]:=simplifica(F[i,j,m]):
od:
od:
if m>0 then
sol:=sum( '(F[0,0,m-k] -2*F[2,0,m-k])*(a*F[4,0,k]+b*F[0,4,k]
+2*c*F[2,2,k]-2*a*F[4,2,k]-2*b*F[0,6,k]
-4*c*F[2,4,k]) ', 'k '=0..m)+a*F[4,0,m]+b*F[0,4,m]
+2*c*F[2,2,m]-2*a*F[6,0,m]-2*b*F[2,4,m]-4*c*F[4,2,m]:
sol:=simplify(sol):
F[2,0,m]:=solve(sol,F[2,0,m]):
F[2,0,m]:=simplifica(F[2,0,m]):
fi:
e[m+1]:=2*a*F[4,2,m]+2*b*F[0,6,m]
+4*c*F[2,4,m]-a*F[4,0,m]-b*F[0,4,m]-2*c*F[2,2,m]:
e[m+1]:=simplifica(e[m+1]):
od:
end:
```

One easily identifies equations (3.151), (3.152), (3.154), (3.131), a coefficient of the Taylor expansion
of equation (3.150), and equation (3.155). Here R stands for the root in equation (3.154).
The procedure below substitutes $81 * (b - a)^2 + 4 * c^2$ for R^2 in order to simplify the equations,
but it does not substitute the value of R and its negative powers because that would make the results
more complicated.

```
simplifica:=proc(f)
local grado,f1,numera,denomi,j,dd2:
dd2:=81*(b-a) ^2+4*c ^2:
f1:=simplify(f):
numera:=numer(f1):denomi:=denom(f1):
grado:=degree(numera):
for j from grado by -1 to 2 do
```

```
if type(j,even) then
numera:=simplify(subs(R ^j=dd2 ^(j/2),numera)):
else
numera:=simplify(subs(R ^j=dd2 ^((j−1)/2)*R,numera)):
fi:
od:
end:
```

Programs for Chapter 4

6) Method of Swenson and Danforth for the Stark effect in hydrogen in paraboloidal coordinates.

```
delta:=proc(i,j) if i=j then 1 else 0 fi end:
trunca:=proc(x) if x<0 then 0 else 1 fi end:
```

Calculation of the perturbation coefficients E_i in terms of the perturbation coefficients A_j according to equations (4.17)–(4.20). Here s and e stand for σ and E, respectively.

```
PT:=proc(p)
local i,j,l,U: global s,e,A,m:
for i from 0 to p−1 do
U[0,i]:=delta(0,i):
U[−1,i]:=1/A[0]*(−e[i]+trunca(i−1)*3*s/4*U[1,i−1]−sum( 'A[l]*U[−1,i−l] ', 'l '=1..i)):
for j from 1 to p−i do
U[j,i]:=simplify(2/((j+1)*e[0])*(j*(m ^2−j ^2)/4*U[j−2,i]
−(j+1/2)*sum( 'A[l]*U[j−1,i−l] ', 'l '=0..i)
−trunca(i−1)*(j+1)/2*sum('e[l]*U[j,i−l] ', 'l '=1..i)
+trunca(i−1)*s*(j+3/2)/4*U[j+1,i−1])):
od:
e[i+1]:=1/(i+1)*(s*U[1,i]/2−2*sum( 'l*A[l]*U[−1,i+1−l] ', 'l '=1..i+1)):
od:
end:
```

Construction of the sets of coefficients E_j^I and E_j^{II} according to the substitutions (4.21) and (4.22), calculation of A_j from $E_j^I - E_j^{II} = 0$, and substitution of the results into E_j^I. Simply execute $PT(p)$ followed by extract(p).

```
extract:=proc(p)
local i,j,j1,E1,E2,k1,k2:
global A,e,n,q,s:
for j from 1 to p do
E1[j]:=subs(e[0]=−A[0] ^2/(2*k1 ^2),s=1,e[j]):
E2[j]:=subs({e[0]=−(1−A[0]) ^2/(2*k2^2),A[0]=1−A[0],s=−1
,seq(A[j1]=−A[j1],j1=1..j)},e[j]):
od:
A[0]:=k1/(k1+k2):
k1:=(n+q)/2:
k2:=(n−q)/2:
A[0]:=simplify(A[0]):
for j from 1 to p do
A[j]:=−subs(A[j]=0,E1[j]−E2[j])/diff(E1[j]−E2[j],A[j]):
```

A[j]:=simplify(A[j]):
e[j]:=E1[j]:
e[j]:=simplify(e[j]):
od:
end:

7) Moment method for the Zeeman effect in hydrogen: case $j = 0$, $| m |= n - 3$.
   trunca:=proc(i) if i<0 then 0 else 1 fi: end:
   delta:=proc(i,j) if i=j then 1 else 0 fi: end:

Calculation of energy and moment coefficients in terms of the unknown $G_{0,1,0} = g$ according to equations (4.63), (4.64), (4.51), (4.66), and (4.65). In particular, notice that *secq* is equation (4.66) and *sec0* stands for the secular equation.

PT3:=proc(p)
local s,t,q,j,m,l,i,k,secq: global G,e,n,sec0,g:
j:=0:m:=n−3:
G[0,1,0]:=g:
for q from 0 to p−1 do
G[0,0,q]:=simplify(1/(2*n−3)*(trunca(q−1)*G[1,4,q−1]
−trunca(q−1)*sum('e[l]*G[0,2,q−l] ', 'l '=1..q))):
for s from 0 to p−q+1 do
i:=n−3+2*s:
for t from 2 to 3*(p−q)+1 do
i:=n−3+2*s:
if s=1 and t=2 then
G[s,t,q]:=simplify((2*n−4)/(2*n−3)*G[0,2,q]−delta(q,0)):
else
G[s,t,q]:=simplify(n/(k−n+1)*((k*(k+1)
−(i+j)*(i+j+1))/2*G[s,t−1,q]
+(i ^2−m ^2)/2*G[s−1,t−1,q]+trunca(q−1)*sum('e[l]*G[s,t+1,q−l] ', 'l '=1..q)
−trunca(q−1)*G[s+1,t+3,q−1])):fi:
od:
od:
secq:=simplify(1/(2*n−3)*sum('(G[0,2,l]/(n*(2*n−3))
+n*(n−2)/2*G[0,0,l]−G[0,1,l])*((2*n−4)*G[1,4,q−l]
−(2*n−3)*G[2,4,q−l]) ', 'l '=0..q)−n*(n−2)/2*G[1,2,q]−G[1,4,q]/(n*(2*n−3))+G[1,3,q]):
if q=0 then
sec0:=secq:
else
assign(solve({secq},G[0,1,q])):fi:
e[q+1]:=simplify((2*n−4)/(2*n−3)*G[1,4,q]−G[2,4,q]):
od:
end:

Substitution of the root (4.67) of the secular equation so that simplifications take place.

prepara:=proc()
global R1,g:
R1:=RootOf(z ^2−16*n ^2+48*n−41,z):
g:=(3−2*n)/(20*n ^2*(n ^2−3*n+2))*(8*n ^2−24*n+13+(2*n−3)*R1):
end:

Simplification of the results by substitution of R for the square root.

simplifica:=proc(f)

```
global R1,R,g:
subs(RootOf(z ^2−16*n ^2+48*n−41,z)=R,simplify(f)):
end:
```

Execute first *PT(p)*, where *p* is a positive integer, then *prepara()*, and finally apply *simplifica(f)* to the chosen perturbation coefficient *f* to be simplified.

8) Moment method for the hydrogen molecular ion

```
trunca:=proc(i) if i<0 then 0 else 1 fi: end:
delta:=proc(i,j) if i=j then 1 else 0 fi: end:
```

Coefficients of the expansion of the potential-energy function according to equation (4.79)

```
coeC:=proc(M)
local i,j: global C:
C[0,0]:=1:
for j from 0 to M−1 do
for i from 0 to j+1 do
C[j+1,i]:=1/(j+1)*((2*j+1)*trunca(i−1)*C[j,i−1]
−j*trunca(j−i−1)*C[j−1,i]):
od:
od:
end:
```

Perturbation corrections to the energy and moments according to equations (4.82)–(4.84).

```
PT2:=proc(p)
local t,q,j,m,l,i,k,u,v: global G,e,n,C:
coeC(p−1):
i:=n−1:
for q from 0 to p−1 do
for j from 0 to p−q−1 do
for t from 0 to p−q−1 do
k:=n+t:
if j=0 and t=0 then G[j,t,q]:=delta(q,0) else
G[j,t,q]:=simplify(n/(k−n+1)*((k*(k+1)
−(i+j)*(i+j+1))/2*G[j,t−1,q]+j*(j−1)/2*G[j−2,t−1,q]
+trunca(q−1)*sum( 'e[l]*G[j,t+1,q−l] ', 'l '=1..q)
+trunca(q−1)*sum('sum( 'C[u,v]*G[j+v,t+u+1,q−u−1] ', 'v '=0..u) ', 'u '=0..q−1))):
fi:
od:
od:
e[q+1]:=−simplify(sum('sum( 'C[u,v]*G[v,u,q−u] ', 'v '=0..u) ', 'u '=0..q)):
od:
end:
```

Programs for Chapter 5

9) Straightforward integration of the perturbation equations for the particle in a box with a perturbation

```
trunca:=proc(x) if x<0 then 0 else 1 fi end:
```

Construction of the perturbation corrections to the eigenfunction by straightforward integration as indicated in equation (5.11). Determination of the perturbation correction to the energy ϵ_j from

$\Phi_j(1) = 0$. Normalization of the approximate eigenfunction order by order according to equation (5.13). $PT(v,p)$ gives the perturbation corrections through order p for the interaction $v(q)$ in terms of $\omega = n\pi$.

```
PT:=proc(v,p)
local i,j,f,norma: global Fi,e,omega,q,q1,C:
Fi[0]:=sqrt(2)*sin(omega*q):
for j from 1 to p do
f[j]:=subs(q=q1,2*v*Fi[j−1]−2*trunca(j−1)*sum( 'e[i]*Fi[j−i] ', 'i '=1..j)):
Fi[j]:=C[j]*sin(omega*q)
+1/omega*int(sin(omega*(q−q1))*f[j],q1=0..q):
e[j]:=simplify(solve(subs(q=1,Fi[j]),e[j]),{sin(omega)=0,cos(omega) ^2=1}):
e[j]:=factor(e[j]):
norma:=simplify(sum('int(Fi[i]*Fi[j−i],q=0..1) ', 'i '=1..j),
{sin(omega)=0,cos(omega) ^2=1}):
C[j]:=solve(norma,C[j]):
Fi[j]:=factor(subs(sin(omega*q)=1,cos(omega*q)=0,Fi[j]))*sin(omega*q)
+factor(subs(sin(omega*q)=0,cos(omega*q)=1,Fi[j]))*cos(omega*q):
od:
end:
```

Programs for Chapter 6

10) Intelligent approximants for the anharmonic oscillator $\hat{H} = \hat{p}^2/2 + \hat{x}^2/2 + \lambda\hat{x}^4$. It is supposed that one has previously calculated the perturbation coefficients E_j

```
delta:=proc(i,j) if i=j then 1 else 0 fi end:
```
Order of the perturbation series required by $A[M, 3M]$
```
PO:=proc(M)M−1+sum('floor((3*M−3*m)/2) ', 'm '=0..M) end:
```
Calculation of the coefficients C_{kj} according to equation (6.103).
```
C:=proc(k,j)
local i: global e,C:
option remember:
if k=0 then delta(k,j) elif
k=1 then e[j]: else
sum( 'C(k−1,i)*e[j−i] ', 'i '=0..j):
fi:
end:
```
Equations (6.101) to be solved for the approximant coefficients A_{mj}.
```
equa:=proc(o,M,N)
local m,j,o1: global e,A:
o1:=min(o,M):
A[0,0]:=1:
expand(sum('sum( 'A[m,j]*C(N−3*m−2*j,o−m) ',
'j '=0..floor((N−3*m)/2)) ', 'm '=0..o1)):
end:
```
Intelligent approximants with unevaluated coefficients A_{mj}
```
apro:=proc(M,N)
```

```
local m,j,m1: global aprox,e,la,A:
A[0,0]:=1:
m1:=min(M,floor(N/3)):
aprox:=sum('sum( 'A[m,j]*la ^m*e ^(N−3*m−2*j) ',
'j '=0..floor((N−3*m)/2)) ', 'm '=0..m1):
end:
```

Calculation of the coefficients A_{mj} and construction of approximants

```
aproxi:=proc(M,N)
local aes,naes,sol: global e,la,aprox,A:
uneva(M,N):
aes:=indets(apro(M,N)) minus {la,e}:
naes:=nops(aes):
print('Perturbation order = ',naes−1):
if e[naes−1]=evaln(e[naes−1]) then ERROR('Execute PT') fi:
sol:=solve({seq(equa(j,M,N),j=0..naes−1)},aes):
assign(sol):
end:
```

Implicit equations (6.109) for $W(g)$

```
strong:=proc(M,N)
local m,j,m1: global apros,W,g,A:
m1:=min(M,floor(N/3)):
apros:=sum('sum( 'A[m,j]*g ^j*W ^(N−3*m−2*j) ',
'j '=0..floor((N−3*m)/2)) ', 'm '=0..m1):
end:
```

Unevaluation of the approximant coefficients for subsequent calculation

```
uneva:=proc(M,N)
local m,j,m1: global aprox,e,la,A:
A[0,0]:=1:
m1:=min(M,floor(N/3)):
for m from 0 to m1 do
for j from 0 to floor((N−3*m)/2) do
A[m,j]:=evaln(A[m,j]):
od:
od:
end:
```

Programs for Chapter 8

11)Transmission coefficient calculated by means of equations (8.21). The arguments of the procedure are the function $F(q) = 2a^2[V(q) - \epsilon]$, the variable q, and the initial point q_0.

```
TMaple:=proc(F,q,q0)
local Nu,De,u0,u1,v0,v1,Du0,Du1,Dv0,Dv1,solu,solv,u,v,equ,eqv:
global T,a,e:
equ:=diff(u(q),q,q)=F*u(q):
eqv:=diff(v(q),q,q)=F*v(q):
solu:=dsolve({equ,u(q0)=1,D(u)(q0)=0},u(q),type=numeric):
```

solv:=dsolve({eqv,v(q0)=0,D(v)(q0)=1},v(q),type=numeric):
u0:=subs(solu(0),u(q)):u1:=subs(solu(1),u(q)):
v0:=subs(solv(0),v(q)):v1:=subs(solv(1),v(q)):
Du0:=subs(solu(0),diff(u(q),q)):
Du1:=subs(solu(1),diff(u(q),q)):
Dv0:=subs(solv(0),diff(v(q),q)):
Dv1:=subs(solv(1),diff(v(q),q)):
Nu=8*a ^2*e*(Du1*v1−u1*Dv1) ^2:
De=(2*a ^2*e*(v1*u0−u1*v0)−Du1*Dv0+Dv1*Du0) ^2
+2*a ^2*e*(Dv1*u0−Du1*v0+u1*Dv0−v1*Du0) ^2:
T:=Nu/De:
end:

12) Calculation of the transmission coefficient by means of the third perturbation method (a^2-power series). The procedure is a straightforward translation of equations (8.21) and (8.73).

PT3:=proc(P)
local j,t,k:
global a,Vc,e,u,v,T,q,Su,Sv,DSu,DSv,u0,u1,v0,v1,Du0,Du1,Dv0,Dv1,Nu,De:
u[0]:=1:v[0]:=q:
for j from 1 to P do
u[j]:=2*a ^2*int((q−t)*subs(q=t,(Vc−e)*u[j−1]),t=0..q):
v[j]:=2*a ^2*int((q−t)*subs(q=t,(Vc−e)*v[j−1]),t=0..q):
od:
Su:=sum('u[j] ', 'j '=0..P):
DSu:=diff(Su,q):
Sv:=sum('v[j] ', 'j '=0..P):
DSv:=diff(Sv,q):
u0:=1:Du0:=0:v0:=0:Dv0:=1:
u1:=subs(q=1,Su):
v1:=subs(q=1,Sv):
Du1:=subs(q=1,DSu):
Dv1:=subs(q=1,DSv):
Nu=8*a ^2*e*(Du1*v1−u1*Dv1) ^2:
De=(2*a ^2*e*(v1*u0−u1*v0)−Du1*Dv0+Dv1*Du0) ^2
+2*a ^2*e*(Dv1*u0−Du1*v0+u1*Dv0−v1*Du0) ^2:
T:=Nu/De:
end:

Programs for Chapter 9

13) Polynomial approximation.

delta:=proc(i,j) if i=j then 1 else 0 fi end:

Coefficients of the perturbation series for q^j obtained according to the recurrence relation (9.13):

Q:=proc(j,k)
local i: global q:
if j=0 then delta(k,0) elif
j=1 then q[k] else

```
sum( 'Q(1,i)*Q(j−1,k−i) ', 'i '=0..k):
fi:
end:
```

Coefficients of the perturbation series for $G(s)$ according to equation (9.12):

```
G:=proc(n)
local j:global q,a:
sum( 'a[j]*Q(j+1,n−j) ', 'j '=1..n):
end:
```

Coefficients of the perturbation series for $G(s)$ in the case of an odd force:

```
Godd:=proc(n)
local j:global q,a:
sum( 'a[j]*Q(2*j+1,n−j) ', 'j '=1..n):
end:
```

Coefficients for the perturbation series for the right-hand side of equation (9.34), where $g[j]$ stands for γ_j:

```
GLP:=proc(j)
local m,temp,gj: global g,s,q:
g[j]:=gj:
temp:=collect(combine(G(j)
−sum('g[m]*diff(q[j−m],s$2) ', 'm '=1..j),trig),[sin,cos]):
g[j]:=solve(coeff(temp,cos(s)),gj):
subs(gj=g[j],temp):
end:
```

Unperturbed trajectory (one can substitute other cases)

```
q[0]:=cos(s):
```

Perturbation corrections through order n to the trajectory for the three cases discussed in Chapter 9: arbitrary force $F(x)$ (*case=anything*), odd force $F(x)$ (*case=odd*), and Lindstedt–Poincaré method for arbitrary $F(x)$ (*case=LP*). *PT(case,n)* gives results in terms of arbitrary coefficients a_j:

```
PT:=proc(case,n)
local j,s1,Gloc: global q,s:
for j from 1 to n do
if case=odd then
Gloc:=Godd(j) elif case=LP then
Gloc:=GLP(j) else
Gloc:=G(j)
fi:
q[j]:=int(sin(s−s1)*subs(s=s1,Gloc),s1=0..s):
q[j]:=collect(simplify(combine(q[j],trig)),[sin,cos]):
od:
end:
```

Taylor series for arbitrary and odd forces. Execute this procedure before *PT(case,N)* when interested in a particular model. It produces the appropriate coefficients a_j; for example, $f = -(g/l)\sin(q)$ for the simple pendulum.

```
force:=proc(f,case,N)
local j,f1,F: global a:
f1[0]:=f:
F[0]:=subs(q=0,f1[0]):
if case=odd then
for j from 1 to 2*N+1 do
```

f1[j]:=diff(f1[j−1],q)/j:
od:
for j from 1 to N+1 do
F[j]:=subs(q=0,f1[2*j−1]):
a[j−1]:=−F[j]/F[1]:
od:
else
for j from 1 to N+1 do
f1[j]:=diff(f1[j−1],q)/j:
od:
for j from 1 to N+1 do
F[j]:=subs(q=0,f1[j]):
a[j−1]:=−F[j]/F[1]:
od:
fi:
end:

Perturbation series for the period by systematic application of the argument leading to equations (9.30) and (9.31). Execute after *PT(case,N)*:

period:=proc(N)
local j,fun:global tau:
fun:=subs(s=s+sum('tau[j]*lambda ^j ', 'j '=1..N),
sum('q[j]*lambda ^j ', 'j '=0..N)):
for j from 1 to N do
tau[j]:=limit(solve(simplify(subs(lambda=0,
diff(fun,lambda$j))),tau[j]),s=2*Pi):
od:
end:

14) Test of Jacobi's identity (9.52).

Construction of a set of $2N$ variables $\{q_j,\ p_j,\ j = 1, 2, \ldots N\}$.

vars:=proc(N)
local i:global q,p:
q:=seq(q.i,i=1..N):p:=seq(p.i,i=1..N):
end:

Definition of the Poisson bracket according to equation (9.42)

c:=proc(a,b,q,p)
local n:
n:=nops(q):
sum('diff(a,op(j,q))*diff(b,op(j,p))−
diff(b,op(j,q))*diff(a,op(j,p)) ', 'j '=1..n):
end:

Jacobi identity for the set of variables chosen above. Execute *Jacobi( )* to obtain an expression that simplifies to zero (*simplify(Jacobi( ))*):

Jacobi:=proc( )
c(A(q,p),c(B(q,p),C(q,p),q,p),q,p)
+c(C(q,p),c(A(q,p),B(q,p),q,p),q,p)
+c(B(q,p),c(C(q,p),A(q,p),q,p),q,p):
end:

15) Secular perturbation theory in operator form.

Poisson bracket $\{a, b\}_{[q,p]}$ for arbitrary functions $a(q, p)$ and $b(q, p)$

```
c:=(a,b,q,p)->  diff(a,q)*diff(b,p)-diff(a,p)*diff(b,q):
```
Trajectory for the harmonic oscillator equation (9.99)
```
q[0]:=q0*cos(w*t)+p0*sin(w*t)/(m*w):
p[0]:=p0*cos(w*t)-m*q0*w*sin(w*t):
```
Calculation of the perturbation correction of order $j + 1$ from the perturbation correction of order j according to equation (9.98), where *h1* stands for H'.
```
plus1:=proc(f,h1)
local s,q0ts,p0ts: global m,w,t,p0,q0:
q0ts:=q0*cos(w*(t-s))+p0*sin(w*(t-s))/(m*w):
p0ts:=p0*cos(w*(t-s))-m*q0*w*sin(w*(t-s)):
int(t=s,[q0=q0ts,p0=p0ts],c(f,h1,q0,p0)),s=0..t):
end:
```
Calculation of the first N perturbation corrections by repeated application of equation (9.98). For example, execute $PT(q^3,N)$ for the cubic perturbation, where N is a positive integer.
```
PT:=proc(h1,N)
local j: global w,t,plus1,q,p:
for j from 1 to N do
q[j]:=combine(expand(plus1(q[j-1],h1)),trig):
q[j]:=collect(q[j],[sin,cos]):
p[j]:=combine(expand(plus1(p[j-1],h1)),trig):
p[j]:=collect(p[j],[sin,cos]):
od:
end:
```
16) Canonical perturbation theory in operator form. Notice that we omit explicit reference to the intermediate complex variables a, b, A, and B which at each step we simply call q and p.
```
c:=(a,b,q,p)->  diff(a,q)*diff(b,p)-diff(a,p)*diff(b,q):
```
Transformation $\hat{U}\mathbf{x}$, equation (9.86)
```
Uq:=(q-p*I)/sqrt(2*m*w):
Up:=sqrt(m*w/2)*(p-q*I):
```
Transformation $\hat{U}^{-1}\mathbf{x}$, equation (9.85)
```
U1q:=sqrt(m*w/2)*(q+I*p/(m*w)):
U1p:=sqrt(m*w/2)*(q*I+p/(m*w)):
```
Transformed harmonic oscillator $\tilde{H}_0$, equation (9.88)
```
H0:=-I*w*q*p:
```
Effect of the operator coefficient $\hat{T}_n$ on a given function f according to equation (9.118)
```
T:=proc(f,n)
local j: global W:
if n=0 then f else
1/n*sum( 'c(W[j],T(f,n-j-1),q,p) ', 'j '=0..n-1):
fi:
end:
```
Effect of the operator coefficient $(\hat{T}^{-1})_n$ on a given function f according to equation (9.120)
```
T1:=proc(f,n)
local j: global W:
if n=0 then f else
-1/n*sum( 'T1(c(W[j],f,q,p),n-j-1) ', 'j '=0..n-1):
fi:
end:
```

Calculation of perturbation corrections $K_{j+1}(q,p)$, $W_j(q,p)$, $Q_{j+1}(q,p)$, and $P_{j+1}(q,p)$, for $j = 0, 1, \ldots n-1$. Notice how we extract the diagonal terms $q^k p^k$ from F_j to construct K_{j+1}, and then calculate W_j according to equation (9.124) in the text.

```
canpt:=proc(H1,n)
local j,f,tmp,L: global Uq,Up,U1q,U1p,K,W,P,Q:
for j from 0 to n−1 do
f:=sum( 'c(W[i],T(H0,j−i),q,p) ', 'i '=0..j−1)+(j+1)*T(H1,j):
tmp:=expand(subs(q=q*L,p=p/L,f)):
K[j+1]:=1/(j+1)*coeff(tmp,L,0):
tmp:=expand(tmp−(j+1)*K[j+1]):
W[j]:=limit(int(subs(L=exp(I*w*t),tmp),t),t=0):
Q[j+1]:=expand(subs([q=U1q,p=U1p],T1(Uq,j+1))):
P[j+1]:=expand(subs([q=U1q,p=U1p],T1(Up,j+1))):
od:
end:
```

For a perturbation $H' = q^M$ we simply execute *canpt(Uq ^M,n)*.

Programs for the Appendixes

17) Laplacian in Curvilinear Coordinates
The reader may easily identify equations (A.7) and (A.8) in the procedure below:

```
with(linalg):
g:=proc(xvar,yvar)
local xvar1,i,j,n,e: global G,invG:
n:=nops(yvar):
xvar1:=vector(xvar):
for i from 1 to n do
e[i]:=map(diff,xvar1,yvar[i]):
od:
G:=array(1..n,1..n):
for i from 1 to n do
for j from 1 to i do
G[i,j]:=simplify(evalm(transpose(e[i])&*e[j])):
G[j,i]:=G[i,j]:
od:
od:
invG:=evalm(1/G):
end:
```

The arguments of the procedure are two ordered lists $xvar =[x_1(y), x_2(y), \ldots x_N(y)]$ and $yvar = [y_1, y_2, \ldots y_N]$, and the program produces the matrix $\mathbf{g} = G$ and its inverse $\mathbf{g}^{-1} = invG$.

Appendix A

Laplacian in Curvilinear Coordinates

In several chapters of this book we make use of the Laplacian in different curvilinear coordinates. Some of them are standard, and one finds the necessary expressions in any book on mathematics or quantum mechanics [202, 203]. But when one is interested in a particular set of coordinates that is not so widely used, one should derive the Laplacian oneself. A general expression for the Laplacian in arbitrary orthogonal curvilinear coordinates is available in many books [202, 203]. However, this is not the case of nonorthogonal coordinates because they are not so frequently required. In addition to this, such scarcely available derivations of the Laplacian in arbitrary curvilinear coordinates are typically awkward, requiring special tensor notions and notation [204]. For all these reasons we believe it worthwhile to show a simple and straightforward (although not rigorous) derivation of the Laplacian in arbitrary curvilinear coordinates. We believe that the discussion below is even simpler than a recent pedagogical treatment of the subject [205]. However, the reader who is not interested in the details of the derivation may go directly to the recipe at the end.

Let $x = \{x_1, x_2, \ldots, x_N\}$ and $y = \{y_1, y_2, \ldots, y_N\}$ be sets of Cartesian and curvilinear coordinates, respectively. The volume element is given by

$$dx_1 dx_2 \ldots dx_N = ||\mathbf{J}|| dy_1 dy_2 \ldots dy_N , \tag{A.1}$$

where $\mathbf{J}$ is the Jacobian matrix with elements [206]

$$\mathbf{J}_{ij} = \frac{\partial x_i}{\partial y_j} , \tag{A.2}$$

$|\mathbf{A}|$ denotes the determinant of a square matrix $\mathbf{A}$, and $|a|$ the absolute value of the scalar a. Taking into account that

$$\frac{\partial x_i}{\partial x_j} = \sum_{k=1}^{N} \frac{\partial x_i}{\partial y_k} \frac{\partial y_k}{\partial x_j} = \delta_{ij} \tag{A.3}$$

we conclude that the matrix elements of the inverse $\mathbf{J}^{-1}$ are

$$\left(\mathbf{J}^{-1}\right)_{ij} = \frac{\partial y_i}{\partial x_j} . \tag{A.4}$$

We want to express the Laplacian operator

$$\nabla^2 = \sum_{i=1}^{N} \frac{\partial^2}{\partial x_i^2} \tag{A.5}$$

in terms of the curvilinear coordinates y.

Any vector in $\mathbf{R}^n$ can be written as a linear combination of the Cartesian unit vectors $\mathbf{c}_i$:

$$\mathbf{r} = x_1\mathbf{c}_1 + x_2\mathbf{c}_2 + \cdots + x_N\mathbf{c}_N , \tag{A.6}$$

so that $\mathbf{c}_i = \partial\mathbf{r}/\partial x_i$. Analogously, we define N curvilinear vectors

$$\mathbf{e}_i = \frac{\partial\mathbf{r}}{\partial y_i}, \ i = 1, 2, \ldots, N . \tag{A.7}$$

Although the Cartesian vectors form an orthonormal basis set with respect to the standard scalar product $\mathbf{c}_i \cdot \mathbf{c}_j = \delta_{ij}$, the curvilinear vectors are not necessarily orthogonal, and we define a metric matrix $\mathbf{g}$ with elements

$$\mathbf{g}_{ij} = \mathbf{e}_i \cdot \mathbf{e}_j, \ i, j = 1, 2, \ldots, N . \tag{A.8}$$

Notice that $\mathbf{g}_{ij}$ is symmetric and

$$\mathbf{g}_{ij} = \sum_{k=1}^{N} \frac{\partial x_k}{\partial y_i}\frac{\partial x_k}{\partial y_j} = \sum_{k=1}^{N} \mathbf{J}_{ki}\left(\mathbf{J}^T\right)_{jk} , \tag{A.9}$$

where $\mathbf{J}^T$ denotes the transpose of the Jacobian matrix. Since $\mathbf{g} = \mathbf{J}^T\mathbf{J}$, then $|\mathbf{g}| = |\mathbf{J}|^2 > 0$, and we can write the volume element as

$$dx_1 dx_2 \ldots dx_N = \sqrt{|\mathbf{g}|}dy_1 dy_2 \ldots dy_N . \tag{A.10}$$

The scalar product of two functions $\Psi(x)$ and $\Phi(x)$ that belong to a quantum-mechanical state space is

$$\int \ldots \int \Psi^*\Phi \, dx_1 dx_2 \ldots dx_N = \int \ldots \int \Psi^*\Phi\sqrt{|\mathbf{g}|} \, dy_1 dy_2 \ldots dy_N . \tag{A.11}$$

Straightforward integration by parts, taking into account that the state functions vanish at the boundaries of the coordinate space, shows that

$$\int \ldots \int \Psi^*\nabla^2\Phi \, dx_1 dx_2 \ldots dx_N = -\int \ldots \int \nabla\Psi^* \cdot \nabla\Phi \, dx_1 dx_2 \ldots dx_N$$

$$= -\int \ldots \int \nabla\Psi^* \cdot \nabla\Phi\sqrt{|\mathbf{g}|} \, dy_1 dy_2 \ldots dy_N , \tag{A.12}$$

where

$$\nabla\Psi^* \cdot \nabla\Phi = \sum_{i=1}^{N} \frac{\partial\Psi^*}{\partial x_i}\frac{\partial\Phi^*}{\partial x_i} = \sum_{i=1}^{N}\sum_{j=1}^{N}\sum_{k=1}^{N} \frac{\partial\Psi^*}{\partial y_j}\frac{\partial\Phi^*}{\partial y_k}\frac{\partial y_j}{\partial x_i}\frac{\partial y_k}{\partial x_i}$$

$$= \sum_{j=1}^{N}\sum_{k=1}^{N} \frac{\partial\Psi^*}{\partial y_j}\frac{\partial\Phi^*}{\partial y_k}\left(\mathbf{J}^{-1}\right)_{ki}\left(\mathbf{J}^T\right)_{ij}^{-1} = \sum_{j=1}^{N}\sum_{k=1}^{N}\left(\mathbf{g}^{-1}\right)_{jk}\frac{\partial\Psi^*}{\partial y_j}\frac{\partial\Phi^*}{\partial y_k} . \tag{A.13}$$

Another integration by parts gives us

$$\int \ldots \int \left(\mathbf{g}^{-1}\right)_{jk}\frac{\partial\Psi^*}{\partial y_j}\frac{\partial\Phi^*}{\partial y_k}\sqrt{|\mathbf{g}|} \, dy_1 dy_2 \ldots dy_N$$

$$= -\int \ldots \int \Psi^*\left[\frac{\partial}{\partial y_j}\left(\mathbf{g}^{-1}\right)_{jk}\sqrt{|\mathbf{g}|}\frac{\partial\Phi^*}{\partial y_k}\right] dy_1 dy_2 \ldots dy_N . \tag{A.14}$$

Finally, comparing both sides of

$$\int \ldots \int \Psi^* \nabla^2 \Phi \sqrt{|\mathbf{g}|} \, dy_1 dy_2 \ldots dy_N$$

$$= \sum_{j=1}^{N} \sum_{k=1}^{N} \int \ldots \int \Psi^* \left[\frac{\partial}{\partial y_j} \left(\mathbf{g}^{-1} \right)_{jk} \sqrt{|\mathbf{g}|} \frac{\partial \Phi^*}{\partial y_k} \right] dy_1 dy_2 \ldots dy_N \,, \qquad (A.15)$$

and taking into account that the state vectors Ψ and Φ are arbitrary, we conclude that

$$\nabla^2 = \frac{1}{\sqrt{|\mathbf{g}|}} \sum_{j=1}^{N} \sum_{k=1}^{N} \frac{\partial}{\partial y_j} \left(\mathbf{g}^{-1} \right)_{jk} \sqrt{|\mathbf{g}|} \frac{\partial}{\partial y_k} \,, \qquad (A.16)$$

which is the desired expression. Present proof of equation (A.16) is not rigorous because it requires functions Ψ and Φ that vanish at the boundaries of the coordinate space, while ∇^2 applies to any twice differentiable function. However, in our opinion this lack of rigor is greatly compensated by the remarkable simplicity of the argument which we hope will satisfy most readers.

When the curvilinear vectors are orthogonal we say that the corresponding coordinates are orthogonal. In this simpler case $\mathbf{g}_{ij} = g_i \delta_{ij}$, $(\mathbf{g}^{-1})_{ij} = \delta_{ij}/g_i$, and $|\mathbf{g}| = g_1 g_2 \ldots g_N$. Most coordinates used in physical applications (Cartesian, spherical, cylindrical, etc.) are orthogonal. In such cases the Laplacian (A.16) reduces to a sum of diagonal terms.

Finally, we give the promised recipe to derive the Laplacian in curvilinear coordinates. It suffices to have the expression of either the direct $x(y)$ or inverse $y(x)$ transformation because $\mathbf{J}(x \rightarrow y) = \mathbf{J}^{-1}(y \rightarrow x)$. For concreteness we assume the former and proceed as follows:

a) Obtain the curvilinear vectors according to equation (A.7).
b) Calculate the metric matrix $\mathbf{g}$ according to equation (A.8).
c) Obtain the determinant and inverse of $\mathbf{g}$.
d) Construct the Laplacian according to equation (A.16).

The reader may convince himself that this procedure is simpler than others. At least, it is easy to write a simple and general Maple program. In the program section we show a short procedure that performs the calculation according to the recipe above.

It is worthwhile to notice that equation (A.16) is valid even for a subset of curvilinear coordinates $\{y_1, y_2, \ldots, y_M\}$, $M < N$. For example, if we consider $x = \{r \sin(\theta) \cos(\phi), r \sin(\theta) \sin(\phi), r \cos(\theta)\}$ and $y = \{\theta, \phi\}$, we obtain

$$\nabla^2 = \frac{1}{r^2 \sin(\theta)} \left[\frac{\partial}{\partial \theta} \sin(\theta) \frac{\partial}{\partial \theta} + \frac{1}{\sin(\theta)} \frac{\partial^2}{\partial \phi^2} \right] = -\frac{\hat{L}^2}{\hbar^2 r^2} \,, \qquad (A.17)$$

where $\hat{L}^2$ is the square of the quantum-mechanical angular momentum. This expression of the Laplacian is suitable for the rigid rotors discussed in Section 5.4 [207].

Appendix B

Ordinary Differential Equations with Constant Coefficients

In several chapters of this book we need the solutions of ordinary differential equations with constant coefficients. Such equations are relevant to many branches of physics and chemistry, and are discussed in most introductory courses on mathematical analysis. In this appendix we develop a simple and straightforward algorithm which generalizes common approaches to that mathematical problem [208, 209].

In order to simplify the notation we write the differential operator $\hat{D} = d/dx$. The starting point of our method is the simple identity

$$\exp(rx)\hat{D}\exp(-rx)Y(x) = \left(\hat{D} - r\right)Y(x) \tag{B.1}$$

that enables us to integrate the first-order differential equation

$$\left(\hat{D} - r\right)Y(x) = f(x) \tag{B.2}$$

very easily:

$$Y(x) = \exp(rx)\left[C + \int^x \exp\left(-rx'\right) f\left(x'\right) dx'\right], \tag{B.3}$$

where C is an arbitrary integration constant.

In order to treat differential equations of any order we define the set of functions

$$Y_s(x) = \prod_{j=1}^{s} \left(\hat{D} - r_j\right) Y(x) = \left(\hat{D} - r_s\right) Y_{s-1}(x), \ s = 1, 2, \ldots, \ Y_0(x) = Y(x), \tag{B.4}$$

where $r_1, r_2, \ldots, r_s$ are arbitrary (in general complex) numbers. Arguing as before we integrate equation (B.4) and obtain Y_{s-1} in terms of Y_s as follows:

$$Y_{s-1}(x) = \exp\left(r_s x\right)\left[C_s + \int^x \exp\left(-r_s x'\right) Y_s\left(x'\right) dx'\right]. \tag{B.5}$$

This simple equation is the main result of this appendix.

A general inhomogeneous ordinary differential equation of order n with constant coefficients a_j is of the form

$$\mathcal{L}\left(\hat{D}\right) Y(x) = f(x), \ \mathcal{L}\left(\hat{D}\right) = \sum_{j=0}^{n} a_j \hat{D}^j, \tag{B.6}$$

where we choose $a_n = 1$ without loss of generality. We can factorize the differential operator $\mathcal{L}(\hat{D})$ as

$$\mathcal{L}\left(\hat{D}\right) = \prod_{j=1}^{n} \left(\hat{D} - r_j\right), \tag{B.7}$$

where $r_1, r_2, \ldots, r_n$ are the roots of the characteristic equation $\mathcal{L}(r) = 0$. According to the definition given in equation (B.4) we have $Y_n(x) = f(x)$, and in order to obtain $Y_0(x) = Y(x)$ we simply apply equation (B.5) for $s = n, n-1, \ldots, 1$. Present algorithm is particularly suitable for the application of computer algebra. We do not show a general Maple program here because we are concerned only with the case $n = 2$ that we discuss in what follows.

It is sufficient for our purposes to consider a differential equation of second order

$$Y''(x) + a_1 Y'(x) + a_0 Y(x) = f(x) \tag{B.8}$$

that leads to a quadratic characteristic equation

$$r^2 + a_1 r + a_0 = 0, \tag{B.9}$$

which we easily solve to obtain its two roots r_1 and r_2. Straightforward application of the general recipe outlined above gives us

$$
\begin{aligned}
Y(x) =\ & C_1 \exp(r_1 x) + C_2 \exp(r_1 x) \int^x \exp\left[(r_2 - r_1) x'\right] dx' \\
& + \exp(r_1 x) \int^x \int^{x'} \exp\left[(r_2 - r_1) x' - r_2 x''\right] f(x'') \, dx'' \, dx'.
\end{aligned} \tag{B.10}
$$

Integration by parts enables us to reduce the double integral in equation (B.10) to a single one. To this end it is convenient to consider the cases of equal and different roots separately.

When $r_1 = r_2$, we easily rewrite equation (B.10) as

$$Y(x) = (C_1 + C_2 x) \exp(r_1 x) + \int^x (x - x') \exp\left[r_1 (x - x')\right] f(x') \, dx'. \tag{B.11}$$

On the other hand, when $r_1 \neq r_2$ we have

$$
\begin{aligned}
Y(x) =\ & C_1 \exp(r_1 x) + \frac{C_2}{r_2 - r_1} \exp(r_2 x) \\
& + \frac{1}{r_2 - r_1} \int^x \left\{ \exp\left[r_2 (x - x')\right] - \exp\left[r_1 (x - x')\right] \right\} f(x') \, dx'.
\end{aligned} \tag{B.12}
$$

In some chapters of this book we face an example of the latter case given by $a_1 = 0$ and $a_0 = \omega^2$. Because the roots of equation (B.9) are $r_1 = -r_2 = i\omega$ (we choose $\omega > 0$ without loss of generality), we rewrite equation (B.12) as

$$Y(x) = C \sin(\omega x) + C' \cos(\omega x) + \frac{1}{\omega} \int^x \sin\left[\omega (x - x')\right] f(x') \, dx', \tag{B.13}$$

where the constants of integration C and C' are related to C_1 and C_2 in a straightforward way.

Appendix C

Canonical Transformations

In this appendix we give a brief account of canonical transformations [48] that considerably facilitate the discussion of several subjects covered by this book. In particular we are interested in canonical transformations of the form

$$\hat{B}_A(\alpha) = \hat{U}_A^{-1}(\alpha)\hat{B}\hat{U}_A(\alpha), \quad \hat{U}_A(\alpha) = \exp\left(-\alpha\hat{A}\right) , \tag{C.1}$$

where $\hat{A}$ and $\hat{B}$ are two linear operators. Notice that

$$\hat{B}_A(0) = \hat{B} . \tag{C.2}$$

One easily proves that canonical transformations preserve commutators; that is to say

$$\left[\hat{B}, \hat{C}\right] = \hat{D} \Rightarrow \left[\hat{B}_A, \hat{C}_A\right] = \hat{D}_A . \tag{C.3}$$

In particular, if $\hat{A}$ is antihermitian $\hat{A}^\dagger = -\hat{A}$ and α is real, then $\hat{U}_A$ is unitary $\hat{U}_A^\dagger = \hat{U}_A^{-1}$.

Many equations regarding canonical transformations take considerably simpler forms in terms of superoperators [48]. For example, if we define the superoperator $\widehat{\hat{A}}$ as

$$\widehat{\hat{A}}\hat{B} = \left[\hat{A}, \hat{B}\right] \tag{C.4}$$

we easily prove that

$$\frac{d^n}{d\alpha^n}\hat{B}_A = \widehat{\hat{A}}^n\hat{B}_A = \hat{U}_A^{-1}\widehat{\hat{A}}^n\hat{B}\hat{U}_A \tag{C.5}$$

and can formally write

$$\hat{B}_A = \exp\left(\alpha\widehat{\hat{A}}\right)\hat{B} . \tag{C.6}$$

By repeated application of the rule

$$\hat{U}_A^{-1}\hat{B}^n\hat{U}_A = \hat{U}_A^{-1}\hat{B}\hat{U}_A\hat{U}_A^{-1}\hat{B}^{n-1}\hat{U}_A \tag{C.7}$$

we conclude that

$$\hat{U}_A^{-1}\hat{B}^n\hat{U}_A = \left(\hat{U}_A^{-1}\hat{B}\hat{U}_A\right)^n = \hat{B}_A^n . \tag{C.8}$$

Operator differential equations like (C.5) with the initial condition (C.2) are suitable for obtaining explicit expressions of canonical transformations. In what follows we consider a few simple cases that are useful in this book.

1) If $[\hat{A}, \hat{B}] = a$, where a is a scalar, then

$$\hat{B}_A = \hat{B} + a\alpha \ . \tag{C.9}$$

2) If $[\hat{A}, \hat{B}] = b\hat{B}$, where b is a scalar, then

$$\hat{B}_A(\alpha) = \exp(b\alpha)\hat{B} \ . \tag{C.10}$$

In particular, notice that

$$\hat{B}_A(\pi i/b) = \hat{U}_A^{-1}(\pi i/b)\hat{B}\hat{U}_A(\pi i/b) = -\hat{B} \ . \tag{C.11}$$

3) If $\widehat{A}^2 B = \omega^2 \hat{B}$, where ω is a constant, then

$$\hat{B}_A(\alpha) = \cosh(\omega\alpha)\hat{B} + \frac{\sinh(\omega\alpha)}{\omega}\widehat{A}\hat{B} \ . \tag{C.12}$$

When $\omega^2 < 0$ we rewrite this equation in a more convenient form:

$$\hat{B}_A(\alpha) = \cos(|\omega|\alpha)\hat{B} + \frac{\sin(|\omega|\alpha)}{|\omega|}\widehat{A}\hat{B} \ . \tag{C.13}$$

If we apply $\hat{U}^{-1}$ to the Schrödinger equation

$$\hat{H}\Psi = E\Psi \tag{C.14}$$

from the left, we obtain

$$\hat{U}^{-1}\hat{H}\Psi = \hat{U}^{-1}\hat{H}\hat{U}\hat{U}^{-1}\Psi = E\hat{U}^{-1}\Psi \ . \tag{C.15}$$

If $\hat{H}$ is invariant under the canonical transformation

$$\hat{U}^{-1}\hat{H}\hat{U} = \hat{H} \ , \tag{C.16}$$

then $\hat{U}^{-1}\Psi$ is an eigenfunction of $\hat{H}$ with eigenvalue E. If this eigenvalue is not degenerate then $\hat{U}^{-1}\Psi \propto \Psi$.

A particularly useful canonical transformation is the so-called scaling or dilatation. Consider dimensionless coordinate and momentum operators $\hat{x}$ and $\hat{p}$, respectively, which satisfy $[\hat{x}, \hat{p}] = i$, and construct the unitary operator

$$\hat{U}_A = \exp\left(-\alpha\hat{A}\right), \quad A = \frac{i}{2}\left(\hat{x}\hat{p} + \hat{p}\hat{x}\right) \ , \tag{C.17}$$

where α is a real parameter. Taking into account that $[\hat{A}, \hat{x}] = \hat{x}$, and $[\hat{A}, \hat{p}] = -\hat{p}$, then we conclude from Case 2 above that

$$\hat{x}_A = e^\alpha \hat{x}, \quad \hat{p}_A = e^{-\alpha}\hat{p} \ . \tag{C.18}$$

Moreover, if $V(x)$ is an analytic function of x at $x = 0$, and we apply the result in equation (C.8) to the Taylor series of $V(x)$ around $x = 0$, we conclude that

$$\hat{U}_A^\dagger V\left(\hat{x}\right)\hat{U}_A = V\left(\hat{x}_A\right) \ . \tag{C.19}$$

Consequently, the scaling transformation of the Hamiltonian operator

$$\hat{H} = \frac{\hat{p}^2}{2} + V(\hat{x}) \tag{C.20}$$

reads

$$\hat{H}_A = \frac{\hat{p}_A^2}{2} + V(\hat{x}_A) \ . \tag{C.21}$$

An interesting particular case is given by $\alpha = i\pi$ because the scaling transformation simply changes the sign of the operators: $\hat{x}_A = -\hat{x}$, $\hat{p}_A = -\hat{p}$.

As an illustrative example consider the anharmonic oscillator

$$\hat{H}(a, b, \lambda) = a\,\hat{p}^2 + b\,\hat{x}^2 + \lambda\,\hat{x}^k \ , \tag{C.22}$$

where a, b, λ, and k are chosen so that this operator supports bound-state eigenvalues $E(a, b, \lambda)$. First of all notice that $E(a, b, \lambda) = cE(a/c, b/c, \lambda/c)$. It follows from the results above that

$$\hat{U}_A^\dagger \hat{H}(a, b, \lambda)\hat{U}_A = \hat{H}\left(a\,e^{-2\alpha}, b\,e^{2\alpha}, \lambda\,e^{\alpha k}\right) = e^{-2\alpha}\hat{H}\left(a, b\,e^{4\alpha}, \lambda\,e^{\alpha(k+2)}\right) \ ; \tag{C.23}$$

consequently,

$$E(a, b, \lambda) = E\left(a\,e^{-2\alpha}, b\,e^{2\alpha}, \lambda\,e^{\alpha k}\right) = e^{-2\alpha}E\left(a, b\,e^{4\alpha}, \lambda\,e^{\alpha(k+2)}\right) \ . \tag{C.24}$$

This argument due to Symanzik [210] proved useful in the study of the analytic properties of the eigenvalues of anharmonic oscillators [111].

On choosing $e^{2\alpha} = \lambda^{-2/(k+2)}$ equation (C.24) becomes

$$E(a, b, \lambda) = \lambda^{2/(k+2)}E\left(a, b\,\lambda^{-4/(k+2)}, 1\right) \ , \tag{C.25}$$

which suggests that the eigenvalues of the anharmonic oscillator can be expanded as

$$E(a, b, \lambda) = \lambda^{2/(k+2)}\sum_{j=0}^{\infty}e_j\lambda^{-4j/(k+2)} \ . \tag{C.26}$$

It has been proved that this series already exists for k even and exhibits finite convergence radius [111]. The leading coefficient e_0 is an eigenvalue of the anharmonic oscillator $\hat{H}(a, 0, 1) = a\hat{p}^2 + \hat{x}^k$.

The scaling transformation also proves useful to relate the eigenvalues of anharmonic oscillators with different parameters. For example, starting from $E(1/2, 0, 1) = E(1, 0, 2e^{\alpha(k+2)})/(2e^{2\alpha})$, and choosing $e^{2\alpha} = 2^{-2/(k+2)}$, we prove that $E(1/2, 0, 1) = 2^{-k/(k+2)}E(1, 0, 1)$.

If the Hamiltonian operator $\hat{H}(\lambda)$ depends on a parameter λ, its eigenfunctions and eigenvalues will also depend on λ $\hat{H}(\lambda)\Psi(\lambda) = E(\lambda)\Psi(\lambda)$. Suppose that $\hat{H}(0)$ supports discrete states and that there is a canonical transformation such that

$$\hat{U}^{-1}\hat{H}(\lambda)\hat{U} = \hat{H}(-\lambda) \ . \tag{C.27}$$

It follows from $\hat{H}(-\lambda)\Psi(-\lambda) = E(-\lambda)\Psi(-\lambda)$ and equation (C.27) that

$$\hat{H}(\lambda)\hat{U}\Psi(-\lambda) = E(-\lambda)\hat{U}\Psi(-\lambda) \ . \tag{C.28}$$

This equation tells us that $E_j(-\lambda) = E_k(\lambda)$ for some pair of quantum numbers j and k; in particular, $E_j(0) = E_k(0)$. If the spectrum of $\hat{H}(0)$ is nondegenerate we conclude that $j = k$, and the Taylor expansion of $E_j(\lambda)$ about $\lambda = 0$ will have only even terms:

$$E_j(\lambda) = \sum_{i=0}^{\infty}E_{j,2i}\lambda^{2i} \ . \tag{C.29}$$

Throughout this book we show several quantum-mechanical problems that exhibit such a feature.

References

[1] Maple V Release 5.1, Waterloo Maple Inc.

[2] Messiah, A., *Quantum Mechanics,* John Wiley & Sons, New York, 1961, Vol. 1, 73.

[3] Messiah, A., *Quantum Mechanics,* John Wiley & Sons, New York, 1961, Vol. 2, chap. 16.

[4] Adams, B.G., *Algebraic Approach to Simple Quantum Systems,* Springer-Verlag, Berlin, 1994, chap. 7.

[5] Messiah, A., *Quantum Mechanics,* John Wiley & Sons, New York, 1961, Vol. 2, chap. 8.

[6] Messiah, A., *Quantum Mechanics,* John Wiley & Sons, New York, 1961, Vol. 2, chap. 17.

[7] Fernández, F.M. and Castro, E.A., *Algebraic Methods in Quantum Chemistry and Physics,* CRC Press, Boca Raton, FL, 1996, chap. 3.

[8] Fernández, F.M. and Castro, E.A., *Algebraic Methods in Quantum Chemistry and Physics,* CRC Press, Boca Raton, FL, 1996, chap. 9.

[9] Eyring, H., Walter, J., and Kimball, G.E., *Quantum Chemistry,* John Wiley & Sons, New York, 1944, chap. 8.

[10] Dalgarno, A. and Stewart, A.L., On the perturbation theory of small disturbances, *Proc. Roy. Soc. A,* 238, 269, 1956.

[11] Arteca, G.A., Fernández, F.M., and Castro, E.A., *Large Order Perturbation Theory and Summation Methods in Quantum Mechanics,* in Lecture Notes in Chemistry, Vol. 53, Springer-Verlag, Berlin, 1990, chap. 9.

[12] Adams, B.G., *Algebraic Approach to Simple Quantum Systems,* Springer-Verlag, Berlin, 1994, chap. 9.

[13] Arteca, G.A., Fernández, F.M., and Castro, E.A., *Large Order Perturbation Theory and Summation Methods in Quantum Mechanics,* in Lecture Notes in Chemistry, Vol. 53, Springer-Verlag, Berlin, 1990, Appendix H.

[14] Rösner, W., Wunner, G., Herold, H., and Ruder, H., Hydrogen atoms in arbitrary magnetic fields: I. Energy levels and wavefunctions, *J. Phys. B,* 17, 29, 1984.

[15] Price, P.J., Perturbation theory for the one-dimensional wave equation, *Proc. Phys. Soc. London,* 67, 383, 1954.

[16] Dolgov, A.D. and Popov, V.S., Modified perturbation theories for an anharmonic oscillator, *Phys. Lett. B,* 79, 403, 1978.

[17] Au, C.K. and Aharonov, Y., Logarithmic perturbation expansions, *Phys. Rev. A*, 20, 2245, 1979.

[18] Aharonov, Y. and Au, C.K., New approach to perturbation theory, *Phys. Rev. Lett.*, 42, 1582, 1979.

[19] Privman, V., New method of perturbation-theory calculation of the Stark effect for the ground state of hydrogen, *Phys. Rev. A*, 22, 1833, 1980.

[20] Turbiner, A.V., The hydrogen atom in an external magnetic field, *J. Phys. A*, 17, 859, 1984.

[21] Fernández, F.M., Alternative approach to stationary perturbation theory for separable problems, *J. Chem. Phys.*, 97, 8465, 1992.

[22] Fernández, F.M. and Castro E.A., A new implementation of stationary perturbation theory for separable quantum- mechanical problems, *J. Chem. Phys.*, 98, 6392, 1993.

[23] Fernández, F.M., Approach to perturbation theory for box models, *Phys. Rev. A*, 48, 189, 1993.

[24] Adams, B.G., *Algebraic Approach to Simple Quantum Systems*, Springer-Verlag, Berlin, 1994, 72.

[25] Fernández, F.M. and Castro, E.A., *Hypervirial Theorems*, Lecture Notes in Chemistry, Vol. 43, Springer-Verlag, Berlin, 1987, chap. 1.

[26] Swenson, R.J. and Danforth, S.H., Hypervirial and Hellmann-Feynman Theorems applied to anharmonic oscillators, *J. Chem. Phys.*, 37, 1734, 1972.

[27] Killingbeck, J., Perturbation theory without wavefunctions, *Phys. Lett. A*, 65, 87, 1978.

[28] Fernández, F.M. and Castro, E.A., *Hypervirial Theorems*, Lecture Notes in Chemistry, Vol. 43, Springer-Verlag, Berlin, 1987, chap. 6.

[29] Arteca, G.A., Fernández, F.M., and Castro, E.A., *Large Order Perturbation Theory and Summation Methods in Quantum Mechanics*, in Lecture Notes in Chemistry, Vol. 53, Springer-Verlag, Berlin, 1990, 54.

[30] Fernández, F.M. and Castro, E.A., *Algebraic Methods in Quantum Chemistry and Physics*, CRC Press, Boca Raton, FL, 1996, chap. 3, 82.

[31] Eyring, H., Walter, J., and Kimball, G.E., *Quantum Chemistry*, John Wiley & Sons, New York, 1944, chap. 8, 47.

[32] Fernández, F.M. and Castro, E.A., *Hypervirial Theorems*, Lecture Notes in Chemistry, Vol. 43 (Part B), Springer-Verlag, Berlin, 1987.

[33] Fernández, F.M. and Castro, E.A., *Algebraic Methods in Quantum Chemistry and Physics*, CRC Press, Boca Raton, FL, 1996, chap. 5.

[34] Blankenbecler, R., DeGrand, T., and Sugar, L.R., Moment method for eigenvalues and expectation values, *Phys. Rev. D*, 23, 1055, 1980.

[35] Ader, J.P., Moment Method and the Schrödinger equation in the large N limit, *Phys. Lett. A*, 97, 178, 1983.

[36] Fernández, F.M. and Morales, J.A., Perturbation theory without wave function for the Zeeman effect in hydrogen, *Phys. Rev. A*, 46, 318, 1992.

[37] Fernández, F.M., Moment-method perturbation theory for the hydrogen atom in parallel electric and magnetic fields and in inhomogeneous electric fields, *Int. J. Quantum Chem.*, 26, 117, 1992.

[38] Fernández, F.M. and Ogilvie, J.F., Perturbation theory by the moment method applied to coupled anharmonic oscillators, *Phys. Lett. A*, 178, 11, 1993.

[39] Radicioni, M.D., Diaz, C.G., and Fernández, F.M., Renormalized perturbation theory by the moment method for degenerate states: anharmonic oscillators, *Int. J. Quantum Chem.*, 66, 261, 1998.

[40] Eyring, H., Walter, J., and Kimball, G.E., *Quantum Chemistry*, John Wiley & Sons, New York, 1944, chap. 6.

[41] Killingbeck, J. and Jones, M.N., The perturbed two-dimensional oscillator, *J. Phys. A*, 19, 705, 1986.

[42] Witwit, M.R.M., The eigenvalues of the Schrödinger equation for spherically symmetric states for various types of potentials in two, three, and N dimensions, by using perturbative and non-perturbative methods, *J. Phys. A*, 24, 4535, 1991.

[43] Radicioni, M.D., Diaz, C.G., and Fernández, F.M., Application of Perturbation Theory to Coupled Morse Oscillators, *J. Molec. Struct. (THEOCHEM)*, 488, 37, 1999.

[44] Banwell, C.N. and Primas, H., On the analysis of high-resolution nuclear magnetic resonance spectra, I. Methods of calculating N.M.R. spectra., *Mol. Phys.*, 6, 225, 1963.

[45] Primas, H., Eine verallgemeinerte störungstheorie für quantenmechanische mehrteilchenprobleme, *Helv. Phys. Acta*, 34, 331, 1961.

[46] Primas, H., Generalized perturbation theory in operator form, *Rev. Mod. Phys.*, 35, 710, 1963.

[47] Fernández, F.M. and Castro, E.A., *Algebraic Methods in Quantum Chemistry and Physics*, CRC Press, Boca Raton, FL, 1996, 109.

[48] Fernández, F.M. and Castro, E.A., *Algebraic Methods in Quantum Chemistry and Physics*, CRC Press, Boca Raton, FL, 1996, chaps. 1 and 2.

[49] Fernández, F.M. and Castro, E.A., *Algebraic Methods in Quantum Chemistry and Physics*, CRC Press, Boca Raton, FL, 1996, chap. 3.

[50] Van Vleck, J.H., On σ-type doubling and electron spin in the spectra of diatomic molecules, *Phys. Rev.*, 33, 467, 1929.

[51] Shavitt, I., Quasidegenerate perturbation theories. A canonical Van Vleck formalism and its relationship to other approaches, *J. Chem. Phys.*, 73, 5711, 1980.

[52] Sibert III, E.L., Theoretical studies of vibrationally excited polyatomic molecules using canonical Van Vleck perturbation theory, *J. Chem. Phys.*, 88, 4378, 1988.

[53] Eyring, H., Walter, J., and Kimball, G.E., *Quantum Chemistry*, John Wiley & Sons, New York, 1944, 190.

[54] Eyring, H., Walter, J., and Kimball, G.E., *Quantum Chemistry*, John Wiley & Sons, New York, 1944, 203.

[55] Austin, E.J., Perturbation theory and Padé approximants for a hydrogen atom in an electric field, *Mol. Phys.*, 40, 893, 1980.

[56] Lai, C.S., Calculation of the Stark effect in hydrogen atoms by using the hypervirial relations, *Phys. Lett.*, 83A, 322, 1981.

[57] Adams, B.G., *Algebraic Approach to Simple Quantum Systems*, Springer-Verlag, Berlin, 1994, chap. 8.

[58] Fernández, F.M., Large-order perturbation theory without a wavefunction for the LoSurdo-Stark effect in hydrogen, *J. Phys. A*, 25, 495, 1992.

[59] Fernández, F.M. and Morales, J.A., Perturbation theory without wavefunction for the Zeeman effect in hydrogen, *Phys. Rev. A*, 46, 318, 1992.

[60] Fernández, F.M., Moment-method perturbation theory for the hydrogen atom in parallel electric and magnetic fields and in inhomogeneous electric fields, *Int. J. Quantum Chem.*, 26, 117, 1992.

[61] Eyring, H., Walter, J., and Kimball, G.E., *Quantum Chemistry*, John Wiley & Sons, New York, 1944, 201.

[62] Eyring, H., Walter, J., and Kimball, G.E., *Quantum Chemistry*, John Wiley & Sons, New York, 1944, 369.

[63] Damburg, R.J. and Propin, R.Kh., On asymptotic expansions of electronic terms of the molecular ion H_2^+, *J. Phys. B*, 1, 681, 1968.

[64] Morgan III, J.D. and Simon, B., Behavior of molecular potential energy curves for large nuclear separations, *Int. J. Quantum Chem.*, 17, 1143, 1980.

[65] Čížek, J., Damburg, R.J., Graffi, S., Grecchi, V., Harrell II, E.M., Harris, J.G., Nakai, S., Paldus, J., Propin, R.Kh., and Silverstone, H.J., $1/R$ expansion for H_2^+: calculation of exponentially small terms and asymptotics, *Phys. Rev. A*, 33, 12, 1986.

[66] Harrell, E.M., Double wells, *Commun. Math. Phys.*, 75, 239, 1980.

[67] Lapidus, R.I., One-dimensional model of a diatomic ion, *Amer. J. Phys.*, 38, 905, 1970.

[68] Chandrasekhar, S., Dinamical friction. II. The rate of escape of stars from clusters and the evidence for the operation of dinamical friction, *Astrophys. J.*, 97, 263, 1943.

[69] Corson, E.M. and Kaplan, I, The oscillator concept in the theory of solids, *Phys. Rev.*, 71, 130, 1947.

[70] Suryan, B., Bounded linear harmonic oscillator and phase transitions of second order, *Phys. Rev.*, 71, 1947, 741.

[71] Michels, A., De Boer, J., and Bijl, A., Remarks concerning molecular interaction and their influence on the polarisbility, *Physica*, 4, 981, 1937.

[72] Rabinovitch, A. and Zak, J., Electrons in crystals in a finite-range electric field, *Phys. Rev. B*, 4, 2358, 1971.

[73] Lukes, T., Ringwood, G.A., and Suprapto, B., A particle in a box in the presence of an electric field and applications to disordered systems, *Physica A*, 84, 421, 1976.

[74] Bastard, G., Mendez, E.E., Chang, L.L., and Esaki, L., Variational calculations on a quantum well in an electric field, *Phys. Rev. B*, 28, 3241, 1983.

[75] Dingle, R.B., Some magnetic properties of metals. IV. Properties of small systems of electrons. *Proc. R. Soc. London Ser. A*, 212, 47, 1952.

[76] Fernández, F.M. and Castro, E.A., Hypervirial treatment of multidimensional isotropic bounded oscillators, *Phys. Rev. A*, 24, 2883, 1981.

[77] Fernández, F.M. and Castro, E.A., Hypervirial-perturbational treatment of the bounded hydrogen atom, *J. Math. Phys.*, 23, 1103, 1982.

[78] Townes, C.H. and Schawlow, A.L., *Microwave Spectroscopy*, McGraw-Hill, New York, 1955.

[79] Wollrab, J.E., *Rotational Spectra and Molecular Structure*, Academic Press, New York, 1967.

[80] Abramowitz, M. and Stegun, I.A., *Handbook of Mathematical Functions*, Dover, New York, 1972, 358, 437.

[81] Fernández, F.M., Alternative approach to stationary perturbation theory for separable problems, *J. Chem. Phys.*, 97, 8465, 1992.

[82] Fernández, F.M. and Castro, E.A., *Algebraic Methods in Quantum Chemistry and Physics*, CRC Press, Boca Raton, FL, 1996, 249.

[83] Abramowitz, M. and Stegun, I.A., *Handbook of Mathematical Functions*, Dover, New York, 1972, 409.

[84] Hughes, H.K., The electric resonance method of radiofrequency spectroscopy. The moment of inertia and electric dipole moment of CsF, *Phys. Rev.*, 72, 614, 1947.

[85] Scharpen, L.H., Muenter, J.S., and Laurie, V.W., Determination of polarizability anisotropy of OCS by microwave spectroscopy, *J. Chem. Phys.*, 46, 2431, 1967.

[86] Røeggen, J., Polynomial approximations to the Stark perturbed rotational energy levels of the rigid symmetric top rotor, *Atomic Data*, 4, 289, 1972.

[87] Wijnberg, L., Fourth order perturbation theoretic Stark energy levels in linear rotors, *J. Chem. Phys.*, 60, 4632, 1974.

[88] Propin, R., On the levels of a diatomic rigid polar molecule in a uniform electric field, *J. Phys. B*, 11, 4179, 1978.

[89] Fernández, F.M., Energy of a perturbed rigid rotor, *J. Math. Chem.*, 18, 197, 1995 (and references therein).

[90] Peter, M. and Strandberg, M.W.P., High-field Stark effect in linear rotors, *J. Chem. Phys.*, 26, 1657, 1957.

[91] Propin, R., Stark effect for the rotational levels of a diatomic polar molecule in a strong field, *J. Phys. B*, 11, 257, 1978.

[92] Cohen, M. and Feldmann, T., Rayleigh-Schrödinger perturbation theory with a non-Hermitian perturbation, *J. Phys. B*, 15, 2563, 1982.

[93] Maluendes, S.A., Fernández, F.M., and Castro, E.A., Rotational energies of polar symmetric-top molecules in high electric fields, *J. Molec. Spectrosc.*, 100, 24, 1983.

[94] Maluendes, S.A., Fernández, F.M., and Castro, E.A., Rotational energy of the hydrogen molecular ion in a magnetic field, *Phys. Rev. A*, 28, 2059, 1983.

[95] Maluendes, S.A., Fernández, F.M., Mesón, A.M., and Castro, E.A., Calculation of rotational energies of molecules in electric and magnetic fields by perturbation theory, *Phys. Rev. A*, 30, 2227, 1984.

[96] Apostol, T.M., *Calculus*, Vol. 1, 2nd ed., Blaisdell, Waltham, 1969, chap. 10.

[97] Irving, J. and Mullineux, N., *Mathematics in Physics and Engineering,* Academic Press, New York, 1959, 438.

[98] Bender, C.M. and Orszag, S.A., *Advanced Mathematical Methods for Scientists and Engineers,* McGraw-Hill, New York, 1978, chap. 3.

[99] Fernández, F.M., Arteca, F.M., and Castro, E.A., Shifted ratio method for calculation of critical parameters from powers series expansions, *Physica A,* 137, 639, 1986.

[100] Arteca, F.M., Fernández, F.M., and Castro, E.A., Study of eigenvalue singularities from perturbation series: application to two-electron atoms, *J. Chem. Phys.,* 84, 1624, 1986.

[101] Arteca, G.A., Fernández, F.M., and Castro, E.A., A new method of analysis of critical-point singularities from power series expansions: application to eigenvalue problems and virial series, *J. Chem. Phys.,* 85, 6713, 1986.

[102] Fernández, F.M., Arteca, G.A., and Castro, E.A., Convergence radii of the Rayleigh-Schrödinger perturbation series for the bounded oscillators, *J. Phys. A,* 20, 6303, 1987.

[103] Fernández, F.M., Arteca, G.A., and Castro, E.A., Singular points from Taylor series, *J. Math. Phys.,* 28, 323, 1987.

[104] Bender, C.M. and Orszag, S.A., *Advanced Mathematical Methods for Scientists and Engineers,* McGraw-Hill, New York, 1978, 350.

[105] Reed, M. and Simon, B., *Methods of Modern Mathematical Physics, IV. Analysis of Operators,* Academic, New York, 1978.

[106] Robinson, P.D., H_2^+: a problem in perturbation theory, *Proc. Phys. Soc.,* 78, 537, 1961.

[107] Claverie, P., Study of the convergence radius of the Rayleigh–Schrödinger perturbation series for the delta-function model of H_2^+, *Int. J. Quantum Chem.,* 3, 349, 1969.

[108] Certain, P.R. and Byers Brown, W., Branch point singularities in the energy of the delta-function model of one-electron diatoms, *Int. J. Quantum Chem.,* 6, 131, 1972.

[109] Ahlrichs, R. and Claverie, P., Convergence radii for the perturbation expansions of the energy and of the wave function: the case of the delta function model H_2^+, *Int. J. Quantum Chem.,* 6, 1001, 1972.

[110] Simon, B., Large orders and summability of eigenvalue perturbation theory: a mathematical overview, *Int. J. Quantum Chem.,* 21, 3, 1982 (and references therein).

[111] Simon, B., Coupling constant analyticity for the anharmonic oscillator, *Ann. Phys. (N.Y.),* 58, 76, 1970.

[112] Kroto, H.W., *Molecular Rotation Spectra,* Dover, New York, 1992.

[113] Fernández, F.M. and Castro, E.A., Convergence radii of the Rayleigh-Schrödinger perturbation series for a diatomic rigid polar molecule in a uniform electric field, *Phys. Lett. A,* 10, 215, 1985.

[114] Maluendes, S.A., Fernández, F.M., and Castro, E.A., Convergent renormalized perturbation series for the Stark rotational energies of diatomic molecules, *J. Chem. Phys.,* 83, 4599, 1985.

[115] Bender, C.M. and Wu, T.T., Anharmonic oscillator, *Phys. Rev.,* 184, 1231, 1969.

[116] Fernández, F.M. and Guardiola, R., Accurate eigenvalues and eigenfunctions for quantum-mechanical anharmonic oscillators. *J. Phys. A,* 26, 7169, 1993.

[117] Bessis, N. and Bessis, G., Open perturbation and the Riccati equation: algebraic determination of the quartic anharmonic oscillator energies and eigenfunctions, *J. Math. Phys.*, 38, 5483, 1997.

[118] Killingbeck, J., Renormalized perturbation series, *J. Phys. A*, 14, 1005, 1981.

[119] Stevenson, P.M., Optimized perturbation theory, *Phys. Rev. D*, 23, 2916, 1981.

[120] Banerjee, K., Bhatnagar, S.P., Choudhry, V., and Kanwal, S.S., The anharmonic oscillator, *Proc. R. Soc. London, A*, 360, 575, 1978.

[121] Fernández, F.M., Renormalized perturbation series and the semiclassical limit of quantum mechanics, *J. Math. Phys.*, 36, 3922, 1995.

[122] Guida, R., Konishi, K., and Suzuki, H., Improved convergence proof of the delta expansion and order dependent mappings, *Ann. Phys. (N.Y.)*, 249, 109, 1996.

[123] Austin, E.J., Further applications of the renormalised series technique, *J. Phys. A*, 17, 367, 1984.

[124] Weniger, E.J., Čížek, J., and Vinette, F., Very accurate summation for the infinite coupling limit of the perturbation series expansions of anharmonic oscillators, *Phys. Lett. A*, 156, 169, 1991.

[125] Guardiola, R., Solis M.A., and Ros, J., Strong-coupling expansion for the anharmonic oscillators, *Nuovo Cim.*, 107B, 713, 1992.

[126] Weniger, E.J., Čížek, J. and Vinette, F., The summation of the ordinary and renormalized perturbation series for the ground state energy of the quartic, sextic, and octic anharmonic oscillators using nonlinear sequence of transformations, *J. Math. Phys.*, 34, 571, 1993.

[127] Janke, W. and Kleinert, H., Convergent strong-coupling expansions from divergent weak-coupling perturbation theory, *Phys. Rev. Lett.*, 75, 2787, 1995.

[128] Feranchuk, I.D. and Komarov, L.I., The operator method of the approximate solution of the Schrödinger equation, *Phys. Lett. A*, 88, 211, 1982.

[129] Feranchuk, I.D. and Komarov, L.I., The operator method of the approximate description of the quantum and classical systems, *J. Phys. A*, 17, 3111, 1984.

[130] Fernández, F.M. and Castro, E.A., Comment on the operator method and perturbational solution of the Schrödinger equation, *Phys. Lett. A*, 91, 339, 1982.

[131] Fernández, F.M., Mesón, A.M., and Castro, E.A., On the convergence of the operator method perturbation series, *Phys. Lett. A*, 104, 401, 1984.

[132] Fernández, F.M., Mesón, A.M., and Castro, E.A., Energy eigenvalues from a modified operator method, *Phys. Lett. A*, 111, 104, 1985.

[133] Arteca, G.A., Fernández, F.M., and Castro, E.A., Energies of the $\alpha x^3 + \beta x^4$ anharmonic oscillator from a modified operator method., *Molec. Phys.*, 58, 365, 1986.

[134] Arteca, G.A., Fernández, F.M., and Castro, E.A., Further comments on the modified operator method. *Phys. Lett. A*, 119, 149, 1986.

[135] Bender, C.M. and Orszag, S.A., *Advanced Mathematical Methods for Scientists and Engineers,* McGraw-Hill, New York, 1978, chap. 8.

[136] Arteca, G.A., Fernández, F.M., and Castro, E.A., *Large Order Perturbation Theory and Summation Methods in Quantum Mechanics,* in Lecture Notes in Chemistry, Vol. 53, Springer-Verlag, Berlin, 1990, chap. 5.

[137] Weniger, E.J., Nonlinear sequence transformations for the acceleration of convergence and the summation of divergent series, *Comp. Phys. Rep.,* 10, 189, 1989.

[138] Sergeev, A.V. and Goodson, D.Z., Summation of asymptotic expansions of multiple-valued functions using algebraic approximants: application to anharmonic oscillators, *J. Phys. A,* 31, 4301, 1998.

[139] Weniger, E.J., Performance of superconvergent perturbation theory, *Phys. Rev. A,* 56, 5165, 1997.

[140] Fernández, F.M. and Tipping, R.H., Analytical expressions for the energies of anharmonic oscillators, *Can. J. Phys.,* in press.

[141] Fernández, F.M., Strong coupling expansion for anharmonic oscillators and perturbed Coulomb potentials, *Phys. Lett. A,* 166, 173, 1992.

[142] Weniger, E.J., A convergent renormalized strong coupling perturbation expansion for the ground state energy of the quartic, sextic and octic anharmonic oscillator, *Ann. Phys.,* 246, 133, 1996.

[143] Weniger, E.J., Construction of the strong coupling expansion for the ground state energy of the quartic, sextic and octic anharmonic oscillator via a renormalized strong coupling expansion, *Phys. Rev. Lett.,* 77, 2859, 1996.

[144] Fernández, F.M. and Guardiola, R., The strong coupling expansion for anharmonic oscillators, *J. Phys. A,* 30, 7187, 1997.

[145] Fernández, F.M., Fast and accurate method for the summation of divergent series, unpublished.

[146] Shanley, P.E., Spectral singularities of the quartic anharmonic oscillator, *Phys. Lett. A,* 117, 161, 1986.

[147] Shanley, P.E., Spectral properties of the scaled quartic anharmonic oscillator, *Ann. Phys.,* 186, 292, 1988.

[148] Chatterjee, A., Large-N expansions in quantum mechanics, atomic physics and some $O(N)$ invariant systems, *Phys. Rep.,* 6, 249, 1990.

[149] Messiah, A., *Quantum Mechanics,* John Wiley & Sons, New York, 1961, Vol. 1, 231.

[150] ter Haar, D., *Problems in Quantum Mechanics,* Pion, London, 3rd ed., 1975, 361.

[151] ter Haar, D., *Problems in Quantum Mechanics,* Pion, London, 3rd ed., 1975, 91.

[152] Simon, B., The bound state of weakly coupled Schrödinger operators in one and two dimensions, *Ann. Phys., (N.Y.),* 97, 279, 1976.

[153] Blankenbecler, R., Goldberger, M.L., and Simon, B., The bound states of weakly coupled long-range one-dimensional quantum Hamiltonians, *Ann. Phys.,* 108, 69, 1977.

[154] Patil, S.H., T-matrix analysis of one-dimensional weakly coupled bound states, *Phys. Rev. A,* 22, 1655, 1980.

[155] Patil, S.H., Scattering amplitudes and bound state energies for one-dimensional, weak potentials, *Amer. J. Phys.,* 67, 616, 1999.

[156] Mateo, F.G., Zuñiga, J., Requena, A., and Hidalgo, A., Energy eigenvalues for Lennard-Jones potentials using the hypervirial perturbative method, *J. Phys. B,* 23, 2771, 1990.

[157] ter Haar, D., *Problems in Quantum Mechanics,* Pion, London, 3rd ed., 1975, 357.

[158] Fernández, F.M. and Castro, E.A., *Algebraic Methods in Quantum Chemistry and Physics,* CRC Press, Boca Raton, FL, 1996, 94.

[159] Pilar, F.L., *Elementary Quantum Chemistry,* McGraw-Hill, New York, 1968, 414.

[160] Ogilvie, J.F., *The Vibrational and Rotational Spectrometry of Diatomic Molecules,* Academic Press, San Diego, 1998, 87.

[161] Fernández, F.M., Maluendes, S.A. and Castro, E.A., Comment on "$1/N$ expansion for a Mie-type potential," *Phys. Rev. D,* 36, 650, 1987.

[162] Fernández, F.M., Quantization condition for bound and quasibound states, *J. Phys. A,* 29, 3167, 1996.

[163] Kirschner, S.M. and Le Roy, R.J., On the application, breakdown, and near-dissociation behavior of the higher-order JWKB quantization condition, *J. Chem. Phys.,* 68, 3139, 1978.

[164] Imbo, T., Pagnamenta, A., and Sukhatme, U., Energy eigenstates of spherically symmetric potentials using the shifted $1/N$ expansion, *Phys. Rev. D,* 29, 1669, 1984.

[165] Maluendes, S.A., Fernández, F.M., Mesón, A.M., and Castro, E.A., Large-order shifted $1/N$ expansions, *Phys. Rev. D,* 24, 1835, 1986.

[166] Maluendes, S.A., Fernández, F.M., and Castro, E.A., Modified $1/N$ expansion, *Phys. Lett. A,* 124, 215, 1987.

[167] Maluendes, S.A., Fernández, F.M., and Castro, E.A., Modified large-N expansion, *Phys. Rev. A,* 36, 1452, 1987.

[168] Born, M. and Huang, K., *Dynamical Theory of Crystals and Lattices,* Oxford, New York, 1959, 166.

[169] Born, M. and Huang, K., *Dynamical Theory of Crystals and Lattices,* Oxford, New York, 1959, 406.

[170] Fernández, F.M., Corrections to the Born-Oppenheimer approximation by means of perturbation theory, *Phys. Rev. A,* 50, 2953, 1994.

[171] Whitton, W.N. and Connor, J.N.L., Wronskian analysis of resonance tunnelling reactions, *Mol. Phys.,* 26, 1511, 1973.

[172] Siegert, A.J.F., On the derivation of the dispersion formula for nuclear reactions, *Phys. Rev.,* 56, 750, 1939.

[173] James, P.B., Integral equation formulation of one-dimensional quantum mechanics, *Amer. J. Phys.,* 38, 1319, 1970.

[174] Messiah, A., *Quantum Mechanics,* John Wiley & Sons, New York, 1961, Vol. 2, 808.

[175] ter Haar, D., *Problems in Quantum Mechanics,* Pion, London, 3rd ed., 1975, 11.

[176] Moiseyev, N., Certain, P.R., and Weinhold, F., Resonance properties of complex-rotated hamiltonians, *Mol. Phys.,* 36, 1613, 1978.

[177] Fernández, F.M., Rayleigh-Schrödinger perturbation theory for resonance tunneling, *J. Chem. Phys.,* 105, 10444, 1996.

[178] Eckart, C., The penetration of a potential barrier by electrons, *Phys. Rev.,* 35, 1303, 1930.

[179] Goldstein, H., *Classical Mechanics,* 2nd ed., Addison-Wesley, Reading, MA, 1980, chap. 11.

[180] Nayfeh, A.H., *Introduction to Perturbation Techniques,* John Wiley & Sons, New York, 1981.

[181] Fernández, F.M., Perturbation theory in classical mechanics, *Eur. J. Phys.,* 18, 436, 1997.

[182] Nayfeh, A.H., *Introduction to Perturbation Techniques,* John Wiley & Sons, New York, 1981, chap. 4.

[183] Spiegel, M.R., *Theory and Problems of Theoretical Mechanics,* Schaum, New York, 1967, 90.

[184] Goldstein, H., *Classical Mechanics,* 2nd ed., Addison-Wesley, Reading, MA, 1980, chap. 8.

[185] Goldstein, H., *Classical Mechanics,* 2nd ed., Addison-Wesley, Reading, MA, 1980, chap. 9.

[186] Deprit, A., Canonical transformations depending on a small parameter, *Cel. Mech.,* 1, 12, 1969.

[187] Dewar, R.L., Renormalized canonical perturbation theory for stochastic propagators, *J. Phys. A,* 9, 2043, 1976.

[188] Dragt, A.J. and Finn, J.M., Lie series and invariant functions for analytic sympletic maps, *J. Math. Phys.,* 17, 2215, 1976.

[189] Cary, J.R., Lie transform perturbation theory for Hamiltonian systems, *Phys. Rep.,* 79, 131, 1981.

[190] Oteo, J.A. and Ros, J., The Magnus expansion for classical Hamiltonian systems, *J. Phys. A,* 24, 5751, 1991.

[191] Casas, F., Oteo, J.A. and Ros, J., *J. Phys. A,* 24, 4037, 1991.

[192] Fernández, F.M., Construction of invariants by perturbation theory, *Phys. Lett. A,* 242, 4, 1998.

[193] Eckardt, B., Birkhoff-Gustavson normal form in classical and quantum mechanics, *J. Phys. A,* 19, 1961, 1986.

[194] Ali, M.K., The quantum normal form and its equivalents, *J. Math. Phys.,* 26, 2565, 1985.

[195] Ali, M.K., Wood, W.R., and Devitt, J.S., On the summation of the Birkhoff-Gustavson normal form of an anharmonic oscillator, *J. Math. Phys.,* 27, 1806, 1986.

[196] Goldstein, H., *Classical Mechanics,* 2nd ed., Addison-Wesley, Reading, MA, 1980, chap. 10.

[197] Carhart, R.A., Canonical perturbation theory for nonlinear quantum oscillators and fields, *J. Math. Phys.,* 12, 1748, 1971.

[198] Swimm, R.T. and Delos, J.B., Semiclassical calculations of vibrational energy levels for nonseparable systems using the Birkhoff-Gustavson normal form, *J. Chem. Phys.,* 71, 1706, 1979.

[199] Varandas, A.J.C. and Mil'nikov, G.V., Incorporation of tunneling effects in classical trajectories via a method of canonical transformations, *Chem. Phys. Lett.,* 259, 605, 1996.

[200] McRae, S.M. and Vrscay, E.R., Canonical perturbation expansions to large order from classical hypervirial and Hellmann-Feynman theorems, *J. Math. Phys.,* 33, 3004, 1992.

[201] Goldstein, H., *Classical Mechanics, Section 10-7,* 2nd ed., Addison-Wesley, Reading, MA, 1980.

[202] Apostol, T.M., *Calculus,* Vol. 2, 2nd ed., Blaisdell, Waltham, 1969, 293.

[203] Eyring, H., Walter, J., and Kimball, G.E., *Quantum Chemistry,* John Wiley & Sons, New York, 1944, 363.

[204] Spiegel, M.R., *Theory and Problems of Vector Analysis,* Schaum's Outline Series, McGraw-Hill, New York, 1959, chap. 8.

[205] Kjaergaard, H.G. and Mortensen, O.S., The quantum mechanical Hamiltonian in curvilinear coordinates: a simple derivation, *Amer. J. Phys.,* 58, 344, 1990.

[206] Apostol, T.M., *Calculus,* Vol. 2, 2nd ed., Blaisdell, Waltham, 1969, 408.

[207] Eyring, H., Walter, J., and Kimball, G.E., *Quantum Chemistry,* John Wiley & Sons, New York, 1944, 39, 72.

[208] Powles, J.G., Teaching the prototype differential equation of physics, *Eur. J. Phys.,* 11, 323, 1990.

[209] Apostol, T.M., *Calculus,* Vol. 2, 2nd ed., Blaisdell, Waltham, 1969, 163.

[210] Symanzik, K. (unpublished)

Index

Printed in the United States
by Baker & Taylor Publisher Services